PHENOMENOLOGICAL THERMODYNAMICS

With Applications To Chemistry

PHENOMENOLOGICAL THERMODYNAMICS

With Applications To Chemistry

JOSEPH de HEER

Professor Emeritus of Chemistry
Department of Chemistry
University of Colorado, Boulder

PRENTICE-HALL, INC. / Englewood Cliffs NJ 07632

Library of Congress Cataloging in Publication Data

de Heer, Joseph (date)
Phenomenological thermodynamics with applications to chemistry.

Bibliography: p.
Includes index.
1. Thermodynamics. I. Title.
QD504.D4 1986 541.3′69 85-6516
ISBN 0-13-662172-4

Editorial/production supervision and
interior design: Nicholas Romanelli
Cover design: Edsel Enterprises
Manufacturing buyer: John Hall

Printed in the United States of America

10 9 8 7 6 5 4 3 2 1

ISBN 0-13-662172-4

Prentice-Hall International (UK) Limited, *London*
Prentice-Hall of Australia Pty. Limited, *Sydney*
Prentice-Hall Canada Inc., *Toronto*
Prentice-Hall Hispanoameracana, S.A., *Mexico*
Prentice-Hall of India Private Limited, *New Delhi*
Prentice-Hall of Japan, Inc., *Tokyo*
Prentice-Hall of Southeast Asia Pte. Ltd., *Singapore*
Editora Prentice-Hall do Brasil, Ltda., *Rio de Janeiro*
Whitehall Books Limited, *Wellington, New Zealand*

Contents

Part I. Basic Concepts and Definitions; Mathematical Techniques

Part II. The Laws of Thermodynamics; Some Basic Applications

Chapter 18 Two-Phase Equilibria in One-Component Systems 202

Chapter 19 Chemical Equilibrium 220

Chapter 20 The Principle of Le Chatelier 240

Chapter 21 The Phase Rule and Related Topics 253

Preface

The most explicit title appropriate for this book would be *Lectures on Phenomenological Equilibrium Thermodynamics, with Special Emphasis on Applications to Chemistry.* Although this lengthy title does not appear on the cover, it serves as the proper heading under which to outline what the author of this monograph *does* and *does not* intend to accomplish.

First, this book is an outgrowth of the author's *lectures* on the subject, both on an introductory and on a more advanced level. Since these presentations were, in general, well received by the students involved, a continuous effort has been made to preserve the characteristics of lectures as much as possible. This has lead to some obvious, as well as to some unexpected, difficulties. Obviously, while frequent repetitions with a proper verbal emphasis are a powerful lecture tool, printing costs limit repetitions (in particular as far as mathematical formalisms are concerned), and the verbal emphasis can at best be substituted by an imaginative use of italics. The latter will not please those grammatical purists who consider themselves disciples of Fowler. It came as a surprise to the author that in a few instances the lecture style threatened to impart an undesirable flavor to the text. For example, a Socratic-type dialogue (in the derivation of the Phase Rule), which had proven to be very effective in the classroom, came across as "arrogant" to several helpful reviewers of the manuscript. Every effort has been made to eliminate such unintended effects.

Second, this is a book on *phenomenological* thermodynamics only. (In the first chapter the author explains why he prefers this adjective to *classical,* so frequently used in this context.) The argument as to whether the phenomenological and the statistical approaches should be combined will undoubtedly go on forever. Although the author sides with those who think they should *not* be combined, he wishes to avoid a "dogmatic" attitude regarding this matter; if an occasional corpuscular "picture" can make the phenomenological framework considerably easier to understand, it is included.

Of course, criteria for the occurrence of spontaneous (irreversible, "natural") processes are as important to likely readers as are the conditions for equilibrium (and stability). This book is restricted to *equilibrium* thermodynamics only insofar as it excludes the so-called "thermodynamics of irreversible processes," in the sense of Onsager, Prigogine, de Groot, and others.

This newer branch of thermodynamics will be alluded to occasionally but requires, in the author's opinion, a separate monograph.

Since the author is a physical chemist, and the book is an outgrowth of lectures given primarily, if not exclusively, to chemistry and chemical engineering students, problems in *chemistry* dominate the applications discussed in some detail. On the other hand, several sections have been added which traditionally have had more appeal to physicists than to chemists. These include sections 10.3 and 14.2 on the Joule–Kelvin effect, 14.3 on adiabatic demagnetization, as well as 17.1 and 17.2 on the equilibrium and stability of uniform and nonuniform gases. Beginning chemistry students may wish to skip these sections, although to many chemists a study of one or more of these subjects should be a rewarding experience.

Even with all the restrictions described above, the area of thermodynamics and its applications is so vast that it cannot possibly be contained in a single volume. Several opinions about this were expressed by editors and reviewers of the manuscript in its initial stages, but ultimately the author is solely responsible for the decisions regarding what to include and what to leave out, knowing full well that *any* such decisions would leave *some* potential readers dissatisfied. Throughout the book a special effort has been made to emphasize a *critical* discussion, in particular of the basic concepts on which the development of the subject is based. In the light of the fact that some consider phenomenological equilibrium thermodynamics a "closed" science. ("There is nothing in it that was not known already to Clausius and Gibbs" is the extreme expression of this viewpoint), it is incredible that many of the concepts are still highly controversial and sometimes definitely not yet understood. In some instances the author hopes to be able to shed some light on such topics, but in other cases he shall have to admit that the issues concerned are as yet unresolved. If admitting the latter status is to be interpreted as a weakness, so be it. The one thing the author does not wish to be accused of, in this context, is to sweep these intricate problems routinely under the proverbial rug.

Throughout the book, a large number of paragraphs (usually rather brief, but occasionally of considerable length), appear in small print. These contain clarifications for, elaborations upon, or digressions from the main body of the text. It is left up to the individual readers to decide which of these comments are helpful to them and which ones do not reflect their primary interests.

The contents of this book divides logically into four parts. Part I presents the basic concepts, definitions, and mathematical techniques. For the beginning student, these eight chapters may be too much to swallow. Consequently, on an introductory level, parts of many chapters could be dealt with more superficially. Chapter 7 may be skipped in its entirety (if one has no intention to study Carathéodory's approach to the Second Law), and some of the material in Chapter 8 could be postponed. On the other hand, the better we master the mathematical techniques, the greater the reward in the sense of a preparation for the study of the main subject as it will be presented in the rest of the book.

In Part II the laws of thermodynamics are developed. This is one of the very few treatises on thermodynamics which gives *both* the "traditional" *and* the "axiomatic" (Carathéodory) approach. Readers who wish to confine themselves to the former can skip section 9.3 as well as all of Chapter 13. Chapters 10 and 14 are entirely devoted to some basic applications. Those primarily interested in applications to chemistry may wish to skip sections 10.3 and 14.2 on the Joule–Kelvin effect, as well as section 14.3 on adiabatic demagnetization.

Part III, on equilibrium and stability, contains the major features involved in the applications to chemistry. Again there is considerable flexibility in what one should, or should not, include at various levels of study. In introductory courses, some obvious candidates for omission are Chapter 17, sections 18.3 and 18.4, 20.3, and 21.4, and selected parts of Chapter 22. In choosing the subject matter for this part of the book, the author's strong affinity for the thermodynamics of phase equilibria is very apparent. The sections concerned attempt to give both a rather novel discussion of many features that have been "hidden" in the literature for a long time, as well as an initial presentation (as far as textbooks are concerned) of a number of recently published new ideas.

Part IV, dealing with some of the most important model systems, is obviously even more "open ended" than any of the preceding parts. The subject matter covers a wide range, from the derivation of the familiar equations for the colligative properties in dilute solutions (section 26.3) to a discussion of such esoteric topics as retrograde and barotropic phenomena (section 27.3). Obviously, one more time, readers and teachers can make a selection commensurate with their special interests. It is the fervent hope of the author that a study of the material *presented* will make the vast literature on topics *omitted* in this book much more accessible to readers.

In the appropriate chapters, readers will find some Exercises and Problems. The former contain primarily derivations and questions of a qualitative nature, which are woven into the text. The latter are mostly numerical applications and are given at the end of the relevant chapters. In both categories, some assignments are brief and very simple, others are much more challenging. Teachers "adopting" this book will undoubtedly wish to augment the various types of assignments with their own favorite examples. In selecting *units*, the author strongly feels that we should use those with which most of us are comfortable and which best express the magnitude they are trying to convey. There is no reason why we should not use an SI unit such as the joule, in lieu of the "rival" calorie, throughout most of this book. On the other hand, the author does not use SI units dogmatically; e.g., he balks at replacing atmospheres by pascals. Some relevant tables of units and conversion factors appear at the end of this book. Several other decisions had to be made regarding controversial issues involving notations and sign conventions. These are justified as they are encountered for the first time in the text.

In acknowledging all the help the author has received, directly and indirectly, he should first give credit to his teachers. As a sophomore chemical engineering student at Delft Institute of Technology, he had to take his first course on the subject in the *physics* department. This course was extremely hard to follow, and so was the textbook used, *De Beide Hoofdwetten der Thermodynamika*, by G. L. de Haas-Lorentz. It took many years to appreciate fully the approach taken, which was stimulated primarily by the writings of T. Ehrenfest-Afanassjewa. A few years later, this initial exposure was followed by a series of lectures on *chemical* thermodynamics by F. E. C. Scheffer, a brilliant classroom teacher. As a former pupil of Bakhuis-Roozeboom and van der Waals, Scheffer's interest was concentrated first and foremost in the area of chemical phase theory. This is reflected in some of the preferences of the present author, already referred to above.

In the summer of 1967, after having taught thermodynamics for fifteen years at various levels of sophistication, the author had the opportunity to attend a set of illuminating lectures on the subject by G. E. Uhlenbeck. These followed largely the methodology of *his* teacher, H. A. Lorentz, but were strongly flavored by Uhlenbeck's own penetrating insights. The present book has profited from the writings of many others, in the form of both articles and monographs. The indebtedness concerned is acknowledged in the relevant sections. At the end of several chapters, an annotated list of references is provided.

At the beginning of this preface we mentioned that this book is an outgrowth of the author's lectures on the subject. Thus he is indebted in the first place to many students, for their general encouragement as well as for their criticism of specific items. Good students are often the best teachers! The completion of this project owes much to Dr. Arlan D. Norman, who, as Chairman of the Chemistry Department of the University of Colorado at Boulder, lived up to his expressed conviction that writing a book of this nature is a valuable scholarly activity, which justifies *some* relief of teaching duties. Many colleagues, inside as well as outside this department, have offered advice on specific points. Their input is acknowledged in the appropriate chapters. In this context, a special word of thanks goes out to J. T. Hynes and S. J. Strickler, whose frequent counsel was invaluable.

This manuscript has been reviewed, and "rereviewed," by more peers than the author is able to recount. All this started with the critical reading of several "sample chapters" at the request of six potential publishers, and it finally ended with a third set of reviews on behalf of Prentice-Hall. Since most of the persons involved have remained anonymous (occasionally, the names of those who reviewed the same section of the manuscript at the same time were released as a group*), it is only possible to convey to them a feeling of "collective gratitude." William E. Palke and John S. Winn deserve a special word of thanks for having read the *entire* manuscript *twice.* Since his services were solicited by the author, Bill Palke was the only reviewer who never had the opportunity to hide under the cloak of anonymity. By leveling more valuable criticism at the manuscript than all other reviewers taken together, he proved that lack of anonymity need not be a liability to being effective as a critic! Finally, a special word of thanks to Ellsworth G. Mason for his advice on problems of style and grammar.

From the above it should be crystal clear that the author has had the benefit of much advice and help. Writing this book has not only confirmed to him the controversial nature and intricate complexity of phenomenological equilibrium thermodynamics, but also his fallibility in mastering this subject. The latter condition alone should be held responsible for any remaining errors.

The author expresses his gratitude to the editors of Prentice-Hall, in particular Elizabeth G. Perry, the former chemistry editor, who showed enough confidence in him to offer a publication contract against the background of a highly competitive market. Her successor, Nancy L. Forsyth, coordinated the final phases of the reviewing process in a most efficient manner, and "launched" the manuscript into production.

Gail Maxwell typed the "sample" chapters for the initial reviews, and the first version of Part I of the manuscript. She also deserves credit for all the illustrations, in the sense that the professional draftsperson, later engaged by Prentice-Hall, had to do little but retrace her drawings. Vicky Nelson did a superb job in typing the rest of the manuscript, and bore with me through the tedious chore of making revisions and corrections.

Last, but not least, a special word of thanks to Nicholas Romanelli, who supervised the entire production process for Prentice-Hall. Thanks to his expertise, a very complex manuscript has been transformed into a book with a far better aesthetic appearance than the author ever thought possible.

Boulder, Colorado, April 1, 1985 Joseph de Heer

---

* The following anecdotal point may give the teachers among the readers some food for thought. One of the scientists, whose name was revealed as one of those who was asked to evaluate the sample chapters, got his first instruction in thermodynamics from . . . the author!

PHENOMENOLOGICAL THERMODYNAMICS

With Applications To Chemistry

Part I

Basic Concepts and Definitions; Mathematical Techniques

Chapter 1

Thermodynamics as a Phenomenological Theory

This chapter introduces the essence of the author's approach to the subject. Alternative points of view are possible, depending on one's opinions about the structure of scientific theories.

1.1 A Few Remarks about the Philosophy of Science

According to one point of view, the philosophy of science addresses four topics: *location, structure, validation, and assessment.*[*] In analyzing the *location* issue, one raises questions of domain, such as: Is mathematics (or sociology) a science? Or, with special reference to the topic of this book, one may ask whether the laws of thermodynamics can be applied to, e.g., biology or economics. The *structure* of thermodynamics constitutes the main topic of section 1.2. *Validation* deals with the justification of the methods used in science, the "inductive approach" in particular. This often leads one into what is frequently called "metaphysics," although the usage of the latter term is ambiguous. An elaboration would fall entirely outside the scope of this introduction. *Assessment* goes beyond these "logical" aspects of science; in stressing its "relevance" one is concerned with what Philip Frank has called its "sociological" aspects. To give a contemporary example, Berry has linked the need for waste recycling to thermodynamic concepts.[†] It should come as no surprise that questions of *assessment* are frequently tangled up with those of *location.* (See also section 1.3.)

---

* Peter J. Caws, *Scientific Research*, September 1967, p. 81.

† R. Stephen Berry, "The Option for Survival," *Bulletin of the Atomic Scientists*, May 1971, p. 22.

1.2 The Structure of Phenomenological Theories, with Special Reference to Thermodynamics

The development of what is commonly called a "law" can best be analyzed with the aid of the flowchart shown below. We see that *phenomena* (associated with observations, hence with experimentation) are located at the beginning and at the end of a "cycle." They are tied together by a *logical*

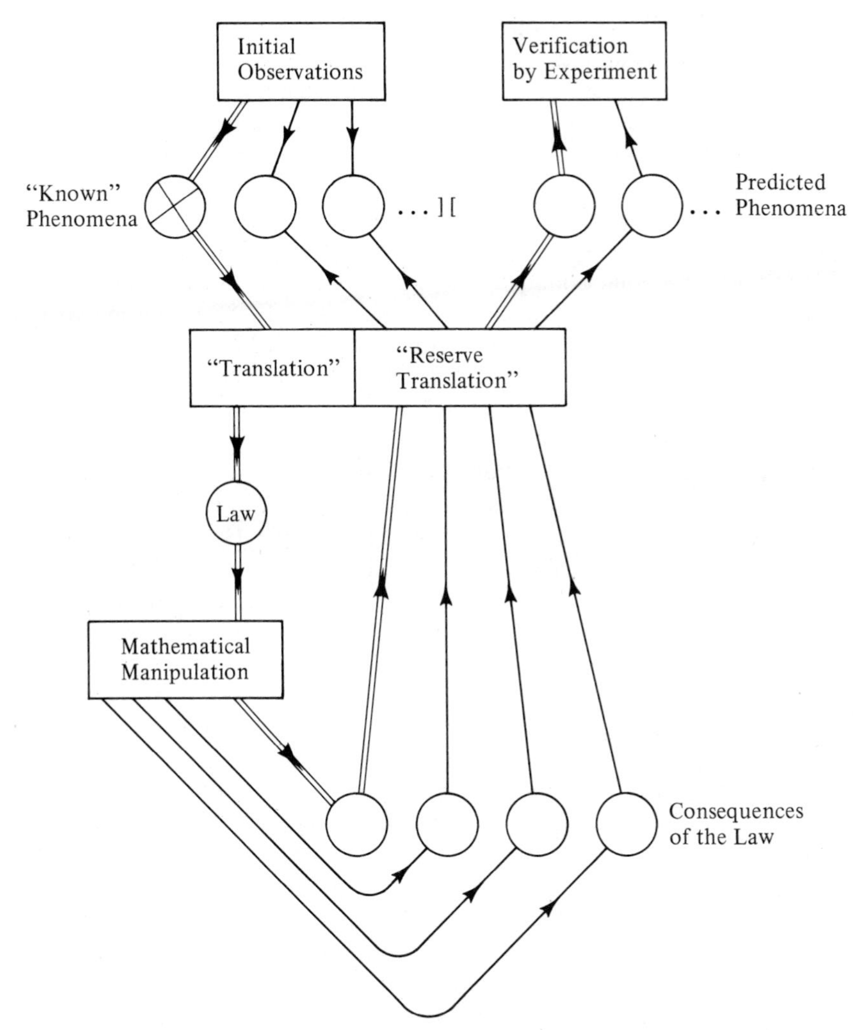

An example of the logical analysis, loosely called a "cycle," is indicated by double lines. For an illustration, see the text.

analysis, whence the designation *phenomenological* theory. The *law* occupies a central position; it appears as a "translation" of some particular result of our initial observations (*that* known phenomenon crossed in the diagram) into a convenient, i.e., mathematical, language. Sometimes the translation process looks misleadingly simple, the law becomes "obvious," and its entire origin may be obscured. This is particularly true in some formulations of the *First* Law of thermodynamics. However, for the *Second* Law, the translation process is lengthy as well as complex, and is commonly referred to as a "derivation." In this case there is the danger that the student gets lost in the details and is left up in the air to what exactly is "derived" from what. Once the law is formulated, it allows for manipulations, which lead to a variety of "consequences." These can be "translated in reverse," and the rest of the flowchart speaks for itself. For every law the entire pattern is repeated.

By way of illustration, let us put the First Law of thermodynamics within the framework of this flowchart. We shall do so in a very preliminary fashion, foregoing the rigor we shall impose when these matters come up in detail in Chapter 9. At this stage we need only a very elementary knowledge of the subject (e.g., that obtained in a general chemistry course). The "initial observations" are the experiments of Joule (1843–1848), which led to the realization that work and heat are "equivalent." This is our "known" phenomenon, which plays a vital part in the analysis, the "privileged" one, crossed in the diagram. "Translation" leads to the powerful mathematical form $\Delta\mathbf{E} = Q \pm W$ (+ or − depending on one's sign convention), the First Law. "Mathematical manipulation" leads to many "consequences of the law," among which $Q_p = \Delta\mathbf{H}$. "Reverse translation" ("interpretation," if one prefers) shows that the last equality justifies, for example, Hess's "Law of Constant Heat Summation" (1840), which had been discovered, prior to Joule's experiments, in several cases (other "known phenomena"), but also led to the prediction of many other thermochemical data ("predicted phenomena"), which could be verified by experiment.

Two points require some elaboration. First the question arises of how one picks the "privileged" phenomenon that will generate the law. As our flowchart suggests, the choice is not necessarily unique. It is tied up with one's preference regarding both the translation process ("derivation") to follow *and* the resulting mathematical formulation. This flexibility will become quite evident in Chapters 11 through 13, dealing with the development of the Second Law of thermodynamics. How the "founding fathers" arrived at their choices is quite a different story. As McGlashan has put it: "The history of thermodynamics is in fact a much more difficult subject than thermodynamics itself, and much less well understood . . . ."* (A good case can be

---

* M. L. McGlashan, "The Use and Misuse of the Laws of Thermodynamics," *J. Chem. Educ.* **43**, 226 (1966).

made for the contention that the *Second* Law was already being developed *before* the *First* Law was understood.) But no matter how the original choice was arrived at, as time has passed the privileged observation has become associated in our minds with some very basic attributes, and has subsequently been referred to as a "fundamental fact of experience" or even a "principle" (see, e.g., section 11.2). The net result is that its empirical nature is often forgotten. The laws of thermodynamics are *not* derived from "first principles" (whether spelled with a capital P or not), and, in this sense, do not "explain" anything.

This brings us to the second point. The preceding discussion should have made it crystal clear that, through its laws, *thermodynamics correlates phenomena*, and that is all! Obviously, this ought to convey a strong *aesthetic appeal*, in the sense that initially unrelated observations now appear "tied together"; they are, loosely speaking, different aspects of the same thing. One of the early triumphs of thermodynamics was the realization that Hess's law of constant heat summation is not a separate (fundamental) fact of experience, but a consequence of the First Law (see above and section 10.4). But such aesthetic appeal, by itself, would not make thermodynamics a compulsory subject for most engineers. As McGlashan put it,* the *practical* importance of correlating phenomena lies in the fact that some quantities are easier to measure than others. Thus the same Law of Constant Heat Summation allows one to obtain heats of reaction that are hard to obtain directly by the optimal use of those that are readily accessible. Readers will recall all the exercises involved from their glorious days in general chemistry courses.

1.3 Controversies: Axiomatization and Extrapolation

The approach embodied in the preceding section, although perhaps presented in a somewhat unorthodox way, is in fact a "traditional" one, sometimes referred to as "classical" or "historical." The use of "classical" can be misleading in this context, for reasons to be given in section 1.5. As already hinted above, "historical" seems even more out of place, since any *systematic* presentations will certainly *not* parallel the *historical* development. The designation "traditional" approach is the least objectionable if only because it preceded the alternative put forward by Carathéodory. After observing that (phenomenological) thermodynamics is a closed science (see also section 1.5), he realized that an "axiomatic" approach had to be feasible. Since, as noted above, the "privileged" known observation of our flowchart had already become a "principle," it was but a small step to turn it into an "axiom." By

* See footnote, p. 4.

ingeniously coupling such axioms with an appropriate set of definitions, a consistent scheme could be developed which, at least on the surface, looks quite different from the "traditional" pattern. Pushed to the limit, a corresponding treatise on (phenomenological) thermodynamics appears to be closer to a branch of symbolic logic than to a physical science firmly rooted in experimentation. Although we shall disregard such extremes and admit some bias toward the traditional approach, it would be foolish to reject completely the magnificent contributions of Carathéodory and his followers (notably Buchdahl), since they have greatly enriched the subject, both conceptually and analytically. Therefore, although our treatment remains essentially "traditional," we shall occasionally draw attention to the "axiomatic"* alternative, with which every serious student of the subject should have *some* familiarity. Readers have the option to skip this material (contained primarily in Chapter 7, section 11.3, and Chapter 13); this will not impair their ability to understand the various applications to be given in the second half of the book.

The judgment on the value of certain "extrapolations" is even more controversial and brings us right back to the philosophical issues of *location* and *assessment* (section 1.1). As outlined above (section 1.2), the author takes the view that experimentation stands at the beginning and at the end of the development of phenomenological thermodynamics. Here experimentation is implied to involve only (macroscopic) laboratory systems. It is true that in this context the meaning of "laboratory" may be stretched to include, e.g., a pilot plant or even the total physical plant involved in a certain industrial process. But it is quite a jump to extend the validity of the maximum theorem of the entropy (section 15.1) to "the world" or "the universe." On a different level, applications to problems in biology, although certainly not out of the question, should be handled with the utmost care (cf. the discussion of *intrinsically irreversible processes* in section 14.4). Sometimes the phenomenological formalism of thermodynamics, coupled with arguments based on its statistical interpretation, is even applied to socioeconomic issues. Although the author has made every effort to keep an open mind in these matters, he must admit that such extrapolations frequently leave him with an uneasy feeling at best.

1.4 Systems and Surroundings

One difficulty that may arise in making such extrapolations is the need for a precise definition and parameterization of a thermodynamic *system*, as set

---

* Buchdahl stresses that his adaptation of this approach, superficial appearances notwithstanding, is *not* axiomatic. See reference 3 at the end of the chapter.

against its *surroundings* (environment). But even in "ordinary" applications, the latter in principle comprises everything that is not the former, so that it becomes tempting to drag the entire world into the discussion once more and replace "the surroundings" by "the rest of the universe." However, it is not difficult to realize that all that is relevant in this context is the condition of a reasonably close portion of this environment. What should be considered "reasonably close" will be quite evident in any specific case. What goes on simultaneously in distant parts of the universe is *irrelevant*, and the information concerned is not available to us instantaneously. "Ordinary" thermodynamics deals with "here-and-now" phenomena.

Extending the surroundings outward may be ambiguous, but its demarcation against the system is usually straightforward. From a practical point of view, however, we shall occasionally find ourselves faced with a nontrivial choice. For example, if we have a chemical reaction taking place in a vessel inside a thermostat, which in turn is located inside a laboratory, there are two possibilities: Sometimes it may be convenient to incorporate the thermostat as part of the system, whereas on other occasions it may be more useful to consider it as part of the surroundings.

The systems themselves may be classified in a variety of ways, which are usually not mutually exclusive. Several such classifications, e.g., on the basis of the number of phases or the number of components, will be introduced in later chapters as the need arises. It is useful, however, to have one important aspect of nomenclature at our disposal immediately. This characterizes systems with respect to their possible interactions with the surroundings. If exchanges of both matter and energy are possible, the system concerned is called *open.* If such an exchange is confined to energy (hence excludes matter), the system is designated as *closed.* Finally, an *isolated* system can exchange neither energy, nor matter with its surroundings.

In this entire book, open systems will be touched on only in a peripheral fashion. This is not because they are of no importance; most biological ones fall into this category. Some warning concerning applications of thermodynamics to biological systems has already been given. Most meaningful discussions in this area (as in certain other areas, such as the study of continuous reactions in chemical industry) would force us into the realm of the *thermodynamics of irreversible processes* (sometimes called the *thermodynamics of the steady state*). This brings in so many new facets that it would be best to treat it elsewhere. The interested reader will have no difficulty locating monographs on this challenging subject.* The present book will concentrate on what in *this* context has been called "equilibrium thermodynamics," although this designation may suggest a much more severe restriction than is

* See, e.g., reference 4 at the end of the chapter.

actually implied. Irreversible processes will be considered frequently, in particular in conjunction with the Second Law.

1.5 The Power and Limitations of Phenomenological Theories, Thermodynamics in Particular; Supplementary Approaches

Let us preface this final section of this introductory chapter with another reminder that *phenomenological theories deal with observable properties of macroscopic systems.* As such they completely disregard the atomistic structure of matter. This has great advantages! For example, when the classical mechanics of atoms and molecules had to be abandoned in favor of the quantum mechanics of same, the laws of (phenomenological) thermodynamics remained unchanged. Parenthetically, this is one of the reasons the author dislikes the designation *classical* thermodynamics in lieu of *phenomenological* thermodynamics. But by the same token, this points toward the weakness and limitations of the phenomenological approach. The very general relationships that link macroscopic properties typically contain "material constants," which have to be found by other means. Thus Ohm's Law involves a material constant, the electrical resistance, which is the ratio of the potential difference across the ends of a conductor, and the current caused to flow through it. But no *phenomenological* theory will tell us why different substances have different resistances, why (to go to the extremes) copper is an excellent conductor and glass is an insulator. For examples "closer to home," thermodynamics gives us some very general relations for $(C_p - C_v)$ of any substance (see section 10.1), but it does not tell us why either of them (C_p or C_v) is quite different for He and for CO_2 vapor under the same circumstances.

Hence it becomes clear that phenomenological theories have to be supplemented with *corpuscular* ones, which deal exactly with the microscopic, atomistic substructures neglected in the former. It is equally obvious that since bulk matter contains a large number of particles, these corpuscular theories are of necessity statistical in nature. Thus *statistical mechanics* forms a bridge between the classical or quantum mechanics of atoms and molecules on the one hand and phenomenological thermodynamics on the other hand. For many decades educators have argued the pros and cons of teaching these disciplines jointly or separately. There are strong arguments both ways, but the title of this book suggests that we shall confine ourselves to the phenomenological approach. However, we need not be "dogmatic" about this; if, occasionally, a qualitative or semiquantitative corpuscular argument can greatly clarify a difficult issue, we shall not hesitate to incorporate it.

Fortunately, in bringing the very general equations of (phenomenological) thermodynamics to life, there are alternatives to the explicit introduction of statistical mechanics. Obviously, we can make optimal use of experimental data. In this way a thermodynamic relation between $(C_p - C_v)$, "fed"* with an experimental C_v, will give us C_p. This is entirely in line with one of the important practical goals of phenomenological thermodynamics, as mentioned toward the end of section 1.2; we have obtained the (often more important) quantity C_p from the (at least in the gas phase) easier to measure C_v. Another powerful way to utilize the very general equations under consideration is to "feed" them with the results of the manipulation of *models*. Since these play such an important part in the development of thermodynamics, the entire second chapter is devoted to this subject.

REFERENCES

1. P. W. Bridgman, *The Nature of Some of Our Physical Concepts*, Philosophical Library, New York, 1952. Introduction to "operationalism" in science, with an occasional reference to thermodynamics.
2. P. W. Bridgman, *The Nature of Thermodynamics*, Harvard University Press, Cambridge, Mass., 1941. A classic contribution.
3. H. A. Buchdahl, *The Concepts of Classical Thermodynamics*, Cambridge University Press, Cambridge, 1973. By many considered the definitive adaptation of the "axiomatic" approach† originated by Born and Carathéodory. The first few chapters contain much material of quite general interest.
4. I. Prigogine, *Introduction to the Thermodynamics of Irreversible Processes*, 3rd ed., Wiley-Interscience, New York, 1967 (an updated and expanded version of a monograph originally published by Charles C. Thomas, Springfield, Ill., in 1955). A concise, readable introduction for those who are interested in this relatively new field, frequently mentioned in the present book. Considerable knowledge of "equilibrium thermodynamics" is desirable.

---

* As G. E. Uhlenbeck used to say in his lectures: "We have to give these equations something to eat!"

† See, however, the remark in the footnote on p. 6.

Chapter 2

The Use of Models

(With an Excursion into the Ideal Gas Concept)

2.1 Models and Their Definition

In many branches of the sciences, both social and physical, the "real" system is too complex to be described quantitatively (i.e., mathematically). Therefore, we abstract from reality and, instead, introduce a "model" system. The latter cannot be realized experimentally; it is a theoretical concept, *defined* with two requisites in mind:

1. It must be simple enough to be mathematically manageable.
2. It must be close enough to "reality" to be useful.

Obviously, these two requirements, if not mutually exclusive, will normally interfere with each other. The closer the model is to reality, the more complex are its characteristics and the more difficult it will be to manipulate mathematically. Hence the definition of each model necessarily embodies a compromise. Such a definition can be given in a number of ways.

(a) A *corpuscular definition* should be straightforward and unambiguous. When subjected to the techniques of either kinetic-molecular theories or statistical mechanics, the model thus introduced should yield, at least in principle, an *explicit* expression for any desired macroscopic property. As mentioned in Chapter 1, this book will not incorporate such manipulations, and therefore the corpuscular description of a model is of no *direct* use to us. *Indirectly*, however, it can still play an important role. Frequently, it gives us a "picture" which is not only aesthetically pleasing but may also suggest to what extent various real systems *approach* a given abstract model. This, in turn, may suggest an *empirical* (perhaps we should call it "quasi-empirical") method of obtaining the properties of interest, namely through the appropriate extrapolation of experimental data.

An explicit equation for a particular property is often more then we really need. Let us consider an example, the broader meaning and importance of which may very well escape readers at this state, but which nevertheless clearly illustrates the point to be made. For the chemical potential of a component i in an ideal mixture of ideal gases, we may write

$$\mu_i = \mu_i^{\ominus}(T) + RT \ln |p_i|, \qquad (2\text{-}1)$$

in which $|p_i|$ is the (dimensionless value of the) partial pressure of i, and $u_i^{\ominus}$ is the corresponding chemical potential in the proper "standard state." Statistical mechanics should give us, at least in principle, an *explicit expression* for $\mu_i^{\ominus}$. However, the mere existence of the *functional relationship*, expressed by Eq. (2-1), stressing that $\mu_i^{\ominus}$ is a temperature function only, is normally all that we need to take cognizance of in phenomenological thermodynamics. Among other things, Eq. (2-1), *as such*, will allow us to derive the most important features of the thermodynamic equilibrium constant associated with a chemical reaction in an ideal mixture of ideal gases. The "origin" and other implications of Eq. (2-1) are discussed thoroughly in Chapters 19 and 23.

(b) The foregoing considerations lead us to look for a *phenomenological definition* of a model. This consists of the *minimum* number of phenomenological statements from which all relevant functional relationships may be derived. How many such statements are necessary and sufficient in any given case often depends on what laws and concepts of thermodynamics are already at our disposal. This should come as no surprise for, as emphasized in Chapter 1, it is entirely within the nature of this discipline to *interrelate* certain, at first sight *unrelated*, phenomenological statements. For the same reason, this type of defining process is not necessarily unique. Finally, we stress that not all authors employ the same terminology. For example, some use "ideal gas mixture" for what we call an "ideal mixture of ideal gases," while others join us in restricting the former designation to a gaseous system that is ideal only in the Lewis and Randall sense (Chapter 24). Needless to say, the readers should approach the extensive literature on this subject with the utmost care. Handled properly, the phenomenologically defined models can provide us with a very powerful means to bring the general thermodynamic equations to life, in the sense discussed in section 1.5.

We discuss the most interesting model systems in the last part of this book when a maximum of thermodynamic formalism will be at our disposal. We end the present chapter with a preliminary consideration of a very familiar model in the physical sciences, the single-component ideal gas. This plays an important part in the *development* of our subject. At the same time, this gives us the opportunity to illustrate several of the more general features mentioned above.

2.2 An Important Example: The Single-Component Ideal Gas

The ideal, or perfect,* gas (the qualification "single-component" will be implied throughout this section) is probably one of the best known and most widely used models in the physical sciences. In this section we assume that the reader already has a pragmatic familiarity with the temperature concept, which we examine critically in Chapter 3. As much as possible, our presentation will be couched in terms of an *empirical temperature*, t, which does not require access to the (First and) Second Law of thermodynamics.

(a) Corpuscular Discussion

A real gas consists of a very large number of atoms or molecules, confined in some container of volume $\mathbf{V}$, and engaged in translational, vibrational, and rotational motion. The particles interact with each other, and with the walls of the container, so that there are very complex changes in all these types of motion. Such a system cannot be treated mathematically in an exact way. This is the typical situation in which one introduces models, and the ideal gas is the simplest one. In giving its definition, it is customary to consider translational motion separately from the internal molecular motions.

(i) Concentrating first on the *translational motion*, the model is usually defined as consisting of noninteracting mass points† which undergo only elastic collisions with the walls of the container. Classical kinetic-molecular theory, first given by D. Bernoulli in 1738, then gives the famous formula

$$P\mathbf{V} = \tfrac{1}{3}Nm\overline{u^2}, \tag{2-2}$$

where P is the pressure of the gas, N the total number of particles, each of mass m, with $\overline{u^2}$ as their mean-square speed. Equation (2-2) is sometimes called the *mechanical* equation of state. A more or less simplified derivation is found in many textbooks on physical—or even general—chemistry. We

* We shall join the vast majority of physicists and chemists who use these terms *interchangeably*. A few authors do otherwise. For example, Ira N. Levine, in his *Physical Chemistry* (McGraw-Hill, New York, 1978), initially uses "ideal gas" and "perfect gas" to designate *different* models, but on p. 114 he proves that they are, in fact, identical. See Exercise 2-1.

† Since, barring charged particles, the range of all molecular forces (even the so-called "long-range" ones) is small, the model is strictly valid if the mean free path (between interparticle collisions) is very large compared to (1) the diameter of the particles and (2) the dimensions of the container. It may remain valid to a good approximation if only the *second* condition is relaxed.

are more interested in the corresponding relation between pressure, volume, and *temperature*, the *thermal* equation of state, usually just referred to as *the* equation of state for the model. In terms of the empirical temperature this may be written

$$P\mathbf{V} = nA(t), \tag{2-3}$$

where n is the number of moles of our gas and $A(t)$ is a temperature function only. The conversion of Eq. (2-3) to the more familiar form, in terms of an *absolute temperature*, T,

$$P\mathbf{V} = nRT, \tag{2-4}$$

is discussed in section 3.3. The link between the mechanical equation of state, on the one hand, and the thermal equations of state, on the other hand, becomes obvious *if* we can show that the average translational kinetic energy is a function of t only:

$$\tfrac{1}{2}m\overline{u^2} = a(t) \tag{2-5a}$$

or, more explicitly, in terms of T:

$$\tfrac{1}{2}m\overline{u^2} = \tfrac{3}{2}kT. \tag{2-5b}$$

The Boltzmann constant, k, is the quotient of the universal gas constant, R, and Avogadro's number, N_{Av}:

$$k = \frac{R}{N_{\mathrm{Av}}} \quad \text{while} \quad n = \frac{N}{N_{\mathrm{Av}}}.$$

A discussion of the origin of Eqs. (2-5) falls beyond the scope of this book; the reader is referred to any text on statistical mechanics. It should be stressed that the existence of t (and T) is, once more, accepted beforehand; temperature, unlike pressure, cannot be *introduced* on a purely corpuscular level.

(ii) For a monatomic gas, such as He or Ne, the translational energy is all we have to be concerned with. For diatomic and polyatomic gases, we obviously must consider *other forms of energy* as well, but all results obtained under (i) above are still assumed to be valid. This implies that, in dealing with translations, we still treat the molecules involved as if they are noninteracting mass points, notwithstanding the fact that we must allow for a molecular substructure if we turn our attention to vibrations and rotations. The last two forms of motion must also be taken into account if we wish to consider such important properties as C_p and C_v, the molar heat capacities of the gas at constant pressure and volume, respectively. This requires the superposition of *independent* models, normally the harmonic oscillator and the rigid rotor. Once more the reader has to be referred to textbooks on statistical mechanics. The most important result is that *C_p and C_v turn out to be functions of the temperature.* They become constants (the familiar multiples of R) only if the temperature is high enough for the classical limit to be valid, and this would place an unnecessary constraint on the applicability of the model.

The "dualistic" aspect of the introduction of the model, outlined above, carries over into a phenomenological framework, where it has given rise to some confusion. It also turns up in an "empirical" discussion, to which we next turn our attention.

(*b*) "*Empirical*" *Discussion*

(i) The most general equation of state for a real gas is the *virial expansion.* It can be cast in a variety of forms, one of which is

$$PV = A(t) + B(t)P + C(t)P^2 + \dots. \tag{2-6}$$

Since the ideal gas model (noninteracting mass points) should be approached as $P \to 0$ when $V \equiv (\mathbf{V}/n) \to \infty$, Eq. (2-6) immediately reduces to

$$P\mathbf{V} = nA(t). \tag{2-3}$$

This embodies a number of empirical equations of historical significance (which were obtained from measurements at relatively low pressures and were valid within the accuracy available at the time). In current terminology,* they pertain to

$\mathbf{V}$ as $f(n)$ at constant P and t (Avogadro's Law),
$\mathbf{V}$ as $f(P)$ at constant n and t (Boyle's Law), and
$\mathbf{V}$ as $f(t)$ at constant n and P (Charles' Law).

While the *virial coefficients* $B(t)$, $C(t)$, etc., in Eq. (2-6) are different for different gases, Gay-Lussac discovered in the early part of the nineteenth century that $A(t)$ is a universal function of t. As we shall see in section 3.3, this allows us to use Eq. (2-3) as the basis for the introduction of the *ideal gas temperature.*

(ii) There is another important piece of empirical information. When a gas expands into a vacuum, under isolated conditions, a temperature change, Δt, will in general occur. As the initial gas pressure, P, is lowered, Δt becomes smaller and smaller. As $P \to 0$ (ideal gas limit), $\Delta t \to 0$. This is known as Joule's Law, because in 1843 he found, within the accuracy allowed by his instruments, Δt to be negligibly small at initial pressures up to 22 atm. As we shall see in section 10.3, Joule's result implies that the total internal energy *of an ideal gas* is a function of the temperature only.

---

* The mole concept was not available when these laws were first formulated.

(c) *Phenomenological Definitions*

In this context we have to distinguish two cases:

Case I. Suppose we have developed the First, but not yet the Second, Law of thermodynamics.

At this stage the change in internal energy, $\Delta\mathbf{E}$, is well defined, but the absolute temperature, T, has not yet been introduced. The ideal gas must still be defined by *two* statements:

$$P\mathbf{V} = nA(t) \tag{2-3}$$

and

$$\mathbf{E} = nE(t), \tag{2-7}$$

where we have also used the fact that $\mathbf{E}$ is an *extensive* quantity (see section 4.2). For reasons suggested in subsection (b) above, these two statements are sometimes referred to as Boyle's Law and Joule's Law, respectively. Equation (2-7) is also occasionally called a *caloric equation of state.* Unlike $A(t)$, $E(t)$ is different for different gases. Note that neither $A(t)$ nor $E(t)$ is given *explicitly* (compare the general discussion in section 2.1).

Case II. Suppose that both the First and Second Laws of thermodynamics are at our disposal.

At this stage we shall also have introduced the absolute temperature, T, and the equation of state reads (see also section 3.3)

$$P\mathbf{V} = nRT. \tag{2-4}$$

In this situation Eq. (2-4) *alone* suffices to define the ideal gas, since the Second Law has provided us with the very powerful relation (section 14.1):

$$\left(\frac{\partial \mathbf{E}}{\partial \mathbf{V}}\right)_T = T\left(\frac{\partial P}{\partial T}\right)_\mathbf{V} - P, \tag{2-8}$$

which, in turn, allows us to *derive* the fact that $\mathbf{E}$ is a function of the temperature only. Taking into account the extensive character of $\mathbf{E}$, Eq. (2-7) thus appears as a *consequence* of Eq. (2-4).

Exercise 2-1

(a) Use Eq. (2-8) to prove, from Eq. (2-4), that for an ideal gas

$$\left(\frac{\partial \mathbf{E}}{\partial \mathbf{V}}\right)_T = 0.$$

(b) Convince yourself that this implies that $\mathbf{E}$ is a function of the temperature only.

(c) Use Eq. (2-8) to show that for any equation of state of the form

$$P = Tf(\mathbf{V}), \tag{2-9}$$

the result $\mathbf{E} = \mathbf{E}(T)$ is obtained. This shows that whereas $\mathbf{E} = \mathbf{E}(T)$ follows from Eq. (2-4), the reverse is *not* true; Eq. (2-4) is just one special case of the general form (2-9). Another example is provided by the equation of state for the *elastic hard-sphere model* (see section 10.2):

$$\mathbf{P}\mathbf{V} = nRT + \mathbf{B}P, \tag{2-10a}$$

or for 1 mole:

$$PV = RT + BP. \tag{2-10b}$$

Note: Levine (see footnote, p. 12) initially calls an *ideal* gas one that obeys Eq. (2-4) *only*, whereas a *perfect* gas obeys Eqs. (2-4) *and* (2-7). An analysis corresponding to parts (a) and (b) above is subsequently taken as the proof that the two models are, in fact, identical.

Chapter 3

Temperature, Heat, and Heat Capacity

3.1 Introduction

The development of the concepts of temperature, heat, and heat capacity occupies many lengthy chapters in the history and philosophy of science.* Some of the controversies involved have raged for centuries and are still the subject of heated discussions.

On an introductory level, readers with a working knowledge of these three concepts, and with little or no interest in the axiomatic approach, especially to the First Law, may wish to confine their attention to sections 3.2(a) and 3.3. They should familiarize themselves with the notions of *empirical* temperature, *ideal gas* temperature, and *absolute* temperature, as presented in those parts of this chapter.

For those readers who are seriously interested in the conceptual development of thermodynamics, section 3.2(c) contains a critical discussion of the so-called Zeroth Law, while section 3.4 emphasizes one of the lesser known ways to introduce heat capacities and (quantity of) heat *in that order.* The entire presentation is brief and makes no claim to completeness. Nevertheless, it ought to be quite helpful in assessing the alternative approaches to the First Law, which are discussed in Chapter 9.

3.2 The Empirical Temperature and the Zeroth Law of Thermodynamics

(a) Although historians of science are not sure who first measured temperatures, we do know that Galileo designed a *barothermoscope* toward the end of the sixteenth century. During the three centuries that followed, there

* See, e.g., "The Early Development of the Concepts of Temperature and Heat," by Duane Roller, Case 3 of the *Harvard Case Histories in Experimental Sciences*, J. B. Conant, ed., Harvard University Press, Cambridge, Mass., 1957, Vol. I, p. 119.

was an ongoing debate about what the *true* temperature is and about what constitutes a *good* thermometer, until it was finally recognized that such questions made no sense. A critical discussion of the temperature concept was not given until the latter part of the nineteenth century, when it caught the attention of several prominent physicists, notably Mach, Maxwell, and Poincaré. In 1872, Maxwell made the definition of temperature contingent upon what he considered to be an experimental result, the *transitivity* of thermal equilibrium:

> If two bodies are in thermal equilibrium with a third body, they are in thermal equilibrium with each other. (3-1)

This became accepted as the *basis* for the construction of a test body, the *thermometer*, which would associate a *number*, the *empirical temperature*, *t*, with any equilibrium state of any macroscopic system. The "transitivity principle" could then be expressed as:

$$\text{If } t_A = t_B \text{ and } t_B = t_C, \text{ then } t_A = t_C. \tag{3-2}$$

In 1931, R. H. Fowler "sanctified" the statement (3-1), and those equivalent to it, by the designation *Zeroth Law of Thermodynamics.* We shall return to it in subsection (c).

(b) In this way, the temperature *concept* can be provided with a much firmer basis. At the same time, there remains a considerable degree of arbitrariness in how to construct a thermometer and how to assign numerical values to the empirical temperatures introduced in this fashion. First, one *arbitrarily* chooses a "thermoscopic substance," with an associated "thermoscopic property." In addition, one has to *arbitrarily* define a "temperature scale." To give a very familiar example, we can choose mercury as the substance,* its volume (at constant pressure) as the property, and the centigrade (Celsius) system as the scale. Calibration may be achieved by assigning the "ice point" and "steam point" of water the numbers 0 and 100, respectively, and using a linear interpolation, so that intermediate temperatures can be obtained from

$$t = \frac{\mathbf{V}_t - \mathbf{V}_0}{\mathbf{V}_{100} - \mathbf{V}_0} \times 100. \tag{3-3}$$

The thermometer is marked accordingly, and can be made usable above 100 and below 0 by extrapolation of this scale beyond these two calibration points. Of course, boiling and melting of the mercury limit this process. The main trouble with these well-known "liquid expansion thermometers" is that different substances yield different values of the temperature (except,

* Strictly speaking, one should also specify the material out of which such a thermometer is made, normally glass.

of course, at the calibration points). For example, when a mercury thermometer indicates 50.0°C, a carbon disulfide one would read 49.5°C. Discrepancies of the order of a degree (centigrade) are not at all unusual! As we shall see in section 3.3, this has led to the development of the ideal gas thermometer. For a description of other types of thermometers, alternative temperature scales, and adopted standards, the reader is referred to appropriate monographs on thermometry; this material falls outside the scope of this book.

(c) We now return to the Zeroth Law and subject it to closer scrutiny. Three main points have to be raised.

In the first place we assumed that we know the meaning of "thermal equilibrium." This is a relatively straightforward concept if one is willing to accept the notion of (flow of) heat. It is indeed possible to give an operational definition of the latter without using the temperature; we shall return to this issue briefly in section 3.4. However, in the "axiomatic" method of Born and Carathéodory, one adopts the Zeroth Law while postponing the introduction of the concept of heat until the development of the *First* Law (section 9.3). In this case the definition of "thermal equilibrium" is far from trivial. The usual procedure is to distinguish initially between *adiabatic* and *nonadiabatic* (or *diathermal*) *walls*. Assuming, for simplicity, that all such walls are impermeable to matter, the former are characterized by the property that they can transmit a disturbance only by *mechanical* means. Such a definition (several variations are found) is purely formal.* Surely, *in practice*, one would ascertain whether a partition is adiabatic or not by checking if *heat* can flow across it! But if one accepts these definitions of adiabatic and diathermal, one can proceed by considering two bodies to be in "thermal contact" if they either touch directly (e.g., two pieces of metal or two immiscible liquids) or are separated by a diathermal wall (e.g., in the case of two miscible fluids). When, after a certain time interval following the initiation of such contact, no more observable macroscopic changes occur, the bodies concerned are said to be "in thermal equilibrium."

The second point is that even when we accept the notion of thermal equilibrium, the temperature concept appears to have a precise meaning only in equilibrium states. Different authors resolve this difficulty in different ways. Some refer t to the *surroundings*. However, we all know that when we carry out typical chemical reactions in the laboratory, the temperature *inside* the reaction vessel can usually be measured accurately. Hence this author prefers to join those who look toward molecular considerations for a justification of this obvious extension of the temperature concept to many nonequilibrium states. In brief, as long as many nonreactive collisions occur,

---

* In Bridgman's terminology (see references 1 and 2 of Chapter 1), it is "operational" only in a paper-and-pencil sense.

as opposed to reactive ones, the energy distribution of the molecules over translational, vibrational, and rotational states will be so close to equilibrium that a single statistical parameter, *the* temperature, suffices to specify the system.

This is no longer necessarily the case if the chemical reaction is very fast as, for example, in flash photolysis experiments under adiabatic conditions, in flame reactions, or in shocktube processes. Here each form of energy may be associated with a different temperature. If we are only interested in specifying initial and final states, a reference to the temperature of the surroundings may be relevant. Otherwise, such reactions are excluded from our discussions.

The third and last point concerns the question of whether the transitivity, expressed by Eq. (3-1), can justifiably be called a "law of thermodynamics." Redlich was the first one to point out that this "principle" does not reflect what we have called earlier (section 1.2) a fundamental fact of experience.* He illustrates this by a number of examples; we shall consider only one of these:

Let system A be zeolite (molecular sieves), system B water, and system C mesitylene. The sieves attract water, liberating a heat of adsorption which will largely disperse into the water. There is no such interaction between the sieves and mesitylene. Hence there can be thermal equilibrium between A and C, as well as between B and C, but not between A and B.

When faced with such a situation, we take it for granted that some *special* interaction occurs. Whether this is easily recognized and understood (as in this example) or not, the fact remains that the transitivity principle has no strict empirical basis. Recently, Bergthorsson placed the emphasis somewhat differently† in the sense that whenever the transitivity principle cannot be verified experimentally there *must* be a "new" type of interaction. Thus *the Zeroth Law can never be refuted*; rather than designating it as a phenomenological law, Redlich calls it "a guideline for checking our description of nature," a "rule of order." Obviously, a further discussion would lead us too deeply into the realm of the philosophy of science. The interested reader may consult the papers by Redlich and Bergthorsson, in which many supplementary references can be found.

As a postscript, the author feels that there is no essential difference between calling the "transitivity principle" the Zeroth Law and designating the "principle of the conservation of energy" as the First Law. Whenever we find a violation of the latter, we promptly look for new forms of energy. We return to this issue in section 9.2.

---

* Otto Redlich, "The So-Called Zeroth Law of Thermodynamics," *J. Chem. Educ.* **47**, 740 (1970); see also *Rev. Mod. Phys.* **40**, 559 (1968).

† Bjørn Bergthorsson, "Temperature, Transitivity, and the Zeroth Law," *Am. J. Phys.* **45**, 270 (1977).

3.3 The Ideal Gas Temperature

In the seventeenth and eighteenth centuries most thermometers were of the familiar liquid expansion type. As already mentioned, their serious drawback was the fact that *different liquids* gave *different temperatures* (except, of course, at the calibration points). Early in the nineteenth century, Gay-Lussac showed that at low (constant) pressures, $\mathbf{V}$ is the *same function* of t *for all gases.* This result is embodied in the equation of state

$$P\mathbf{V} = nA(t) \qquad (2\text{-}3)$$

discussed extensively in Chapter 2. In 1840, Regnault pointed out that this could be the basis for the definition of an ideal gas scale (also known as the Avogadro scale), and the measurement of an ideal gas temperature, θ.

Two points should be mentioned briefly. In the first place, with the ideal gas thermometer one actually measures $P(t)$ at constant $\mathbf{V}$, rather than $\mathbf{V}(t)$ at constant P, since this is the simplest experimental procedure. Second, when later during the century this type of instrument became a *standard thermometer*, one was forced to introduce corrections for the nonideality of the gas involved.

As a first step he adopted a *linear scale*, by requiring that

$$A(t) = R\theta, \qquad (3\text{-}4)$$

so that the equation of state becomes

$$P\mathbf{V} = nR\theta. \qquad (3\text{-}5)$$

This does not yet define θ uniquely, since we can multiply it by an arbitrary constant, provided that we divide the *gas constant*, R, by the same. Therefore, the additional stipulation was made:

$$\theta_{\substack{\text{normal boiling}\\ \text{point of water}}} - \theta_{\substack{\text{normal freezing}\\ \text{point of water}}} = 100, \qquad (3\text{-}6)$$

so that the *size* of a degree on the ideal gas scale became the same as that in the centigrade system. Thus everything is fixed, as least in principle.

In 1954, the International Committee on Weights and Measures chose

$$\theta_{\substack{\text{triple point}\\ \text{of water}}} = 273.16,$$

which makes

$$\theta_{\substack{\text{normal freezing}\\ \text{point of water}}} = 273.15.$$

The gas constant then becomes

$$\begin{aligned} R = 0.082058\ \ell\ \text{atm deg}^{-1}\ \text{mole}^{-1} &= 1.9872\ \text{cal deg}^{-1}\ \text{mole}^{-1} \\ &= 8.3143\ \text{J deg}^{-1}\ \text{mole}^{-1}. \end{aligned}$$

In this way we can define an *ideal gas temperature*, θ, without recourse to the First and Second Laws of thermodynamics. On the other hand, the introduction of the *absolute, thermodynamic, or Kelvin temperature*,* T, is intimately tied to the development of the Second Law. This will be discussed in section 12.2 within the "traditional" framework, and in section 13.2 along the lines of the "axiomatic" method. Subsequently, in either scheme, we shall *prove* that

$$T \equiv \theta; \tag{3-7}$$

that is, *the two temperatures are identical.* Purists would therefore insist on couching the presentation of the first eleven chapters entirely in terms of t and θ. We shall follow most textbooks in *anticipating* the result (3-7), so that we can use T, more or less exclusively, from now on. Only in Chapters 12 and 13 will it become necessary to reintroduce briefly t and θ.

3.4 Heat Capacities and (Quantity of) Heat

Following most authors, "quantity of heat"† will henceforth be called, simply, "heat" and will be denoted by Q. If one adopts the axiomatic formulation, Q is introduced within the framework of the First Law. This procedure is discussed critically in section 9.3. With the temperature concept *and* the notion of heat at our disposal, the definition of *heat capacities* becomes trivial.

In the "traditional" approach there are a number of options. Most authors will define Q first. This, in turn, can be achieved in several ways, some of which do not even involve the temperature explicitly. Thus Q can be determined by means of a phase transition, e.g., from the volume change in an "ice calorimeter." Or, utilizing Joule's law of electricity, it is related to the effect of an electric current passing through a known resistance over a given time interval.

The latter procedure is consistent with the modern *definition* of the calorie, in terms of the (international) joule:

$$1 \text{ (gram) calorie} \equiv 1 \text{ cal} \equiv 4.184 \text{ J } \textit{exactly}.$$

This definition utilizes the equivalence of heat and work and thus, in a way, anticipates the First Law of thermodynamics. "Classically," the calorie was defined as the amount of heat needed to raise the temperature of 1 g of water from 14.5°C to 15.5°C.

---

* Most authors use these terms interchangeably, but some make rather subtle distinctions.

† Some prefer "heat flow."

There is an interesting, though little known alternative, in which heat *capacities* are defined first.* Consider two homogeneous substances, A and B, which are brought into thermal contact with each other while being effectively isolated from everything else. Let us denote by $\delta T_{i(j)}$ the temperature change of substance i when brought into contact with substance j for a (fixed) short period of time. In general, $\delta T_{\text{A(B)}} \neq \delta T_{\text{B(A)}}$, and we define the *ratio of heat capacities*, $\mathbf{C}_\text{B}/\mathbf{C}_\text{A}$, by

$$\frac{\delta T_{\text{A(B)}}}{\delta T_{\text{B(A)}}} = -\frac{\mathbf{C}_\text{B}}{\mathbf{C}_\text{A}}. \tag{3-8}$$

Since $\delta T_{\text{A(B)}}$ and $\delta T_{\text{B(A)}}$ have opposite signs, the ratio $\mathbf{C}_\text{B}/\mathbf{C}_\text{A}$ is always positive. For three bodies, A, B, and D, one finds empirically, for the three pairs of thermal interactions,

$$-\frac{\delta T_{\text{A(D)}}}{\delta T_{\text{D(A)}}} = \frac{\delta T_{\text{A(B)}}}{\delta T_{\text{B(A)}}} \frac{\delta T_{\text{B(D)}}}{\delta T_{\text{D(B)}}}. \tag{3-9}$$

The existence of such a cycle relation indicates that, in general, $\mathbf{C}_i$ is a property of substance i as such, irrespective of the body j with which it is brought into contact. At the same time, all $\mathbf{C}_i$ turn out to be functions of the (initial) temperature.

Those readers who studied section 3.2 in its entirety will not fail to note that, once more, we assume that we know what is meant by "thermal contact," and that the validity of Eq. (3-9) is contingent upon the absence of *those* "special" interactions which also would have caused the violation of the principle of transitivity [Eq. (3-1)].

Since the $\mathbf{C}_i$ turn out to be proportional to the amount of i present, we define

$$\text{the } molar^\dagger \text{ } heat\ capacity\text{:} \quad C_i \equiv \frac{\mathbf{C}_i}{n_i} \tag{3-10a}$$

and

$$\text{the } specific\ heat\text{:} \quad C_i' \equiv \frac{\mathbf{C}_i}{w_i}. \tag{3-10b}$$

---

* This method was brought to the attention of the author by G. E. Uhlenbeck. He suggests that it can be traced back to Joseph Black's *Lectures in the Elements of Chemistry*, published in 1803 by J. Robinson. (See also the article by D. Roller referred to in the footnote on p. 17.) This procedure was also adopted by P. S. Epstein in his *Textbook of Thermodynamics*, Wiley, New York, 1907, p. 25.

† While there is a clear distinction between *molality* and *molarity* (cf. section 8.1), the terms *molal* and *molar* are used interchangeably through most of the literature. We shall use *molar* exclusively.

Here n_i and w_i denote the number of moles and the mass in grams of substance i, respectively. Up to this point, only *ratios* of heat capacities have been defined. To proceed beyond this, we have to select a particular one as a reference on which to base the desired unit. Classically, one sets C'_{H_2O} at 15°C, 1 atm, arbitrarily equal to 1 cal deg^{-1} g^{-1}. Once we have values of the heat capacities at our disposal, the last step defines the heat transferred in the process as

$$\delta Q_A \equiv \mathbf{C}_A \, \delta T_{A(B)} \quad \text{and} \quad \delta Q_B \equiv \mathbf{C}_B \, \delta T_{B(A)}. \tag{3-11}$$

Equation (3-8) then yields the obvious result,

$$\delta Q_A + \delta Q_B = 0. \tag{3-12}$$

Since all $\mathbf{C}_i$ are positive,* an increase in T_i is associated with a positive δQ_i. This is consistent with the thermodynamic sign convention in which Q is taken to be the heat absorbed *by* the system *from* the environment.

PROBLEMS

Problem 3-1

An ethanol thermometer is calibrated against an ideal gas thermometer at $t = 0°C$ ($\theta = 273.15°$; ice point) and at $t = 50°C$ ($\theta = 323.15°$). In this range ethanol exhibits a thermal expansion according to

$$\frac{1}{V}\left(\frac{\partial V}{\partial \theta}\right)_P = 4.4543 \times 10^{-3} - 2.6556 \times 10^{-5}\theta + 5.158 \times 10^{-8}\theta^2.$$

What is the reading, t, on the ethanol thermometer when $\theta = 298.15°$ (corresponding to 25°C)?

Problem 3-2

(a) How much heat is required to raise the temperature of the air in an otherwise empty room (dimensions 8m × 4m × 3m) from 295 K to 305 K? You may assume that the air behaves as an ideal gas. Over the temperature range involved, the average heat capacity, at atmospheric pressure, may be taken as 21 J K^{-1} mole^{-1}.

(b) How many minutes will it take a 2-kW electric heater to provide this heat?

---

* This is not only intuitively clear (when a system *absorbs* heat, its temperature will *increase*), but it may also be proven explicitly for stable systems (see Chapter 17).

Chapter 4

Systems, States, and Variables; Work

4.1 Some Additional Means of Classifying Systems

The idea of a thermodynamic *system*, as separated from its surroundings, has already been introduced in section 1.4, in which we also became familiar with the distinction between *open*, *closed*, and *isolated* systems. We now consider two additional useful ways of classification.

First, when a system contains a single phase, we shall call it *homogeneous*; multiphase systems are *heterogeneous*.* In most cases it is relatively easy to determine the *number of* phases, although it is much more difficult to give a precise definition of the concept of a phase as such. One possibility is: "Phases are macroscopic regions, exhibiting different properties (such as density, refractive index, etc.), separated by sharp boundaries ('phase boundaries')." Or, "Phases are the different parts into which the system, at least in principle, can be divided *mechanically*." In this context, *mechanical* has a special meaning; it refers to procedures such as sorting out different crystals under a microscope, dividing immiscible liquids by means of a separatory funnel, separating a solid from a liquid by filtration, and so on. On the other hand, the separation of gases by means of effusion is considered a *physical* process, and separations by means of a sequence of *chemical* reactions are, of course, not "mechanical" either. The proper application of our, or any alternative, definitions leads to the following simple rules:

1. There are as many solid phases as there are *different*† crystalline substances; mixed crystals ("solid solutions") constitute a single phase.
2. In general, there is only one liquid phase. An exception arises when we encounter partial miscibility (separation into layers).
3. There is only one gas phase.

---

* The reader should be aware that some authors may assign somewhat different meanings to these terms. See also the section in small print below.

† *Different* crystals of the *same* substance of the *same* crystalline modification should be counted as a *single* phase.

A gas in a long vertical cylinder, in which certain properties (such as its density) vary *continuously* by virtue of its position in a gravitational field, still constitutes a single phase, hence is homogeneous in our terminology. We shall refer to such a system as *nonuniform*, to distinguish it from the idealized gaseous phase, in which the molecules are distributed uniformly throughout a containing vessel. Although this terminology is shared by several authors, alternatives do exist.

There are a few other special situations which may cause problems, for example, the counting of phases in the critical region of a fluid. Such cases will be dealt with as the need arises [see section 21.3(f)].

Second, systems are sometimes classified according to the number of (independent) *components*. Thus we distinguish *one-component* or *single-component* systems* (compare the heading of section 2.2), *two-component* or *binary* systems, *three-component* or *ternary* systems, and so on. In the early parts of this book, our purpose is adequately served by equating the various components with different chemical substances. A precise definition and a critical analysis of this concept are given in Chapter 21, where we discuss the "Phase Rule" and related topics.

4.2 States and Variables; Intensive and Extensive Properties

In phenomenological thermodynamics, the *state* of a system refers to the totality of its macroscopic, measurable, properties. It is *fixed* by a *limited number* of these, called the *state variables*, in the sense that when *these* re-assume a former set of variables, the system is again in the state associated with this set.† Usually, the selection of state variables is not unique; the existence of phenomenological relations, such as *equations of state*, gives us a certain flexibility in that it allows us to *change* variables if and when this becomes desirable. This can be accomplished by the use of standard mathematical techniques, which are summarized in section 6.3.

The separation of properties (or variables) into *extensive* and *intensive* is fairly common, although it occasionally leads to controversies. To illustrate the issues involved, we first confine our attention to a *homogeneous*, *uniform*, system, as defined in section 4.1, at equilibrium. By means of *imaginary* boundaries,‡ we arbitrarily subdivide the system into a number of macroscopic parts. Properties such as pressure, temperature, and (partial) *molar*

---

* It is unfortunate that these are occasionally called "pure" systems.

† This implies that the state of a system is not affected by its history. Phenomena such as *hysteresis*, associated with ferromagnetic materials, will not be considered in this book. Also, fluctuations are disregarded.

‡ Which should not be considered as creating *real* surfaces.

heat capacities that are the same in each of these regions, throughout the system, are called *intensive*. On the other hand, those designated as *extensive*, e.g., volume, energy, and *total* heat capacity, are obtained by summing the corresponding values over all parts.

Frequently, extensive properties are defined as those that are proportional to the *amount* (or *mass*) of the substance(s) involved. This characterization, although usually equivalent to the one given above, may go astray unless it is further qualified. For example, if we pour a fixed amount of water from a bottle onto a large flat tray, its surface area, normally considered an extensive property, increases drastically.

When we turn our attention to heterogeneous systems, some additional features arise. At equilibrium, *some* intensive properties, e.g., pressure and temperature, are still the same throughout the entire system, just as in the homogeneous case. But *other* intensive properties, such as (partial) molar volume and (partial) molar heat capacity, are (in general) different for different phases. Griffiths and Wheeler* have proposed to give different names to these two classes of intensive variables; they call the former *fields* and the latter *densities*. One may argue about the appropriateness of these *names*, but the distinction can be quite helpful, as we shall see in later chapters.

Evidently, the designation of variables as intensive or extensive may, in some cases, lead us into some difficulties. Since the distinction is never *essential*, some authors tend to deemphasize its use. Nevertheless, this author feels that the various terms introduced above can be quite *useful* in a large number of instances, where all controversies can be avoided. Hence readers should be familiar with this terminology.

4.3 Work; Introductory Remarks, with a Preliminary Discussion of Reversibility

Work, W, is defined in mechanics as the scalar product of the force, $\vec{F}$, exerted by the system on the surroundings, and the displacement of the boundary between them. For an infinitesimal displacement, $d\vec{r}$,

$$dW = \vec{F} \cdot d\vec{r}, \tag{4-1}$$

and for a macroscopic change between an initial state, 1, and a final state, 2, along a particular path, a, the total work becomes the line integral:

$$W = \int_{\substack{1 \\ (a)}}^{2} \vec{F} \cdot d\vec{r}. \tag{4-2}$$

* R. B. Griffiths and J. C. Wheeler, *Phys. Rev.* **A2**, 1047 (1970).

In mechanics, there is usually little argument about what $\vec{F}$ and $d\vec{r}$ pertain to, but unfortunately, as soon as we apply this standard definition to situations in thermodynamics, several controversies arise. These cover the entire spectrum, from fairly trivial points regarding *notation*, through internationally debated disagreements on *sign conventions*, to very subtle issues pertaining to both the *definition* and the *evaluation* of work in important special situations.

First, since the value of the integral (4-2) is *path dependent*, the notation dW on the left-hand side of Eq. (4-1) might be misleading. Integration does *not* yield ΔW ($\equiv W_2 - W_1$); in the language of section 6.2, *W is not a state function and dW is not an exact differential.* Some authors wish to remind us of these features at all times, by introducing a special symbol, $đW$ or DW. These same writers will also denote infinitesimal amounts of heat by $đQ$ or DQ. The present author joins many others who consider such notations more of a nuisance than an asset. Consequently, we shall continue to write dW and dQ, and the readers will have to remember that these, in general, do not represent exact differentials.

Second, for decades a battle has raged over the *sign convention* pertaining to W. In most books written before 1970, W represents the work done *by the system* (on the surroundings). However, in 1970, the International Union of Pure and Applied Chemistry (known as I.U.P.A.C.) recommended that W should stand for the "reverse," the work done (by the surroundings) *on the system.* Of course "W_{by}" = −"W_{on}."

There are good arguments both ways. Since the sign convention regarding Q is unambiguous (see p. 24), the I.U.P.A.C. scheme has the advantage that a positive W, just as a positive Q, increases the internal energy of the system. Its disadvantages appear in the formulation of what used to be known as *maximum work* theorems, which would now have to be rephrased in terms of *minimum work.* A consultation of several books, (re-)written since 1970, reveals a trend to abide initially by the I.U.P.A.C. convention, but refer to the work done *by* the system (sometimes given a special symbol) as the theorems concerned arise. In consulting the vast literature on thermodynamics, the readers should always be on guard regarding these matters.

After considerable soul-searching, and soliciting the advice of several colleagues and science editors, this author has opted to stay with the convention in which *W stands for the work done by the system* (on the environment) throughout.

To illustrate a third and much more substantial issue, we shall turn our attention to the familiar type of work associated with the expansion of a gas in a cylinder, frequently referred to as $P,\mathbf{V}$ work. Since the pressure is defined as the force per unit area, the reader will have no difficulty in formally converting Eqs. (4-1) and (4-2) to the well-known expressions

$$dW = P\,d\mathbf{V} \tag{4-3}$$

and

$$W = \int_1^2 P\,d\mathbf{V}, \tag{4-4}$$

respectively. Henceforth the path dependence of the work integrals will be implied.

Exercise 4-1

Derive Eqs. (4-3) and (4-4) from (4-1) and (4-2).

When an expansion occurs *reversibly*, P is well defined and the evaluation of W is straightforward, at least in principle. However, as we shall soon discover, the corresponding analysis of an *irreversible* expansion is no trivial matter. To explain this we shall have to give a very preliminary definition of reversibility, as it pertains to phenomenological thermodynamics. A comprehensive, critical discussion of this concept is given in Chapter 5.

In a reversible expansion the driving force is, at all times, only infinitesimally larger than the opposing force,* so that the system performs a maximum amount of work. The gas goes through a continuous sequence of (quasi-) equilibrium states and, by an infinitesimal change in forces, the entire process can be reversed along exactly the same path. *At each stage in this expansion P is completely determined by the equation of state of the gas*, so that the integration in the right-hand side of Eq. (4-4) can, at least in principle, be carried out. In contrast, the expansion of a gas, initially at a high pressure, into a vacuum is a "very irreversible" process, and it is by no means obvious what we should substitute for P in Eqs. (4-3) and (4-4). The difficulties concerned are discussed in detail in section 4.4.

Finally, in section 4.5 we focus our attention on other forms of work, besides the $P,\mathbf{V}$ type, and we end this chapter with an outline of some generalized formulations.

4.4 The Work Done by a Gas in an Irreversible Expansion

In a large adiabatically enclosed cylinder, a gas G at high pressure is separated from a gas G' at low pressure by a thin piston, p, initially held in place by a set of stops, S. The piston has a mass M_p and is impermeable to heat. Upon withdrawal of the stops, the piston moves without friction through

---

* All frictional forces should be negligible.

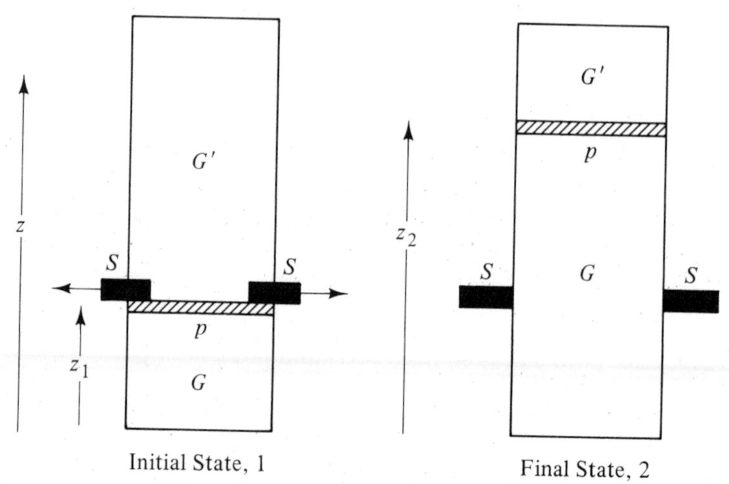

Fig. 1 Irreversible expansion of a gas, G.

the cylinder, until the final equilibrium state is reached (Fig. 1). Let G be our system, while G' *and* p constitute the important parts of the surroundings. As the piston is displaced from z to $z + dz$, the work done by the system is

$$dW = F_{Gp}\,dz, \tag{4-5}$$

in which F_{Gp} is the total force exerted on the piston by the gas G. *By definition*, we have

$$(P_{\text{system}} \equiv)P \equiv \frac{F_{Gp}}{\mathscr{A}}, \tag{4-6}$$

in which $\mathscr{A}$ is the cross-sectional area of the piston. The combination of Eqs. (4-5) and (4-6) yields the old result

$$dW = P\,d\mathbf{V}. \tag{4-3}$$

But while in the limit of a *reversible* process, as we explained in section 4.3, P is determined by the equation of state of gas G, this is no longer so if the expansion proceeds irreversibly. In the latter case Eq. (4-6) is of no help either, since F_{Gp} cannot be evaluated unless we know the details of the hydrodynamic flow pattern of the gas. Thus Eq. (4-3) becomes useless.

In many books this dilemma is apparently resolved by linking the entire process to an *external* pressure. In 1966, Kivelson and Oppenheim* carefully analyzed the procedure involved, for various types of irreversible expansions. In the rest of this section we essentially adapt their methodology to the specific process depicted in Fig. 1. If $F_{G'p}$ denotes the total force exerted

* See reference 5 at the end of the chapter.

on the piston by the gas G', Newton's Second Law for the motion of the piston can be written

$$M_p \frac{d^2 z}{dt^2} \equiv M_p \ddot{z} = F_{Gp} - M_p g - F_{G'p}. \tag{4-7}$$

Solving this equation for F_{Gp} and substituting the result into Eq. (4-5) yields

$$dW = (M_p \ddot{z} + M_p g + F_{G'p})\, dz. \tag{4-8}$$

With the identification

$$(P_{\text{external}} \equiv) P_{\text{ext}} \equiv \frac{M_p g + F_{G'p}}{\mathscr{A}}, \tag{4-9}$$

the external pressure exerted on the system, Eq. (4-8) becomes

$$dW = \left(P_{\text{ext}} + \frac{M_p \ddot{z}}{\mathscr{A}} \right) d\mathbf{V}. \tag{4-10}$$

Unless $\ddot{z} = 0$ (and a constant piston velocity is *not* a common characteristic of an irreversible expansion!), we see that

$$dW \neq P_{\text{ext}}\, d\mathbf{V}. \tag{4-11}$$

Upon integration of Eq. (4-10), we obtain

$$W = \frac{M_p}{\mathscr{A}} \int_1^2 \ddot{z}\, dz + \int_1^2 P_{\text{ext}}\, d\mathbf{V}. \tag{4-12}$$

But if K_p denotes the kinetic energy of the piston, we have

$$M_p \int_1^2 \ddot{z}\, dz = K_{p_2} - K_{p_1}. \tag{4-13}$$

Exercise 4-2

Prove Eq. (4-13).

Since the piston is at rest in both the initial and the final state, the right-hand side of Eq. (4-13) vanishes, so that Eq. (4-12) reduces to

$$W = \int_1^2 P_{\text{ext}}\, d\mathbf{V}, \tag{4-14}$$

even though the *in*equality (4-11) holds! The crucial question is under what circumstances the result (4-14) is any more useful than the "original":

$$W = \int_1^2 P\, d\mathbf{V}. \tag{4-3}$$

For the extreme case $P_{\text{ext}} = 0$ (the gas G pushes a weightless piston into a vacuum), we have a *free expansion* and $W = 0$ (see also the small-print section below). If $P_{\text{ext}} \neq 0$, we distinguish two cases:

Case I

Consider the limit $P_{G'} \to 0$, so that P_{ext} arises entirely from the weight of the frictionless piston. In this situation

$$P_{\text{ext}} = \frac{M_p g}{\mathscr{A}} \tag{4-15}$$

and

$$W = M_p g(z_2 - z_1), \tag{4-16}$$

which is just the increase in potential energy of the piston. Since equilibrium will be reached when

$$M_p g = P_{2,G}\mathscr{A},$$

in which $P_{2,G}$ is the *final* pressure of the gas, a well-defined system property, Eq. (4-16) can be written in the useful alternative form

$$W = P_{2,G}(V_2 - V_1). \tag{4-16$'$}$$

If a gas flows through a nozzle into a vacuum, it performs no work. Of course, this represents a very different experimental setup from the one we are considering. The latter would become equivalent to the former only if we assume (in addition to $P_{G'} \to 0$) that $M_p \to 0$. Unfortunately, a weightless piston is quite unrealistic! Could we get around the complications associated with the weight of the piston by turning the cylinder 90 degrees, i.e., by letting the expansion occur horizontally? Regrettably not; the gas would still have to do work in accelerating the piston, even though the potential energy of the latter would not change. (One could, of course, consider the piston *part of the system*, but then W no longer represents the work done *by the gas* as such.) Ultimately, the frictionless piston would be halted by the wall of the cylinder, or by a second set of stops. Upon impact, the kinetic energy of the piston would be transferred to that wall, or those stops, hence would be dissipated. [It would be possible to consider *both* the piston *and* the second set of stops as *part of our system*, with a further reinterpretation of the quantity W. For the details of the analysis and other elaborations, the readers are referred to the paper by Kivelson and Oppenheim (reference 5 at the end of the chapter).]

Case II

The pressure of the "external gas," G', is not negligible, as we have assumed through most of this section. In this situation, while in Eq. (4-14) we "manipulated F_{Gp} away," P_{ext} contains a term $F_{G'p}$ [see Eq. (4-9)] which we cannot evaluate any easier than F_{Gp} itself. As a result, W still cannot be computed without recourse to hydrodynamics. We reiterate that for reversible changes the situation is relatively simple because at all stages $\ddot{z} \approx 0$, so that the external pressure matches the well-defined system pressure, P, *throughout the process.*

4.5 Other Types of Work; Generalized Formulation

(a) In the preceding sections we focused our attention on $P,\mathbf{V}$ work. In thermodynamics we often encounter other types of work which are of considerable interest. Provided that the relevant changes occur *reversibly*, expressions for dW, analogous to Eq. (4-3), can usually be written down without much trouble. Let us consider a few representative examples:

1. The work done when a body of mass m, raises its height, h, in a gravitational field is

$$dW = -mg\,dh = -\mathscr{G}\,dh, \tag{4-17a}$$

in which $\mathscr{G}$ is the weight of the body [see Problem 4-1(b)].

2. The work done by a liquid, with surface tension γ, when its surface area, $\mathscr{A}$, is extended is

$$dW = -\gamma\,d\mathscr{A}. \tag{4-17b}$$

3. The work done by a wire, subjected to a tension τ, when its length, l, is increased is

$$dW = -\tau\,dl. \tag{4-17c}$$

It should be pointed out that only *elastic* deformations may be present; *plastic* deformations are irreversible in the thermodynamic sense.

4. The work done by a galvanic cell, potential $\mathscr{E}$, when its charge, q, is increased is

$$dW = -\mathscr{E}\,dq. \tag{4-17d}$$

For a discussion of the signs we refer back to section 4.3. It should be clear that whichever of the two possible conventions one adopts, the four terms on the right-hand sides of Eqs. (4-17) must be given the *opposite* sign from the corresponding one in (4-3). For, while a gas *performs* work (on the surroundings) if it expands (i.e., increases its volume), work must be put in (from the outside) to increase the charge of a battery, the length of a wire, etc.

Processes occurring in electric or magnetic fields will not be emphasized in this book. The formulation of the correct expressions for the work performed under these conditions is considerably more difficult than the analysis of the illustrative examples above, and can best be given in the context of specific situations if and when these arise in thermodynamics (see, e.g., section 14.3). For a good general discussion of this subject, the reader is referred to the literature at the end of this chapter, in particular references 6, 7, and 8.

(b) Sometimes it is desirable to have a general notation which covers all possible types of work. Thus we write

$$dW = \sum_i X_i dx_i, \tag{4-18}$$

in which it is understood that the X_i should be given the proper sign. For example, if $x_1 = \mathbf{V}$, then $X_1 = P$ and if $x_2 = \mathscr{A}$, then $X_2 = -\gamma$. In Eq. (4-18) the x_i are *generalized coordinates* (also known as displacement coordinates, work coordinates, or deformation coordinates) and the X_i the corresponding (or "conjugate") *generalized forces*.

Usually, in the terminology of section 4.2, forces are intensive and coordinates extensive variables, but there is no unique correspondence. Redlich* has given an interesting example: In a galvanic cell, the voltage is intensive and the charge extensive. If we arrange two such cells in series, the voltage becomes extensive and the charge intensive.† On the other hand, the voltage is *always* the force and the charge *always* the coordinate.

PROBLEMS

Problem 4-1

(a) Use Eq. (4-3) to prove that for the reversible isothermal expansion of 1 mole of an ideal gas,

$$W = RT \ln \frac{V_2}{V_1}.$$

(b) Compute the work done in joules on a mass of 1.00 kg when it is lifted a distance of 1.00 m in a gravitational field ($g = 9.81$ m sec^{-2}).

(c) What is the maximum height to which this mass can be lifted in such a field through the utilization of the work produced in the reversible expansion of 1 mole of an ideal gas, at 300 K, with $V_2 = 2V_1$?

(d) How long should a current of 1 A flow across a 5-V potential difference to yield the same amount of work?

---

* See reference 3 at the end of the chapter.

† In the Wheeler–Griffith terminology mentioned in section 4.2, the charge now becomes a "density," *not* a "field."

Problem 4-2

Five moles of an ideal gas, initially occupying a volume of 10 ℓ at a pressure of 13 atm, are expanded isothermally to a final pressure of 1 atm:

(i) by suddenly reducing a well-defined external pressure, P_{ext}, from 13 atm to 1 atm;

(ii) by reducing P_{ext} in two stages, first from 13 atm to 7 atm, and then from 7 atm to 1 atm;

(iii) by a similar process occurring in four stages, with intermediate P_{ext} of 10, 7, and 4 atm; and

(iv) reversibly.

(a) Compute W for the four expansions.

(b) Draw a P,**V** diagram and identify the areas representing these amounts of work.

(c) Note (numerically* as well as graphically) that W_{rev} represents the maximum amount of work we can get out of this overall process, and that this amount is approached more and more as the number of stages in the irreversible expansion is increased.

REFERENCES

1. Otto Redlich, "Generalized Coordinates and Forces," *J. Phys. Chem.* **66**, 585 (1962).
2. Otto Redlich, "Fundamental Thermodynamics since Carathéodory," *Rev. Mod. Phys.* **40**, 556 (1968).
3. Otto Redlich, "Intensive and Extensive Properties," *J. Chem. Educ.* **47**, 154 (1970).
4. Otto Redlich, *Thermodynamics*, Elsevier, Amsterdam, 1976, Chapter 1.

References 1 and 3 obviously pertain to the subject matter of this chapter, while 2 and 4 are broader in scope. Redlich's point of view is not necessarily always shared by the author, but his presentation is certainly interesting and thought provoking.

5. Daniel Kivelson and Irwin Oppenheim, "Work in Irreversible Expansions," *J. Chem. Educ.* **43**, 233 (1966). A very thorough analysis of the subject matter of section 4.4.
6. E. A. Guggenheim, *Thermodynamics*, 3rd ed., North-Holland, Amsterdam, 1957, Chapters 11 and 12.
7. A. B. Pippard, *Elements of Classical Thermodynamics*, Cambridge University Press, Cambridge, 1957, pp. 23–28.
8. Richard Becker, *Theory of Heat*, Springer-Verlag, Berlin, 1967, pp. 9–11.

The last three references are recommended in particular for their discussion of work in electric and magnetic fields.

---

* The reader who is in the position to operate a small programmable computer can repeat the calculations with an increased number of intermediate stages, and plot W against this number.

Chapter 5

Reversibility

5.1 Introduction

In this chapter we present a critical discussion of reversible and quasi-static processes, two concepts that are very closely related, though of very different origin. On the one hand, in 1824, Sadi Carnot published his famous "Reflections on the Motive Power of Fire,"* in which, as Ubbelohde so aptly put it, "... the ideas of *reversibility* ... were first introduced almost as a by-product in an engineering argument."† On a very different level, in 1909, the mathematician Carathéodory introduced the concept of a *quasi-static* process, as an integral part of the "axiomatic" formulation of the Second Law.‡ While surveying the relevant definitions which are favored by a number of specialists in the field, the author discovered a striking lack of agreement, which can best be illustrated by the following quotations from three outstanding monographs:§

1. Buchdahl: "... Quasi-static transitions are in fact reversible, but it is by no means obvious that, conversely, all reversible transitions of thermodynamic systems must be quasi-static .... "
2. Callen: "... Every reversible process coincides with a quasi-static one."
3. Kestin: "... All reversible processes are quasi-static but the converse is not true."

Although a generally accepted nomenclature is clearly desirable, this lack of agreement, with its resulting confusion, is not "merely" semantic; there are some underlying "real" problems, to which much of this chapter is devoted.

---

* For a translation in English, see, e.g., J. Kestin, ed., *The Second Law of Thermodynamics*, Dowden, Hutchinson & Ross, Stroudsburg, Pa., 1976, p. 16.

† A. R. Ubbelohde, *Man and Energy*, George Braziller, New York, 1955, p. 142.

‡ C. Carathéodory, *Math. Ann.* **67**, 355 (1909). For the only English translation known to the author, see Kestin, *The Second Law*, p. 229.

§ H. A. Buchdahl, *The Concepts of Classical Thermodynamics*, Cambridge University Press, Cambridge, 1973, p. 12; H. B. Callen, *Thermodynamics*, Wiley, New York, 1963, p. 63; J. Kestin, *A Course in Thermodynamics*, Blaisdell, New York, 1966, Vol. I, p. 134.

As a "preamble" to our brief analysis, the following should be kept in mind. First and foremost, as should already be evident from the preliminary discussion in section 4.3, a reversible change cannot be realized experimentally. It is an idealized concept, a "model" *process*, so to speak, just as an ideal gas is a "model" *system*. Consequently, just as an ideal gas had to be defined, reversibility *has to be defined* and, strictly speaking, no one can claim to have the one and only *correct* definition. These remarks are equally valid for a quasi-static change. The important question is which definitions lead to the most *useful* concepts for the subsequent development of phenomenological thermodynamics, as it will be taken up in the second part of this book. Specialists disagree on how to answer this question, and at least one prominent author in the field, Redlich,* deemphasizes the use of the reversibility concept as much as possible. He considers the notion accessorial rather than essential, but this point of view has, as yet, found few followers. The disagreements on all these related issues have led to considerable controversy, which is likely to remain with us for some time to come.

5.2 Quasi-Static and Infinitely Slow

In anyone's definition, a *quasi-static* process is an idealized one which carries a specific system, at an infinitesimal rate, through a continuous sequence of well-defined states. Most authors, following Carathéodory, add the restriction that work shall be done only by those forces which hold the system in equilibrium. This stipulation excludes all processes in which internal frictions are present whose effects do not go to zero in the infinitely slow limit. Other authors, albeit a minority, do not add such a restriction, so that the concept takes on a very different meaning. Examples of processes which *are* quasi-static in the wider sense, but *not* in the narrower sense, are the discharge of a condenser through a very high electrical resistance, and the flow of heat from a higher to a lower temperature through a piece of material with very large heat resistance. Buchdahl uses the term *pseudo-static* to include these cases, so that quasi-static transformations represent a special class of pseudo-static ones, special in the sense that frictional forces are excluded. Whether or not one likes Buchdahl's *nomenclature*, his *distinction* is a most pertinent one. To illustrate this, let us consider the flow of heat from a temperature T_2 to a temperature T_1 ($T_2 > T_1$), through a wire with a certain heat resistance. In general, this process is neither pseudo-static nor quasi-static. As the heat resistance is made very large, the process becomes infinitely slow and pseudo-static, but it is still *not* quasi-static in the Carathéodory sense. It acquires the latter characteristic only if we go to the limit $T_2 \rightarrow T_1$, irrespective of the magnitude of the heat resistance of the wire.

---

* O. Redlich, *J. Chem. Educ.* **52**, 374 (1975) and private communication.

In the rest of this book we shall rarely have cause to refer to pseudo-static changes that are not quasi-static as well. *If* such a situation arises, the simple designation "infinitely slow" is preferred. As will become clear, *these* processes are *irreversible* in everyone's terminology. From now on, we shall always use the term *quasi-static* in the aforementioned *restricted* sense of this concept, as originally proposed by Carathéodory. The relation between the quasi-static nature of a process and its reversibility is much more complex. This problem is discussed in the next section.

5.3 The Two Faces of Reversibility

(*a*) *Reversibility as Retraceability*

According to this point of view, a reversible process is defined as one that proceeds, in the absence of frictional forces, along a continuous sequence of quasi-equilibrium states, so that it *can return along exactly* the same path.* This implies that at each stage of the reversal, *both* the system *and* the environment reassume their former values, so that upon the completion of the reverse process both are restored to their initial conditions. This definition is favored by most authors, although their wordings may differ slightly. It is also the present author's choice and was, in essence, used in section 4.3, where we discussed certain corollaries regarding the work done in a reversible process. At that time we stressed that in such a change the system, at all stages, faces the largest possible opposing external force, so that it performs a maximum amount of work on the environment.

To characterize this point of view unambiguously, and to distinguish it from the alternative, presented below, it is suggested that the processes concerned could be designated as *retraceable.* It appears that a *retraceable change is always quasi-static in the restricted sense* of section 5.2.

(*b*) *Reversibility as Recoverability*

In this point of view, a process from state A to state B is called reversible if there exists *some* path back from B to A (*not necessarily the one along which the forward process took place*) which restores the initial conditions in *both*

---

* A few authors consider this qualification too strong, and insist that we should allow the reverse process to exhibit *infinitesimal* deviations from the forward path. Such deviations should stay within the experimental limits of accuracy for the characterization of the (macroscopic) changes involved.

the system *and* the environment. Note that in this definition *all reference to intermediate stages is eliminated; the nature of the initial and the final state is decisive.* A relatively small but prominent group of authors has chosen to interpret the reversibility concept in this fashion. Since it is the recoverability of the original situation that counts, not the detailed reversal of the path, Bridgman suggested that such a process should be called *recoverable.**

As far as the author is aware, the first textbooks to give a *critical* discussion of these matters were those of Planck† (1897) and Duhem‡ (1902). In the terminology introduced above, Planck favored recoverability, and Duhem retraceability. In effect, Duhem anticipated Carathéodory's quasi-static processes. This is fascinating territory, as yet insufficiently explored by historians of science.

(c) *Retraceability versus Recoverability*

From an examination of the preceding two definitions, it should be clear that a retraceable process is always recoverable; there is at least one path (the retraceable one) which restores the initial conditions as required. On the other hand, it is by no means evident that recoverable transitions are necessarily retraceable. In the opinion of the author, the current situation may be assessed as follows:

1. No one has *proven* that any recoverable process *must be* retraceable, but at the same time
2. no one has been able to give a single meaningful example of a recoverable process that *is, in fact,* not also retraceable.

Perhaps this unsatisfactory state of affairs is best summed up by Münster, who calls the two concepts "*pragmatically the same, but different from the point of view of logic.*"§ Duhem was right when he called reversibility "one of the most delicate principles in all of thermodynamics."‖ Introductory students may wish to disregard some of these subtleties, and accept the pragmatic equivalence of the two interpretations without further ado. Advanced

---

* P. W. Bridgman, *The Nature of Thermodynamics*, Harvard University Press, Cambridge, Mass., 1941, p. 122.

† M. Planck, *Treatise on Thermodynamics*, 3rd English edition, translated from the 7th German edition by A. Ogg, Dover, New York, 1926, p. 84. (The first German edition was published in 1897.)

‡ P. Duhem, *Thermodynamics and Chemistry*, G. K. Burgess, trans., Wiley, New York, 1903, p. 67. (The first French edition was published in 1902.)

§ A. Münster, *Classical Thermodynamics*, E. S. Halberstadt, trans., Wiley-Interscience, New York, 1970, p. 20.

‖ See footnote ‡ above.

students and researchers in the field should be alerted to these issues, or else they will be thoroughly confused in consulting the vast literature concerned. (See, e.g., the three quotations at the beginning of this chapter.)

Exercise 5-1

Interpret the three statements by Buchdahl, Callen, and Kestin (see section 5.1) in the light of our analysis in sections 5.2 and 5.3. Remember that there are two ways to define quasi-static (section 5.2) *and* two ways to define reversibility (section 5.3).

5.4 A Few Examples; Internal Reversibility; Final Comment

(a) All spontaneous, "natural"* changes are irreversible; reversible processes must be looked upon as idealized limits. This was made clear in Chapter 4 when we discussed different types of expansions of gases. We referred to the flow of a gas into a vacuum as "very irreversible." Within the framework of the present chapter, it is easily seen that this process is indeed neither retraceable nor recoverable. It is not retraceable because the gas will not, *by itself*, retrace its steps, i.e., retrench into its original volume. To restore the initial conditions of the *system*, we must compress it. This involves the expenditure of work, which results in a change in the *environment*, whence the process is not recoverable either. In turning to other examples, we can use the accepted pragmatic equivalence of retraceability and recoverability to choose only one of those characteristics in verifying whether a process is reversible or not. In this context, the author usually favors retraceability as the more *convenient* criterion, but this is a matter of personal preference.

Another obvious example of an irreversible process is the spontaneous mixing of two gases, brought about by punching a hole in a partition between them. It is obvious, because the process cannot return *along the same path*; the two species will not spontaneously retreat into those parts of the system they initially occupied alone. An intriguing question is whether the mixing of two (or more) gases can be carried out reversibly. We discuss this problem in section 23.3.

Exercise 5-2

Design a thought experiment which would restore the system, i.e., the gaseous *mixture*, to its original state of *separated* gaseous species. Check whether this process brings about changes in the environment.

---

* A term favored, in this context, by Max Planck.

(b) A very important and particularly realistic example of a reversible process is a chemical reaction occurring in an appropriate* galvanic cell with electromotive force (EMF), $\mathscr{E}$, subjected to an opposing potential $\mathscr{E}'$, which (e.g., by means of a potentiometer bridge) is made to approach $\mathscr{E}$ arbitrarily closely. Once more (cf. section 4.3) note that the *reversible* discharge, facing a maximum opposing force, delivers a *maximum* amount of work. It should be clear that *at all stages*, an infinitesimal increase in $\mathscr{E}'$ may stop the discharge, and a further infinitesimal increase will force the cell reaction to go in the opposite direction. Thus a reversible process is "under control" at all times. This feature was encountered when we discussed the expansion of gases in Chapter 4. If we go to the other extreme and simply "short" the two electrodes of our cell, a "very irreversible" discharge will take place, associated with (an) erratic chemical reaction(s), such as the disintegration of a shorted car battery. Irreversible processes are "uncontrolled," *dissipative.*

Exercise 5-3

The reaction $H_2 + Cl_2 \rightarrow 2HCl$ is initiated by subjecting a gaseous mixture of H_2 and Cl_2 to a photolytic flash.

(a) Convince yourself that this is an *irreversible* process in the thermodynamic sense (although, of course, there are ways to decompose HCl again into its elements).

(b) Design an experiment in which the same reaction can be carried out *reversibly.*

(c) Phase transitions in one-component systems provide further instructive examples of both reversible and irreversible changes. First consider

$$\text{1 mole (supercooled) water} \xrightarrow[\text{1 atm}]{-10^\circ\text{C}} \text{1 mole ice.} \qquad (5\text{-}1)$$

This process is, of course, irreversible; ice will not melt again *under these exact conditions.*

Exercise 5-4

How would you carry out the same overall process (5-1) in a reversible way?

On the other hand, the transition

$$\text{1 mole ice} \xrightarrow[\text{1 atm}]{0^\circ\text{C}} \text{1 mole water} \qquad (5\text{-}2)$$

can be carried out *reversibly* and is a standard example of a reversible change. Actually, this case is somewhat tricky, since reaction (5-2) can never take place unless the latent heat of melting is supplied. This could be done, for

---

* A critical discussion of the reversibility in galvanic cells is given in section 15.4.

example, by placing the reaction container inside a thermostat with $t > 0°C$, but this results in an irreversible heat flow, hence an irreversible overall process, unless we go to the limit $t \to 0°C$. Once more, the reversible process is approached as the limit of a "real" one.

It should be clear that in the preceding example, the ice–water system can pass through a continuous sequence of *internal* equilibrium states without necessarily being in equilibrium with the thermostat. Guggenheim* has suggested a special nomenclature for such a situation: Although the overall *process is irreversible*, the ice–water transition at 0°C is said to constitute a *reversible change*. Realistically, most of us are unwilling to give up the habit of using the terms *change* and *process* synonymously. Therefore, we reject Guggenheim's *nomenclature*, although his *analysis* is significant. Following several others, the present author suggests referring to such ice–water transitions as *internally reversible* processes.

The distinction introduced in the preceding paragraph is relevant to several other examples. To give just one, an *internally reversible* process may be initiated by the application of an external field when the latter is not regenerated by the reverse process. Since the exact initial conditions are not restored in the environment, the overall process, once more, is irreversible. A few authors have attempted to reintroduce the term *quasi-static* in *this* restricted sense, i.e., synonymous with internally reversible. This can only lead to further confusion.

The examples given in this section suffice to reemphasize the important features of the conceptual framework, which will be needed in subsequent parts of this book. This chapter does not contain an exhaustive compilation of all "personalized" definitions and "unique" terminologies that can be found throughout the literature on phenomenological thermodynamics. We hope that it will provide readers with a guide that will enable them, when the need arises, to find their way through most of the existing semantic labyrinth. We are sure that the last words on the subject of reversibility have not yet been written.

---

* E. A. Guggenheim, *Thermodynamics*, 3rd ed., North-Holland, Amsterdam, 1957, sections 1.14 and 1.15.

Chapter 6

State Functions and Exact Differentials; Mathematical Manipulations

---

6.1 Introduction to Chapters 6 and 7

This book is designed to emphasize the *traditional* approach to phenomenological thermodynamics (see section 1.3) and to present this material with a minimum of mathematical sophistication. The necessary background should have been acquired previously in mathematics courses, for which the following pages are no adequate substitute. Thus Chapter 6 merely summarizes the most important concepts and techniques, without any claim to comprehensiveness or mathematical rigor. After introducing some general formalisms, illustrations and exercises will be couched, as much as possible, in terms of thermodynamically important examples. Those readers who encounter serious problems, have the option to (re-)turn to formal instruction in this area, or to augment the study of this chapter by consulting some of the references given at the end. Mastery of this material is absolutely essential for a proper understanding of what follows.

Chapter 7 is of a quite different nature; it is *not* part of a typical science major's undergraduate curriculum in mathematics. It is specifically concerned with a mathematical preparation to the Carathéodory "derivation" of the Second Law, which in this book is given as an *alternative* approach (see Chapter 13), intended primarily for the more advanced readers. In Chapter 7 the author hopes to take some of the mystery out of "Pfaffian differentials," the "Carathéodory Theorem," and related topics, once more without giving a complete and rigorous treatment of the subject matter involved. Readers who do not intend to study Carathéodory's elegant method may skip this material altogether. On the other hand, those who wish to go beyond the brief discussion in Chapter 7 should consult the literature referred to at the end of that chapter.

6.2 State Functions and Exact Differentials; the Euler Reciprocal Relations

(a) If, for any thermodynamic system, $\mathbf{F}$ is an explicit function of the state variables* $y_1, y_2, \ldots, y_r$, it is called a *state function* or *function of state.*†
Frequently, $\mathbf{F}$ takes the form

$$\mathbf{F}(y_1, y_2, \ldots, y_r) = \mathbf{F}_0 + \mathbf{F}'(y_1, y_2, \ldots, y_r), \tag{6-1}$$

in which $\mathbf{F}_0$ is an arbitrary constant‡ to be fixed by convention. When $\mathbf{F}$ has the form (6-1), only $\Delta\mathbf{F}(\equiv \mathbf{F}_f - \mathbf{F}_i = \mathbf{F}'_f - \mathbf{F}'_i)$ is uniquely determined. Consequently, some prefer to give an alternative definition based on the macroscopic *change* in $\mathbf{F}$: $\mathbf{F}$ is a state function if $\Delta\mathbf{F}$ is completely determined by our (otherwise arbitrary) choice of initial and final states, i and f, hence is path independent. Important examples of such functions are $\mathbf{E}$ (internal energy), $\mathbf{H}$ (enthalpy), and $\mathbf{S}$ (entropy), as well as the "free energy functions," $\mathbf{A}$ and $\mathbf{G}$. In order to *compute* $\Delta\mathbf{F}$, we can always choose the *most convenient* path.

(b) We are assuming throughout that the function $\mathbf{F}$ and its derivatives are "well behaved," in particular that they are always continuous and single valued. With this stipulation, some important *corollaries to our definitions* follow:

1. $d\mathbf{F}$ is an *exact* or *total* differential. Hence the left-hand side of the identity

$$d\mathbf{F} = \sum_{k=1}^{r} \frac{\partial \mathbf{F}}{\partial y_k}\, dy_k \tag{6-2}$$

can be integrated between the states i and f to yield $\Delta\mathbf{F} \equiv \mathbf{F}_f - \mathbf{F}_i$, and obviously $\Delta\mathbf{F}_{i\to f} = -\Delta\mathbf{F}_{f\to i}$.

2. When the system returns to its original state, $\mathbf{F}$ must reassume its former value. Hence, *for any closed path,*

$$\oint d\mathbf{F} = 0. \tag{6-3}$$

3. Let $d\mathbf{F}$ be given in the form

$$d\mathbf{F} = \sum_{k=1}^{r} g_k(y_1, y_2, \ldots, y_r)\, dy_k. \tag{6-4}$$

---

* See section 4.2.

† Some authors use the designation *thermodynamic property* instead.

‡ This is true for several forms of energy. An obvious example is the potential energy of a body, of mass m, in a gravitational field in the z direction, where $\mathscr{V} = \mathscr{V}_0 + mgz$.

Then, since for any well-behaved function

$$\frac{\partial^2 \mathbf{F}}{\partial y_k \, \partial y_l} = \frac{\partial^2 \mathbf{F}}{\partial y_l \, \partial y_k} \qquad \textit{for all } k, l, \tag{6-5}$$

we have, by Eqs. (6-2), (6-4), and (6-5),

$$\frac{\partial g_k}{\partial y_l} = \frac{\partial g_l}{\partial y_k} \qquad \textit{for all } k, l. \tag{6-6}$$

Equations (6-6) are known as the *Euler reciprocal relations.* Without proof, we state that the validity of either Eq. (6-3) *for any closed path,* or Eq. (6-6) *for all k, l,* ensures, conversely, that **F** is a state function.*

(c) To clarify some features of this abstract formalism, let us select a system with two state variables, **V** and T. As we shall see in Chapter 9, the First Law tells us that the internal energy, **E**, is a state function. Hence, by our first and third corollaries, we can write down an expression for the exact differential $d\mathbf{E}$, as a function of the state variables, in the form

$$d\mathbf{E} = g(\mathbf{V}, T)\, dT + h(\mathbf{V}, T)\, d\mathbf{V}. \tag{6-7}$$

The functions g and h depend on the nature of the system at hand. In this example, only *one* Euler relation (6-6) is obtained:

$$\left(\frac{\partial g}{\partial \mathbf{V}}\right)_T = \left(\frac{\partial h}{\partial T}\right)_{\mathbf{V}}. \tag{6-8}$$

In comparison, we frequently come across an expression for dQ, which is *formally* similar to Eq. (6-7):

$$dQ = g'(\mathbf{V}, T)\, dT + h'(\mathbf{V}, T)\, d\mathbf{V}. \tag{6-9}$$

To understand the critical difference, it should be remembered that since Q is *not* a state function, dQ is *not* an exact differential (although we have not adopted a special notation such as $đQ$ or DQ). When properly interpreted, Eq. (6-9) makes perfect sense nevertheless. It tells us that *for a specific infinitesimal change of state* (from given **V**, T to given $\mathbf{V} + d\mathbf{V}$, $T + dT$), g' and h' are well defined, so that *dQ is completely determined.* Moreover, it is perfectly correct to write

$$g'(\mathbf{V}, T) = \left(\frac{\partial Q}{\partial T}\right)_{\mathbf{V}} \tag{6-10a}$$

---

* The mathematically inclined reader may wish to verify this statement. There is also a direct link between Eqs. (6-3) and (6-6) through Stokes' theorem (in the plane).

and

$$h'(\mathbf{V}, T) = \left(\frac{\partial Q}{\partial \mathbf{V}}\right)_T, \qquad (6\text{-}10b)$$

but since there is no state function $Q(\mathbf{V}, T)$, the crucial point is that

$$\left(\frac{\partial g'}{\partial \mathbf{V}}\right)_T \neq \left(\frac{\partial h'}{\partial T}\right)_{\mathbf{V}}. \qquad (6\text{-}11)$$

Exercise 6-1

For a fluid with state variables $\mathbf{V}$ and T, we shall obtain in section 10.1,

$$d\mathbf{E} = \mathbf{C}_v\, dT + \left(\frac{\partial \mathbf{E}}{\partial \mathbf{V}}\right)_T d\mathbf{V}$$

and

$$dQ = \mathbf{C}_v\, dT + \left[P + \left(\frac{\partial \mathbf{E}}{\partial \mathbf{V}}\right)_T\right] d\mathbf{V}.$$

$\mathbf{C}_v$ may be dependent on both $\mathbf{V}$ and T. *Given* the fact that $\mathbf{E}$ is a state function, *prove* that Q is *not*.

The more interesting applications of the Euler reciprocal relations involve the Second Law (see section 14.1).

6.3 Changing Variables and the Use of the Equation of State

(a) The notion of *independent* or *state variables* (these are equivalent in our terminology) has been introduced in section 4.2. It was stressed that the selection of such variables is rarely unique; because of the existence of an *equation of state*, there is a functional relationship between one dependent variable and a set of independent variables. The most common equation of state is a relation between P, $\mathbf{V}$, and T (for a fixed amount of material), but other types do exist and will be encountered later in this book [cf. Eq. (14-24)].

(b) In the examples of the preceding section, $\mathbf{V}$ and T appeared as state variables. Suppose that we wish to switch to the alternative set consisting of P and T. First, the *mere existence* of an equation of state, regardless of its specific form, allows us to write

$$d\mathbf{V} = \left(\frac{\partial \mathbf{V}}{\partial T}\right)_P dT + \left(\frac{\partial \mathbf{V}}{\partial P}\right)_T dP. \qquad (6\text{-}12)$$

Next, substitution of Eq. (6-12) into (6-7) yields

$$d\mathbf{E} = \left[g(\mathbf{V}, T) + h(\mathbf{V}, T) \left(\frac{\partial \mathbf{V}}{\partial T} \right)_P \right] dT + h(\mathbf{V}, T) \left(\frac{\partial \mathbf{V}}{\partial P} \right)_T dP. \qquad (6\text{-}13)$$

To obtain a more explicit result, we have to know (besides g and h) the *specific form* of the equation of state. As an example, take the van der Waals equation,

$$\left(P + \frac{n^2 a}{\mathbf{V}^2} \right) (\mathbf{V} - nb) = nRT, \qquad (6\text{-}14)$$

in which a and b are constants.* It is obviously inconvenient to express $\mathbf{V}$ explicitly in terms of P and T; hence it is difficult to evaluate $(\partial \mathbf{V}/\partial T)_P$ and $(\partial \mathbf{V}/\partial P)_T$ directly. To obtain these two partial differential quotients in such cases, we have several alternative procedures at our disposal. For example, we can make use of identities such as

$$\left(\frac{\partial \mathbf{V}}{\partial P} \right)_T = \frac{1}{(\partial P/\partial \mathbf{V})_T}, \qquad (6\text{-}15)$$

or we can transform the "cycle relationship"

$$\left(\frac{\partial P}{\partial T} \right)_{\mathbf{V}} \left(\frac{\partial T}{\partial \mathbf{V}} \right)_P \left(\frac{\partial \mathbf{V}}{\partial P} \right)_T = -1 \qquad (6\text{-}16)$$

to yield, e.g.,

$$\left(\frac{\partial \mathbf{V}}{\partial P} \right)_T = -\frac{(\partial T/\partial P)_{\mathbf{V}}}{(\partial T/\partial \mathbf{V})_P}. \qquad (6\text{-}17)$$

Exercise 6-2

Derive Eq. (6-16).

Exercise 6-3

Obtain $(\partial \mathbf{V}/\partial P)_T$ for a van der Waals fluid in three ways: (a) use Eq. (6-15), (b) use Eq. (6-17), and (c) differentiate Eq. (6-14) with respect to P at constant T.

(c) The equality (6-15), the equivalence of Eqs. (6-16) and (6-17), and the validity of identities such as

$$\left(\frac{\partial \mathbf{E}}{\partial T} \right)_{\mathbf{V}} \left(\frac{\partial T}{\partial P} \right)_{\mathbf{V}} = \left(\frac{\partial \mathbf{E}}{\partial P} \right)_{\mathbf{V}} \qquad (6\text{-}18)$$

---

* The physical significance of a and b is discussed in section 10.2.

strongly suggest that partial derivatives, in which the same variables are held constant, can be manipulated *as if* they were ordinary fractions. Although this is frequently true, great care should be exercised in this context, particularly with second derivatives. Thus

$$\left(\frac{\partial^2 \mathbf{V}}{\partial P^2}\right)_T \neq \frac{1}{\left(\frac{\partial^2 P}{\partial \mathbf{V}^2}\right)_T}. \tag{6-19}$$

Exercise 6-4

Show that

$$\left(\frac{\partial^2 \mathbf{V}}{\partial P^2}\right)_T = -\frac{(\partial^2 P/\partial \mathbf{V}^2)_T}{[(\partial P/\partial \mathbf{V})_T]^3}.$$

For a proper understanding of these matters, the author refers readers to textbooks on calculus. The essential point is that *differential* quotients (whether "regular" or "partial") are limits of *ordinary* quotients, and that frequently the two operations of manipulating the *latter* quotients and taking the appropriate limits can be reversed.

(d) Another type of relationship that is frequently needed is one between, e.g., $(\partial \mathbf{E}/\partial T)_P$ and $(\partial \mathbf{E}/\partial T)_\mathbf{V}$. From the expression for the total differential of $\mathbf{E}$ in terms of $\mathbf{V}$ and T,

$$d\mathbf{E} = \left(\frac{\partial \mathbf{E}}{\partial T}\right)_\mathbf{V} dT + \left(\frac{\partial \mathbf{E}}{\partial \mathbf{V}}\right)_T d\mathbf{V}, \tag{6-20}$$

one readily obtains

$$\left(\frac{\partial \mathbf{E}}{\partial T}\right)_P = \left(\frac{\partial \mathbf{E}}{\partial T}\right)_\mathbf{V} + \left(\frac{\partial \mathbf{E}}{\partial \mathbf{V}}\right)_T \left(\frac{\partial \mathbf{V}}{\partial T}\right)_P \tag{6-21}$$

and similarly,

$$\left(\frac{\partial \mathbf{E}}{\partial T}\right)_\mathbf{V} = \left(\frac{\partial \mathbf{E}}{\partial T}\right)_P + \left(\frac{\partial \mathbf{E}}{\partial P}\right)_T \left(\frac{\partial P}{\partial T}\right)_\mathbf{V}. \tag{6-22}$$

Exercise 6-5

(a) Verify Eq. (6-21) as indicated.
(b) Derive Eq. (6-22).
(c) If we add Eqs. (6-21) and (6-22), we obtain

$$\left(\frac{\partial \mathbf{E}}{\partial \mathbf{V}}\right)_T \left(\frac{\partial \mathbf{V}}{\partial T}\right)_P = -\left(\frac{\partial \mathbf{E}}{\partial P}\right)_T \left(\frac{\partial P}{\partial T}\right)_\mathbf{V}.$$

Prove this identity "directly," by the use of Eq. (6-16).

REFERENCES

The author does not wish to recommend regular calculus textbooks written for mathematics courses.

There are some books that are specifically intended to be mathematics reviews for scientists and engineers. Besides some very relevant chapters, these will obviously also contain much material that is not needed in the study of phenomenological thermodynamics. A "classic" in this area is

1. H. Margenau and G. M. Murphy, *The Mathematics of Physics and Chemistry*, Van Nostrand, New York, 1943, and many subsequent editions and printings. This book is comprehensive and a good reference, but it is terse and no easy reading for the uninitiated.

Many textbooks on thermodynamics, physical chemistry, etc., contain their own *ad hoc* survey of mathematical methods. The author only wishes to recommend those that are more elaborate than the summary given in the present chapter. One of his favorites is

2. S. M. Blinder, *Advanced Physical Chemistry*, The Macmillan Company, Collier-Macmillan Ltd., Toronto, Canada, 1969; in particular, *Part I: Mathematical Methods in Physical Chemistry*. Unfortunately, this book is out of print, but much of this material can also be found in the following article:
3. S. M. Blinder, "Mathematical Methods in Elementary Thermodynamics," *J. Chem. Educ.* **43**, 85 (1966).

For a rather sophisticated discussion, see also

4. J. Kestin, *A Course in Thermodynamics*, Blaisdell-Ginn, Waltham, Mass., 1966 (currently out of print), Vol. I, Chapter 3: "The Equation of State."

Chapter 7

Mathematical Preparation for the Carathéodory "Derivation" of the Second Law*

7.1 Pfaffian Expressions and Differential Equations; Integrating Factors (Denominators)

An expression of the form

$$dL = \sum_{i=1}^{r} X_i(x_1, x_2, \ldots, x_r)\,dx_i \equiv \sum_{i=1}^{r} X_i\,dx_i, \tag{7-1}$$

in which dL may or may not be an exact differential, is known as a *Pfaffian expression* in the r variables, $x_1, x_2, \ldots, x_r$. Corresponding to this expression we have the *Pfaffian differential equation*

$$dL = \sum_{i=1}^{r} X_i\,dx_i = 0. \tag{7-2}$$

This chapter, although written with a specific scientific purpose in mind, is couched largely in "pure" *mathematical* language. Thus the sums on the right-hand side of Eqs. (7-1) and (7-2) should not *necessarily* be thought of as having a one-to-one correspondence with similar *thermodynamic* expressions seen in Chapter 4. We distinguish three important cases:

Case I. dL is an exact differential.

For this to be true, the "(Euler) reciprocal relations"

$$\frac{\partial X_i}{\partial x_j} = \frac{\partial X_j}{\partial x_i} \tag{7-3}$$

---

* Optional reading. See the remarks in the second paragraph of section 6.1.

must be valid for all $i, j = 1, \ldots, r$. In Case I the expression (7-1) can be written

$$dL = \sum_{i=1}^{r} \frac{\partial L}{\partial x_i} dx_i, \tag{7-4}$$

and the corresponding differential equation (7-2) has solutions

$$L(x_1, x_2, \ldots, x_r) = C, \text{ a constant, acting as a parameter.} \tag{7-5}$$

Equation (7-5) defines a family of nonintersecting, one-parametric surfaces in the r-dimensional space of the x_i.

Theorem I. Let $P(x'_1, x'_2, \ldots, x'_r)$ be on the (specific) surface $L = C_1$. Let R be a point infinitesimally close to P in such a way that the displacements $dx_1, dx_2, \ldots, dx_r$, which take us from P to R, satisfy the Pfaffian equation (7-2). Then R will be on the same surface $L = C_1$.

Proof. We are given that

$$L(x'_1, x'_2, \ldots, x'_r) = C_1. \tag{7-6}$$

A displacement by $dx_1, dx_2, \ldots, dx_r$ in general will take us onto a new surface:

$$L(x'_1 + dx'_1, x'_2 + dx'_2, \ldots, x'_r + dx'_r) = L(x'_1, x'_2, \ldots, x'_r) + \sum_{i=1}^{r} \frac{\partial L}{\partial x_i} dx_i. \tag{7-7}$$

But the last term on the right-hand side of Eq. (7-7) is zero by Eqs. (7-2) and (7-4), since the dx_i's satisfy the Pfaffian equation. Hence

$$L(x'_1 + dx'_1, x'_2 + dx'_2, \ldots, x'_r + dx'_r) = L(x'_1, x'_2, \ldots, x'_r) = C_1. \quad \text{Q.E.D.} \tag{7-8}$$

Although *arbitrary* displacements from P will, in general, take us *off* the surface $L = C_1$ (onto a new surface, $L = C'_1$), Theorem I tells us that *any displacements from P satisfying the Pfaffian equation will keep us on the same surface* ($L = C_1$). Therefore, there are many points infinitesimally close to P that can never be reached from this point by displacements of this *special* type. One says that in this situation there are *many (neighboring) points "inaccessible from P."*

Case II. dL is not an exact differential but has an integrating factor (denominator).

We say that dL has an *integrating factor* $\lambda(x_1, x_2, \ldots, x_r)$ if $d\sigma = \lambda\, dL$ is an exact differential (while dL itself is not). The reciprocal of λ, $\tau(x_1, x_2, \ldots, x_r)$, is known as the corresponding *integrating denominator.* From the discussion of Case I it follows that

$$\lambda\, dL = \frac{dL}{\tau} = d\sigma = \sum_{i=1}^{r} \frac{\partial \sigma}{\partial x_i}\, dx_i, \tag{7-9}$$

and the solutions of the Pfaffian equation $dL = 0$ are identical with those of $d\sigma = 0$. Hence the latter solutions are also of the form

$$\sigma(x_1, x_2, \ldots, x_r) = C, \text{ a constant.} \tag{7-10}$$

This still describes a family of nonintersecting one-parametric surfaces, and Theorem I is still valid (replace L by σ in its formulation).

Theorem II. If a Pfaffian expression has one integrating factor (denominator), it has an infinite number of them.

Proof. Let Σ be an arbitrary function of σ; then the solutions $\Sigma(\sigma) = \Sigma(C)$ will describe the same family of surfaces as Eq. (7-10). We can write

$$d\Sigma = \frac{d\Sigma}{d\sigma}\, d\sigma = \frac{d\Sigma}{d\sigma}\frac{dL}{\tau}. \tag{7-11}$$

This is equivalent to

$$\lambda^*\, dL = \frac{dL}{\tau^*} = d\Sigma, \tag{7-12}$$

in which

$$\lambda^* = \lambda \frac{d\Sigma}{d\sigma}; \tag{7-13a}$$

$$\tau^* = \tau \frac{d\sigma}{d\Sigma}. \tag{7-13b}$$

Thus $\lambda^*(\tau^*)$ is also an integrating factor (denominator) of dL. Since we can form an infinite number of functions Σ, the proof is complete.

Case III. dL is not an exact differential and has no integrating factor.

As we shall see in section 7.2, for *arbitrary* Pfaffian expressions in three or more variables, this situation is by far the most common one. In other words, only in very few cases will such an expression have an integrating factor (denominator). It is extremely important to appreciate this *mathematical* fact, because in the "derivation" of the Second Law of thermodynamics the occurrence of very special *scientific* conditions guarantees that such integrating factors (denominators) *do always exist* (see Chapter 13).

7.2 Pfaffian Expressions in Two and Three Variables; Some Examples

In this section we show that:

(a) A Pfaffian expression in two variables always has an integrating factor (denominator) and hence, by Theorem II, an infinite number of them.
(b) A Pfaffian expression in three variables (and, by obvious generalization, in more than three variables) does not, *in general*, have such factors (denominators).

(a) Write the Pfaffian differential equation in the form

$$dL = X\,dx + Y\,dy = 0. \tag{7-14}$$

Assume that

$$\frac{\partial X}{\partial y} \neq \frac{\partial Y}{\partial x}, \tag{7-15}$$

i.e., dL is not an exact differential. If $\lambda(x, y)$ is an integrating factor, then $d\sigma$, given by

$$d\sigma = \lambda X\,dx + \lambda Y\,dy, \tag{7-16}$$

is an exact differential. Thus λ has to satisfy the single condition

$$\frac{\partial(\lambda X)}{\partial y} = \frac{\partial(\lambda Y)}{\partial x} \tag{7-17}$$

or

$$\lambda\frac{\partial X}{\partial y} + X\frac{\partial \lambda}{\partial y} = \lambda\frac{\partial Y}{\partial x} + Y\frac{\partial \lambda}{\partial x}. \tag{7-18}$$

This equation can always be solved for λ. (In practice this may or may not be an *easy* task. In simple cases the readers will be able to write down possible solutions "by inspection." Elaboration upon such matters falls beyond the scope of this chapter.)

Example 1

$$dL = dx + \frac{x}{y}\,dy = 0. \tag{7-19}$$

It is readily seen that the *inequality* (7-15) holds, hence dL is not an exact differential.

Equation (7-18) yields

$$\lambda \cdot 0 + 1 \cdot \frac{d\lambda}{dy} = \lambda \cdot \frac{1}{y} + \frac{x}{y}\frac{d\lambda}{dx}$$

or

$$\lambda = y\frac{d\lambda}{dy} - x\frac{d\lambda}{dx}. \tag{7-20}$$

Equation (7-20) has many solutions; the simplest one is

$$\lambda = y. \tag{7-21}$$

Correspondingly,

$$d\sigma = y\,dx + x\,dy = d(xy) = 0, \tag{7-22}$$

with solutions

$$\sigma = xy = C, \tag{7-23}$$

a family of equilateral hyperbolas. An alternative solution of (7-20) would be

$$\lambda^* = 2xy^2. \tag{7-24}$$

We can also arrive at this via the reasoning given in conjunction with the proof of Theorem II. Just take

$$\Sigma = \sigma^2. \tag{7-25}$$

Then by (7-13a),

$$\lambda^* = \lambda\frac{d\Sigma}{d\sigma} = y \cdot 2\sigma = 2xy^2.$$

Obviously, the alternative solutions are

$$\Sigma = x^2y^2 = C^* \,(= C^2), \tag{7-26}$$

describing the same curves as in Eq. (7-23).

(b) Let

$$dL = X\,dx + Y\,dy + Z\,dz. \tag{7-27}$$

Assume that at least one of the three reciprocal relations does not hold so that, once more, dL is not an exact differential. If an integrating factor $\lambda(x, y, z)$ exists, we get three relationships of the type (7-18):

$$\lambda\left(\frac{\partial X}{\partial y} - \frac{\partial Y}{\partial x}\right) = Y\frac{\partial \lambda}{\partial x} - X\frac{\partial \lambda}{\partial y}, \tag{7-28a}$$

$$\lambda\left(\frac{\partial Y}{\partial z} - \frac{\partial Z}{\partial y}\right) = Z\frac{\partial \lambda}{\partial y} - Y\frac{\partial \lambda}{\partial z}, \tag{7-28b}$$

and
$$\lambda\left(\frac{\partial Z}{\partial x}-\frac{\partial X}{\partial z}\right)=X\frac{\partial \lambda}{\partial z}-Z\frac{\partial \lambda}{\partial x}. \tag{7-28c}$$

Multiply these three equations by $-Z$, $-X$, and $-Y$, respectively, and add the three resulting equations to yield*

$$\lambda\left[X\left(\frac{\partial Z}{\partial y}-\frac{\partial Y}{\partial z}\right)+Y\left(\frac{\partial X}{\partial z}-\frac{\partial Z}{\partial x}\right)+Z\left(\frac{\partial Y}{\partial x}-\frac{\partial X}{\partial y}\right)\right]=0. \tag{7-29}$$

Since *in general* Eq. (7-29) has only the trivial (and for our purposes irrelevant) solution $\lambda = 0$, this means that the three equations (7-28) cannot be solved simultaneously. Hence it is the *exception* rather than the rule for dL, as given by (7-27), to have an integrating factor (denominator).

Example 2

$$dL = yz\,dx - xz\,dy - y^2\,dz = 0. \tag{7-30}$$

The readers should first check that not all reciprocal relations are satisfied (in fact, *none* of them is!). Next we note that equation (7-29) *does* hold in *this particular* case, with $\lambda \neq 0$, for $-y^2(z+z) + yz(-x+2y) - xz(0-y) = 0$. Without elaboration as to *how* one finds this (since it is not the purpose of this section), we give a possible simultaneous solution to the three equations (7-28) for this example:

$$\lambda = \frac{1}{y^2 z}. \tag{7-31}$$

Hence

$$d\sigma = \frac{1}{y}\,dx - \frac{x}{y^2}\,dy - \frac{1}{z}\,dz = 0 \tag{7-32a}$$

or

$$d\sigma = d\left(\frac{x}{y} - \ln z\right) = 0, \tag{7-32b}$$

with solutions

$$\sigma = \frac{x}{y} - \ln z = C. \tag{7-33}$$

---

* If we consider $\vec{R}(\vec{r})$ a vector with components $X(\vec{r})$, $Y(\vec{r})$, and $Z(\vec{r})$, Eq. (7.29) can be written

$$\lambda\{\vec{R}\cdot \text{Curl}\,\vec{R}\} = 0.$$

7.3 The Carathéodory Theorem

In section 7.1 we found that if a Pfaffian expression has an integrating factor, the solution to the corresponding differential equation is the set of surfaces (7-10), which implies that in the neighborhood of a point P on any one of those surfaces there are points "inaccessible from P." We repeat that the latter statement means that all those displacements from P, restricted by the condition

$$\sum_{i=1}^{r} \frac{\partial \sigma}{\partial x_i} dx_i = 0, \tag{7-34}$$

will never get us off the surface on which P is located. The *Carathéodory Theorem* (also known as the "Theorem of Accessibility") is just the "reverse" of this result:

> If a Pfaffian expression $dL = \sum_{i=1}^{r} X_i dx_i$ has the property that in every *arbitrarily close* neighborhood of any point P there are "inaccessible points," i.e., points that cannot be connected to P along curves which satisfy the equation $dL = 0$, then the Pfaffian expression concerned must have an integrating factor/denominator (and hence, by Theorem II, an infinite number of them).

We end this chapter with the following three remarks:

1. Readers should be aware that there are many equivalent formulations of the theorem, each expressed in a slightly different language.
2. It should be noted that we have only *proven* the "reverse" of the Carathéodory Theorem, not the theorem as such. This reflects our limited aims of the presentation in Chapter 7: ". . . to take some of the mystery out of 'Pfaffian differentials,' the 'Carathéodory Theorem,' and related topics, . . . without giving a complete and rigorous treatment of the subject matter . . . ." (cf. section 6.1). Those who wish to pursue these issues further are referred to the literature (p. 57).
3. Last, but not least, we should clearly distinguish between the purely mathematical "Carathéodory *Theorem*," discussed in this chapter, and the (Restricted) "Carathéodory *Principle*." As we shall see in section 13.1, the latter is a mathematical formulation of a fundamental fact of experience. Thus, in accordance with the discussion in section 1.2, this *principle* has also been designated as the "Carathéodory Statement of the Second Law." The readers should be alerted to the fact that the theorem and the principle are occasionally confused in the literature.*

---

* This has already been pointed out by Born; see reference 2 below, p. 143.

REFERENCES

1. One may go back to "the source" and study Carathéodory's 1909 paper in German, or Kestin's recent translation. (See footnotes, p. 36.)
2. Max Born has discussed this material in an article published in *Z. Physik* **22**, 249 (1921). See also his *Natural Philosophy of Cause and Chance*, based on lectures given at Oxford University in 1948 (Dover, New York, 1964), in particular pp. 38–43 and Appendix V, pp. 143–146.
3. A. Landé, *Handbuch der Physik*, Springer-Verlag, Berlin, 1926, Vol. IX, Chapter 4.
4. A very elaborate proof of the Carathéodory Theorem is given by S. Chandrasekhar in his *Introduction to the Study of Stellar Structure*, University of Chicago Press, Chicago, 1939, Chapter I.
5. The subject matter has also been discussed extensively by H. A. Buchdahl. See, e.g., *Am. J. Phys.* **17**, 44 (1949), and reference 3 of Chapter 1 of this book.
6. See also S. M. Blinder, reference 2 of Chapter 6.

Chapter 8

Partial Molar Quantities and Related Topics; Notation Conventions for (Reaction) Mixtures

8.1 Ways of Expressing the Composition of Mixtures and Solutions

All solutions are mixtures, but only *homogeneous* mixtures (section 4.1) are solutions. If we are interested in *both* the *composition* and the *amount* of our system, we must specify, for example, *all* the masses* w_i $(i = 1, \ldots, r)$, or *all* the mole numbers $n_i(i = 1, \ldots, r)$. Of course, the two are related through

$$n_i = \frac{w_i}{M_i} \qquad (i = 1, \ldots, r), \tag{8-1}$$

in which M_i is the mass of 1 mole of component i. (For an introduction to the notion of components, see the provisional discussion in section 4.1.) We shall write w and n for the *total mass* and the *total number of moles, respectively*:

$$w \equiv \sum_{i=1}^{r} w_i \tag{8-2}$$

and

$$n \equiv \sum_{i=1}^{r} n_i. \tag{8-3}$$

---

* The author does not intend to maintain religiously the distinction between weight and mass. Thus, in accordance with common usage, $|M|$ will usually be called the "molecular *weight*," a dimensionless quantity. We use the symbol w for mass, since m denotes *molality* (see p. 59).

In most instances we shall study those physical or chemical properties that are dependent only on the *composition, not* on the *amount* of the system in question. For this purpose it is sufficient to specify $(r - 1)$ *composition variables.* At this stage it becomes convenient to distinguish between two cases.

Case I. None of the components dominates; the distinction solvent–solute(s) is not relevant.

In this case we want a set of variables which allows us to express conveniently the composition of mixtures over the widest possible range of concentrations* of all components. We can use *weight fractions,*

$$z_i \equiv \frac{w_i}{w} \qquad (i = 1, \ldots, r), \tag{8-4}$$

mole fractions,

$$X_i \equiv \frac{n_i}{n} \qquad (i = 1, \ldots, r), \tag{8-5}$$

weight percents $(=100z_i)$, or *mole percents* $(=100X_i)$. Obviously,

$$\sum_{i=1}^{r} z_i = \sum_{i=1}^{r} X_i = 1, \tag{8-6}$$

whereas the percentages add up to 100. Equation (8-6) confirms that *only* $(r - 1)$ *composition variables are independent.*

Case II. One of the components dominates; the distinction solvent–solute(s) is relevant.

In this situation *we shall always designate the solvent by* 1. Next we specify the composition of the solution by appropriately chosen variables for the $(r - 1)$ solutes i $(=2, \ldots, r)$. In principle we could still deal with any of the four quantities defined in Case I, but only the *mole fractions,* X_i $(i = 2, \ldots, r)$, are widely used in this context. In adopting these (for reasons we do not wish to belabor), some authors would talk about the "*rational*" *basis* for expressing the composition. More frequently, we work with a "*practical*" *basis,* i.e., in terms of the *molarities,* c_i $(i = 2, \ldots, r)$, or the *molalities,* m_i $(i = 2, \ldots, r)$. The molarity is defined as the number of moles of the solute in question *per liter of the solution* and the molality is this number *per kg (1000 g) of the solvent.* The m_i should be preferred over the c_i because the

* In this book the term *concentration* will be used in a general, generic sense. The "concentrations" of elementary chemistry will be referred to as *molarities* (see the text under "Case II" below).

latter are temperature dependent. The X_i, m_i, and c_i are related as follows:

$$X_i = \frac{m_i}{\dfrac{1000}{M_1} + \sum_{k=2}^{r} m_k}$$

$$= \frac{c_i}{\left[(1000\rho - \sum_{k=2}^{r} c_k M_k)/M_1\right] + \sum_{k=2}^{r} c_k} \qquad (i = 2, \ldots, r), \qquad (8\text{-}7)$$

in which ρ is the density of the solution in kg/liter (or g/ml), and M_1 is the weight of one mole of solvent in g/mole.

Exercise 8-1

Derive Eq. (8-7).

It is apparent that in an arbitrary solution there is no simple relation between m_i and c_i. Suppose, however, that we are dealing with a solution that is *infinitely dilute in all solutes.** In this case $\rho \to \rho_1$, i.e., the density of the solution approaches that of the solvent, and Eq. (8-7) reduces to

$$X_i = \frac{m_i}{1000/M_1} = \frac{c_i}{1000\rho_1/M_1} \qquad (i = 2, \ldots, r). \qquad (8\text{-}8)$$

Hence for such an infinitely dilute solution

$$m_i = \frac{c_i}{\rho_1} \qquad (i = 2, \ldots, r). \qquad (8\text{-}9)$$

We see that even for this model $m_i \neq c_i$ in general; the two would become equal only if $\rho_1 = 1$ (e.g., for water at 4°C). Unfortunately, Eqs. (8-8) and/or (8-9) are sometimes used in concentration ranges where their validity is dubious at best.

8.2 Partial Molar† Quantities; the Gibbs–Duhem Equation

Let **J** denote a thermodynamic property of a homogeneous mixture, such as its volume, internal energy, enthalpy, entropy, or free energy, which is an *extensive* quantity (section 4.2) and a *state function* (section 6.2). This same

---

* This is a very important model system to be discussed in detail in the fourth part of this book (Chapter 26).

† Some authors prefer molal to molar; *in this context* the reader may consider the terms interchangeable.

property, **J**, per mole of pure component i will be denoted by $J_i^\bullet$ and called the *molar* volume, *molar* internal energy, etc.

In this section we wish to relate the various thermodynamic properties of solutions to those of its constituent components; *this entire analysis pertains to fixed P and T.* The essential problem is that, except in very few instances, **J** is *not additive in the components.* In other words, *the equation*

$$\mathbf{J} = \sum_{i=1}^{r} n_i J_i^\bullet \tag{8-10}$$

does not hold in general.

A simple example is provided by the volume of ethanol (ethyl alcohol)–water solutions; as most readers are aware, the mixing process results in a volume contraction

$$\mathbf{V} < n_{\text{eth}} V_{\text{eth}}^\bullet + n_{H_2O} V_{H_2O}^\bullet. \tag{8-11}$$

Thus our goal is to find some new quantities, to be denoted by $\bar{J}_i$, the *partial molar* volume, *partial molar* interal energy, etc., such that

$$\mathbf{J} = \sum_{i=1}^{r} n_i \bar{J}_i \tag{8-12}$$

can replace Eq. (8-10) in all possible mixtures of these components. The $\bar{J}_i$ reduce to the $J_i^\bullet$ in those few instances (certain model systems to be discussed in the fourth part of this book) when Eq. (8-10) *is* valid. While, in all cases of interest to us, the various $J_i^\bullet$ may be considered functions of P and T *only*, and while the $\bar{J}_i$ are *formally* written *as if* they were intrinsic properties of some particular component, it is extremely important to realize that these *partial molar quantities are also functions of the composition* of the mixture at hand.

To return to our ethanol–water example (at 25°C, 1 atm), for a solution containing 23% ethanol by weight, $\bar{V}_{\text{eth}} = 53.59$ ml and $\bar{V}_{H_2O} = 18.06$ ml. For a solution containing 68% ethanol by weight, these numbers are 57.70 and 16.90, respectively. This compares with $V_{\text{eth}}^\bullet = 58.37$ ml and $V_{H_2O}^\bullet = 18.02$ ml at the same temperature and pressure.

We claim that the desired new quantities can be defined by

$$\bar{J}_i \equiv \left(\frac{\partial \mathbf{J}}{\partial n_i}\right)_{P,T,n_j} \qquad (i = 1, \ldots, r), \tag{8-13}$$

in which the n_j in the subscript indicates symbolically that *in the differentiation, we should hold* (besides P and T) *all mole numbers, except n_i, constant.* We now have to prove that if the definition (8-13) is adopted, the desired equation (8-12) is indeed true. This can be done in two different ways, following methods independently suggested by J. W. Gibbs (1875) and P. Duhem

(1886). Although these two proofs are related in essence, they differ markedly in where the emphasis is placed, so it is instructive to give them both:

(a) *Duhem's method* makes optimal use of mathematics. One simply observes that all **J** are homogeneous functions in the mole numbers of degree one, and applies Euler's theorem.

$f(x_1, x_2, \ldots, x_r)$ is said to be *homogeneous of degree n* if for any constant, k,

$$f(kx_1, kx_2, \ldots, kx_r) = k^n f(x_1, x_2, \ldots, x_r). \tag{8-14}$$

For such a function *Euler's Theorem* holds:*

$$nf(x_1, x_2, \ldots, x_r) = \sum_{i=1}^{r} x_i \frac{\partial f}{\partial x_i}. \tag{8-15}$$

It is readily seen that for all (extensive) thermodynamic properties **J**,

$$\mathbf{J}(kn_1, kn_2, \ldots, kn_r) = k\mathbf{J}(n_1, n_2, \ldots, n_r). \tag{8-16}$$

For example, taking five times as much of *all* components increases the volume, internal energy, etc., by a factor of 5. Euler's Theorem thus gives

$$\mathbf{J} = \sum_{i=1}^{r} n_i \left(\frac{\partial \mathbf{J}}{\partial n_i}\right)_{P,T,n_j}. \tag{8-17}$$

With the definition (8-13), this indeed leads to the desired result (8-12).

(b) *Gibbs's approach* requires less mathematics, but more insight. Since at constant P and T, **J** is completely determined by the mole numbers, the expression for the total differential (section 6.2) reads

$$d\mathbf{J} = \sum_{i=1}^{r} \left(\frac{\partial \mathbf{J}}{\partial n_i}\right)_{P,T,n_j} dn_i. \tag{8-18}$$

With our definition (8-13) this becomes

$$d\mathbf{J} = \sum_{i=1}^{r} \bar{J}_i \, dn_i. \tag{8-19}$$

Integration of this equation immediately leads to the desired result (8-12). This procedure has often been misunderstood; critics will say that since (at constant P and T) the $\bar{J}_i$ are, *in general*, functions of the n_i, they should not be considered as constants in this integration.† However, in Gibbs's own words‡ we should carry out this integration "supposing the quantity of the

---

* The simple proof is given in many places; see, e.g., S. M. Blinder, reference 2 of Chapter 6, pp. 17–18, or reference 3 of Chapter 6, p. 87.

† In 1962 this procedure was referred to as a frequently occurring *textbook error* in *J. Chem. Educ.* **39**, p. 527.

‡ *The Scientific Papers of J. W. Gibbs*, Dover, New York, 1961, p. 87; a slightly different integration, but the principle involved is the same. For "compound substance" read "mixture."

compound substance considered to vary from zero to any finite value, *its nature and state remaining unchanged*" (the italics are the author's). To put it in another way, we should not think of preparing our solution by starting with the total amount of component 1, then adding 2, then 3, and so on, but we should imagine it being formed by *mixing all components simultaneously in constant (the final) proportions.** The following manipulation may further clarify the procedure. By Eq. (8-5), we can rewrite Eq. (8-19) as

$$d\mathbf{J} = \sum_{i=1}^{r} \bar{J}_i X_i \, dn, \tag{8-20}$$

in which n is the total number of moles [see Eq. (8-3)]. Integrate Eq. (8-20), at constant P and T, *and fixed composition* (all X_i and $\bar{J}_i$ constant!), from zero to an arbitrary final n:

$$\mathbf{J} = \sum_{i=1}^{r} \bar{J}_i X_i \int_0^n dn = \sum_{i=1}^{r} \bar{J}_i X_i n = \sum_{i=1}^{r} \bar{J}_i n_i, \tag{8-12}$$

the desired result. Many times when the author presented this procedure in his lectures, some student asked: "But what if I insisted on preparing my final mixture in some alternative way?" The answer is, of course, that since we know that $\mathbf{J}$ is a *state function*, its final value is independent of the manner in which we prepared our system, hence it is still given by Eq. (8-12). Gibbs very cleverly chose the *most convenient path* to arrive at this result [cf. section 6.2(a)].

Suppose that we ask beginning calculus students, who have *just* learned the technique (and hence have not forgotten it *yet*), to differentiate both sides of Eq. (8-12), disregarding the physicochemical meaning of the various symbols. We hope that they would write the correct general variation:

$$d\mathbf{J} = \sum_{i=1}^{r} n_i d\bar{J}_i + \sum_{i=1}^{r} \bar{J}_i \, dn_i. \tag{8-21}$$

But if we compare this result with Eq. (8-19), we are compelled to conclude that

$$\sum_{i=1}^{r} n_i \, d\bar{J}_i = 0, \tag{8-22}$$

a very powerful relation between the partial molar properties of the various constituents. Several closely related equations appear in the literature:

1. If we divide Eq. (8-22) by n, then by Eq. (8-5),

$$\sum_{i=1}^{r} X_i \, d\bar{J}_i = 0. \tag{8-23}$$

* As E. G. D. Cohen has reminded the author, this is the way "white coffee" is poured in British Railways restaurant cars: The waiter will pour coffee and hot milk simultaneously from two large cans, in the "correct" proportions (he will *not* first pour coffee and then add the milk).

2. If we define *partial specific properties* by

$$J'_i \equiv \left(\frac{\partial \mathbf{J}}{\partial w_i}\right)_{P,T,w_j} \qquad (i = 1, 2, \ldots, r), \tag{8-24}$$

then the analogues to Eqs. (8-22) and (8-23) are

$$\sum_{i=1}^{r} w_i \, dJ'_i = 0 \tag{8-25}$$

and

$$\sum_{i=1}^{r} z_i \, dJ'_i = 0. \tag{8-26}$$

3. For the specific case $\mathbf{J} = \mathbf{G}$ (the Gibbs free energy; see section 15.2), Eqs. (8-22) and (8-23) become

$$\sum_{i=1}^{r} n_i \, d\mu_i = 0 \tag{8-27}$$

and

$$\sum_{i=1}^{r} X_i \, d\mu_i = 0, \tag{8-28}$$

in which we use the *ad hoc* symbol μ_i for the $\bar{G}_i$. This conforms to the notation suggested by Gibbs, long before the introduction of the general concept of partial molar quantities. He called the μ_i *chemical potentials*, for reasons we discuss in Chapter 19. Both Gibbs' notation and his terminology have persisted to the present.

Any or all of Eqs. (8-22), (8-23), and (8-25) through (8-28) are frequently referred to as (the) *Gibbs–Duhem equation(s)*. In fact, Gibbs and Duhem were the first to derive the more general result:

$$\mathbf{S}\, dT - \mathbf{V}\, dP + \sum_i n_i \, d\mu_i = 0. \tag{8-29a}$$

If we define a *mean molar quantity* by*

$$J \equiv \frac{\mathbf{J}}{n}, \tag{8-30}$$

dividing Eq. (8-29a) by n yields

$$S\, dT - V\, dP + \sum_i X_i \, d\mu_i = 0. \tag{8-29b}$$

Some authors think that the designation *Gibbs–Duhem equation(s)* should be restricted to Eq. (8-29a) and/or (8-29b). Obviously, at constant P and T, these reduce to Eqs. (8-27) and (8-28), respectively.

---

* For a single-component system J reduces to the "ordinary" molar property; it is identical with $J_i^{\bullet}$ if i is the substance at hand.

It should be emphasized that there is an important difference in the basis for obtaining these two sets of equations. Clearly, Eqs. (8-27) and (8-28), as special cases of Eqs. (8-22) and (8-23), can be derived without reference to the laws of thermodynamics. On the other hand, Eqs. (8-29) can be derived only if the First and Second Laws are at our disposal (Exercise 19-1). The inclusion of Eqs. (8-29) only serves the purpose of comparison (in connection with an ambiguity in nomenclature, as indicated above); otherwise, they are out of place at this stage of our development.

8.3 Further Discussion of Partial Molar Quantities; Apparent Molar Quantities

Many students have a hard time grasping the concept of partial molar properties and complain that their definition, as given by Eq. (8-13), is far too "abstract." A graphical illustration may clarify some of the most important features involved; to make things easier to visualize we have, in Fig. 2, chosen the volume as an example.

Figure 2a, the trivial linear plot of volume versus mole number for some pure component i, is drawn for comparison purposes only. Every time a mole of i is added, the volume increase is $V_i^\bullet$. Figures 2b–d deal with the addition of the same component i to a solution with initial volume V_o.* In Fig. 2b we have drawn the *exceptional* situation that $\mathbf{V}$ is still linear in n_i: Every time we add 1 mole of i, the volume still increases by the same amount, $\bar{V}_i$. This is only *strictly* true for certain model systems (see Chapter 25 on "ideal mixtures"), for which usually $\bar{V}_i = V_i^\bullet$, i.e., *the volume is additive in the components.* The reader should be warned that certain *other thermodynamic properties* (such as the entropy, $\mathbf{S}$, and the Gibbs free energy, $\mathbf{G}$), for the *same model, do not show such additivity* (section 25.3). Figures 2c and d depict the two possibilities for most real solutions. It is seen that adding the first mole of i leads to a very different volume increase, $\Delta\mathbf{V}_{0\to1}$, than the one in adding, say, the fifth mole, $\Delta\mathbf{V}_{4\to5}$. *Neither of these quantities represents* $V_i^\bullet$ *or even* $\bar{V}_i$! The latter is now dependent on the composition of the solution, which changes in the process of adding 1 mole of i. The definition (8-13) tells us simply that in each point of the $\mathbf{V}$—n_i curve, $\bar{V}_i$ *is given by its slope.*† An alternative interpretation, which many students somehow prefer, gives $\bar{V}_i$ as

---

* It is actually irrelevant whether the initial solution already contains some i; this would simply increase all n_i by the same amount, n_i°.

† In exceptional cases $\bar{V}_i$ (unlike $V_i^\bullet$) *can be negative*! An example is a very dilute solution of $MgSO_4$ in water at 18°C; $\bar{V}_{MgSO_4}$ is negative if the molality is less than about 0.07, then passes through zero and becomes positive at higher concentrations. (See Lewis et al., reference 1 at the end of the chapter, p. 206.)

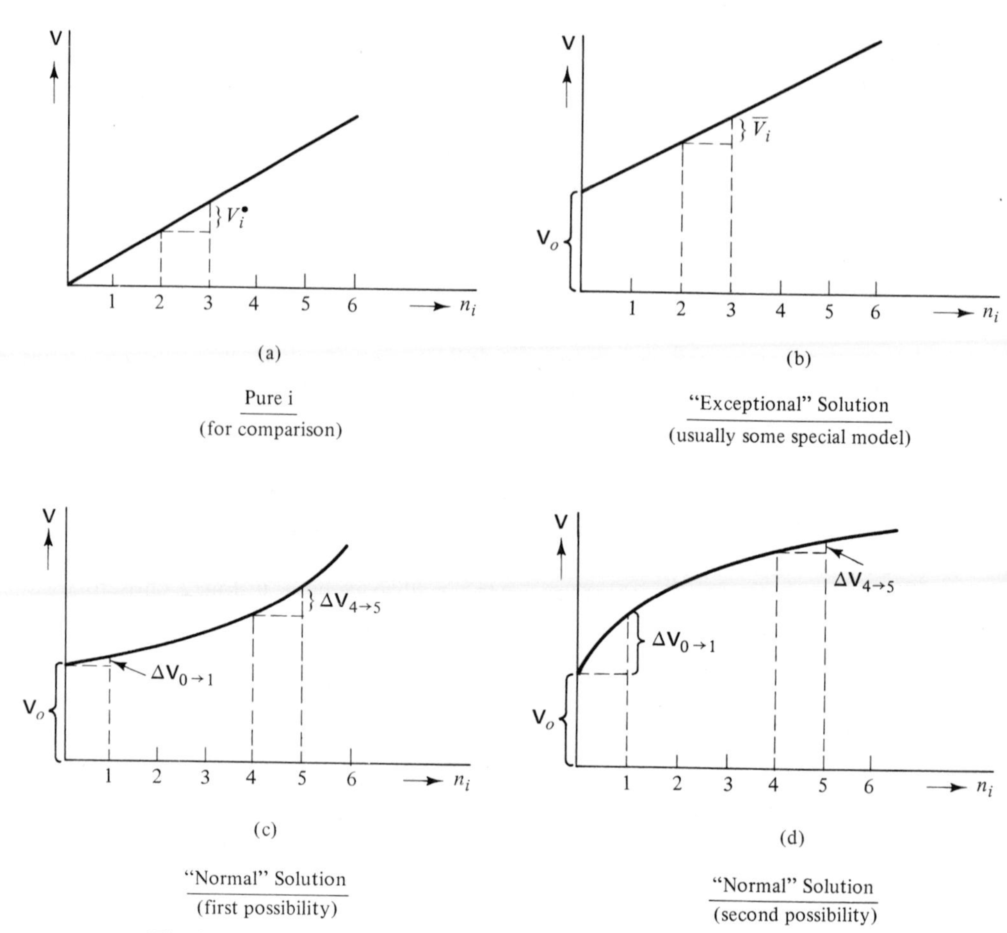

Fig. 2. Illustrating partial molar properties (volume as an example; P and T constant throughout).

the increase in volume if 1 mole of i is added to such a large quantity (strictly speaking, an infinite amount!) of the solution that this addition does not change its composition significantly. Note that we are between the devil and the deep blue sea: We have to define $\bar{V}_i$, and, by obvious generalization, all $\bar{J}_i$, either in terms of infinitesimal changes in composition (using differential quotients) or in terms of additions of 1 mole to an *infinite* amount of solution.

Two final points regarding the $\bar{J}_i$ must be made. First, while the **J** are *extensive* properties, the $\bar{J}_i$, as well as the mean molar quantity, J, of Eq. (8-30), are dependent only on P, T, and *composition*, hence are *intensive*. In the Griffith–Wheeler nomenclature (section 4.2) they are "densities," not "fields."

In addition, the relationships we shall develop in subsequent chapters between many different (extensive) thermodynamic properties also hold between the corresponding partial molar properties, for each component. Two examples will illustrate this assertion. From

$$\mathbf{G} = \mathbf{H} - T\mathbf{S}, \tag{8-31}$$

by differentiation with respect to some n_i, keeping P, T, and all other mole numbers constant, we get

$$\left(\frac{\partial \mathbf{G}}{\partial n_i}\right)_{P,T,n_j} = \left(\frac{\partial \mathbf{H}}{\partial n_i}\right)_{P,T,n_j} - T\left(\frac{\partial \mathbf{S}}{\partial n_i}\right)_{P,T,n_j};$$

hence, by our definition (8-13),

$$\bar{G}_i = \bar{H}_i - T\bar{S}_i. \tag{8-32}$$

Similarly, from

$$\mathbf{C}_p = \left(\frac{\partial \mathbf{H}}{\partial T}\right)_{P,\,\text{all}\,n_k} \tag{8-33}$$

we obtain

$$\bar{C}_{p_i} = \left(\frac{\partial \bar{H}_i}{\partial T}\right)_P, \tag{8-34}$$

due to the circumstance that since $\mathbf{H}$ is a well-behaved (continuous, single-valued) function of *pressure*, temperature, and mole numbers,

$$\left(\frac{\partial^2 \mathbf{H}}{\partial T \partial n_i}\right)_{P,n_j} = \left(\frac{\partial^2 \mathbf{H}}{\partial n_i \partial T}\right)_{P,n_j}.$$

It is obvious that we could not "play this game," without modification, starting with

$$\mathbf{C}_v = \left(\frac{\partial \mathbf{E}}{\partial T}\right)_{\mathbf{V},\,\text{all}\,n_k}, \tag{8-35}$$

in which it is implicit that $\mathbf{C}_v$ and $\mathbf{E}$ are functions of *volume*, temperature, and mole numbers. Differentiation with respect to n_i at constant $\mathbf{V}$, T, and n_j yields, in this case,

$$\left(\frac{\partial \mathbf{C}_v}{\partial n_i}\right)_{\mathbf{V},T,n_j} = \frac{\partial}{\partial T}\left(\frac{\partial \mathbf{E}}{\partial n_i}\right)_{\mathbf{V},T,n_j}. \tag{8-36}$$

Relations such as this one may occasionally be useful, but the terms in this equation are *not* partial molar quantities, since the differentiations are carried out at constant *volume* rather than at constant *pressure*.

The introduction of *apparent molar properties* offers nothing *essentially* new. They are *useful* "intermediate" quantities in only one specific situation, namely, in the calculation of $\bar{J}_2$ from experimental data for a (usually fairly dilute) solution of a single solute, 2, in the solvent, 1. In this situation we define the apparent molar property *for the solute* by*

$$^{\phi}J_2 \equiv \frac{\mathbf{J} - n_1 J_1^{\bullet}}{n_2}. \tag{8-37}$$

No similar quantity is defined for the solvent, since it would not be useful! The name given to the $^{\phi}J_2$ is not difficult to rationalize: If we had n_1 moles of pure solvent, the $\mathbf{J}$ would be $n_1 J_1^{\bullet}$. If we subtract the latter amount from the actual $\mathbf{J}$, the difference is the amount *apparently* due to the solute, which we could call its apparent $\mathbf{J}$. Divide this by the number of moles of solute, n_2, and we get the apparent $\mathbf{J}$ per mole of 2. The reader should readily obtain

$$\bar{J}_2 = n_2 \left(\frac{\partial^{\phi} J_2}{\partial n_2} \right)_{n_1,P,T} + {}^{\phi}J_2 \tag{8-38}$$

and

$$\bar{J}_1 = J_1^{\bullet} + n_2 \left(\frac{\partial^{\phi} J_2}{\partial n_1} \right)_{n_2,P,T} \tag{8-39}$$

Exercise 8-2

Derive Eqs. (8-38) and (8-39).

Equation (8-38) is frequently useful for fairly dilute solutions, for which it is possible to obtain $^{\phi}J_2$ experimentally as a function of n_2, c, or m,† either graphically or analytically. Equation (8-39) is useless in this context; if $\bar{J}_1$ is needed too, we would subsequently compute it from (8-12);

$$\bar{J}_1 = \frac{\mathbf{J} - n_2 \bar{J}_2}{n_1}. \tag{8-40}$$

For completeness sake, we mention that one also defines *apparent specific properties* [compare Eq. (8-24)] by

$$^{\phi}J_2' \equiv \frac{\mathbf{J} - w_1 J_1^{\bullet\prime}}{w_2}. \tag{8-41}$$

---

* As far as the author has been able to ascertain, these quantities were first defined by G. N. Lewis and M. Randall, who also introduced the notation. Unfortunately, in many later publications, the symbol $^{\phi}J_2$ has become ϕJ_2, which is confusing to the uninitiated.

† See Problem 8-2. When a single solute is present, the subscript 2 is often omitted from c and m.

8.4 Some Concepts, Nomenclature, and Notation Conventions for Chemical Reactions

If we write the reaction in the form

$$\nu_A A + \nu_B B + \ldots = \nu_M M + \nu_N N + \ldots, \tag{8-42}$$

then it is customary to refer to A, B, ... as *reactants* and to M, N, ... as *products*. The quantities $\nu_A, \nu_B, \ldots, \nu_M, \nu_N, \ldots$ are called the *stoichiometric coefficients*. These are completely determined except for a common multiplier, the choice of which is arbitrary, unless one agrees, for example, that such coefficients shall be the smallest possible set of integers. Such a convention, desirable as it may be, has not been universally accepted, and we shall not take it for granted in this book. For a *general* discussion of chemical reactions, there are much more convenient, compact, ways to write Eq. (8-42): for example,

$$\sum_{\text{reactants}} \nu_i i = \sum_{\text{products}} \nu_i i \tag{8-42a}$$

or, even simpler,

$$\sum_i \nu_i i = 0. \tag{8-42b}$$

In the last formulation it is implied that each ν_i is negative for a reactant and positive for a product.

A simple example may be helpful at this stage. Equations (8-42) and (8-42a) could represent

$$N_2 + 3H_2 = 2NH_3 \tag{a}$$

or

$$\tfrac{1}{2}N_2 + \tfrac{3}{2}H_2 = NH_3, \tag{a'}$$

while in the form (8-42b) we think of these processes as written

$$2NH_3 - N_2 - 3H_2 = 0 \tag{b}$$

and

$$NH_3 - \tfrac{1}{2}N_2 - \tfrac{3}{2}H_2 = 0, \tag{b'}$$

respectively. Of course, for any equilibrium reaction, we can write the reverse, e.g.,

$$2NH_3 = N_2 + 3H_2. \tag{ã}$$

Now NH_3 is the "reactant" while N_2 and H_2 are the "products." In the form (8-42b) we get

$$N_2 + 3H_2 - 2NH_3 = 0; \tag{b̃}$$

the signs of all ν_i are reversed.

It is often convenient to define a quantity $\Delta\nu$:

$$\Delta\nu \equiv \nu_M + \nu_N + \ldots - \nu_A - \nu_B - \ldots, \tag{8-43}$$

which can also be written as

$$\Delta\nu \equiv \sum_{\text{products}} \nu_i - \sum_{\text{reactants}} \nu_i \tag{8-43a}$$

or

$$\Delta\nu \equiv \sum_i \nu_i. \tag{8-43b}$$

The three definitions (8-43) correspond to the three ways, (8-42), (8-42a), and (8-42b), of writing our chemical reaction.

For the example given in small print above, the various $\Delta\nu$ are -2 for (a), -1 for (a′), -2 for (b), -1 for (b′), $+2$ for ($\tilde{a}$), and $+2$ for ($\tilde{b}$).

Occasionally, we want $\Delta\nu$ to refer to the species in a particular phase only. We shall indicate this by a superscript and write $\Delta\nu^g$, $\Delta\nu^\ell$, etc.

As an example take the reaction

$$2C(s) + O_2(g) = 2CO(g),$$

for which $\Delta\nu = -1$ and $\Delta\nu^g = +1$.

The stoichiometric coefficients, ν_i, of the reaction should not be confused with the mole numbers, n_i, the actual number of moles in the reaction mixture. It can only lead to confusion if we use n_i to denote *both*. The quantity Δn (rarely needed) gives the actual change in the total number of moles of all the substances partaking in the reaction. Thus, for example, if in a given experiment 1.213 moles of N_2 react with 3.639 moles of H_2 to yield 2.426 moles of NH_3, Δn is -2.426, irrespective of how we write the reaction and independent of the associated value of $\Delta\nu$ (see the examples above).

Unless stated otherwise, we shall always assume that a chemical reaction will take place in a *closed* system. Then the *changes* in the various n_i are related to each other as follows:

$$-\frac{dn_A}{\nu_A} = -\frac{dn_B}{\nu_B} = \ldots = \frac{dn_M}{\nu_M} = \frac{dn_N}{\nu_N} = \ldots. \tag{8-44}$$

When the reaction is represented by Eq. (8-42b), these equalities can simply be written

$$\frac{dn_i}{\nu_i} = \frac{dn_j}{\nu_j} \qquad \text{for all } i, j. \tag{8-45}$$

In Eq. (8-45), the proper sign is "automatically" accounted for through the sign convention of the ν_i. The reader should check Eqs. (8-44) and (8-45) for the participants in the ammonia formation, as given in the illustrative examples above.

In 1920, the Belgian physical chemist Th. De Donder introduced the *degree of advancement of a reaction.* For many years this concept was used exclusively by the Belgian school, and it did not become more widely known until Prigogine et al. established its usefulness in the formulation of the *themodynamics of irreversible processes.* In recent decades several authors have recognized that this parameter is useful in "ordinary" *equilibrium thermodynamics* as well. This recognition has been accompanied by the suggestion of alternative names, such as *progress variable* and *extent of reaction.* It has also led to the definition of related but different quantities (such as Zemansky's *degree of reaction*; see below), which has occasionally given rise to some confusion.

If the chemical reaction is written in the form (8-42b),* so that Eq. (8-45) is valid, its degree of advancement, in the original De Donder sense, is defined by

$$d\xi \equiv \frac{dn_i}{\nu_i} \qquad \text{for all } i, \tag{8-46}$$

or, upon integration between $t = 0$ and arbitrary t,

$$\xi = \frac{n_i - n_i^\circ}{\nu_i} \qquad \text{for all } i. \tag{8-47}$$

Here n_i° is the number of moles of i at $t = 0$. Thus defined, ξ is an *extensive variable*, related to the reaction rate, v, by

$$v = \frac{d\xi(t)}{dt}. \tag{8-48}$$

It is important to analyze the range of variables over which ξ extends. Assume that initially, i.e., at $t = 0$, when $\xi = 0$, we have exclusively reactants. Consider two cases:

1. All reactants *are mixed in stoichiometric proportions* and each n_i° equals $|\nu_i|$. If the reaction could go to completion, ξ would reach a theoretical maximum value of unity. In practice this value is rarely attained, since the reaction mixture will reach equilibrium (in the sense that no further macroscopic changes will take place) at $\xi = \xi_e$ $(0 < \xi_e < 1)$. Equilibrium values of the mole numbers will be denoted by n_i^e.
2. The *reactants are not mixed in stoichiometric proportions.* Let there be a single *limiting reactant,* which we shall designate by k. When n_k is reduced to zero, the reaction has gone as far as it possibly can, and ξ has reached its associated maximum value, $\xi_{\max}$. Depending on the

* This is always implied when we use the De Donder scheme.

values of n_k° and the thermodynamic equilibrium constant, the reaction *may be* arrested at $\xi = \xi_e$ $(0 < \xi_e < \xi_{\max})$. Obviously, by Eq. (8-47), with $n_k = 0$,

$$\xi_{\max} = -\frac{n_k^\circ}{\nu_k} = \frac{n_k^\circ}{|\nu_k|} \qquad k = \textit{limiting reactant.*} \tag{8-49}$$

We see that, in Eq. (8-49), $\xi_{\max}$ *is not a constant, but depends on* n_k°. Some authors consider this a disadvantage and, following Zemansky, define a dimensionless *degree of reaction*,† which is ξ/n_k°, an *intensive* quantity. The De Donder definition [Eqs. (8-46) and (8-47)] is the more useful one for several reasons, one of which we shall touch upon below. The others will not be apparent until the concept is applied to the formulation of the thermodynamics of chemical reactions (see, e.g., Chapter 19 and reference 3 given at the end of this chapter).

Finally, we *define* a quantity $\Delta_{P,T}\mathbf{J}$ *of a reaction* (frequently just called $\Delta\mathbf{J}$ of said reaction; the constancy of P and T is implied) by

$$\Delta_{P,T}\mathbf{J} \equiv \Delta\mathbf{J} \equiv \nu_{\mathrm{M}}\bar{J}_{\mathrm{M}} + \nu_{\mathrm{N}}\bar{J}_{\mathrm{N}} + \ldots - \nu_{\mathrm{A}}\bar{J}_{\mathrm{A}} - \nu_{\mathrm{B}}\bar{J}_{\mathrm{B}} \ldots, \tag{8-50a}$$

in which, for the time being, the process involved is represented by the explicit equation (8-42). Of particular interest will be the special cases $\mathbf{J} = \mathbf{H}$ and $\mathbf{J} = \mathbf{G}$, when $\Delta\mathbf{J}$ will give us the heat of the reaction and its free energy change, respectively. As a consequence of the definition of the partial molar quantities discussed in earlier sections of this chapter, $\Delta\mathbf{J}$ can be interpreted quite generally as follows: It is the change in the (extensive) thermodynamic property $\mathbf{J}$ if, at constant P and T, ν_{A} moles of A react completely with ν_{B} moles of B, etc., to yield ν_{M} moles of M and ν_{N} moles of N, etc., in such a large amount of this particular reaction mixture (strictly speaking an infinite amount) that its composition does not change significantly.

Up to the preceding paragraph, $\Delta\mathbf{J}$ has been a shorthand notation for $\mathbf{J}_{\text{final}} - \mathbf{J}_{\text{initial}}$, pertaining to *any process*. Henceforth, when we specifically consider a *chemical reaction*, $\Delta\mathbf{J}$ should be interpreted in the sense of the definition (8-50a). Used in this *particular* fashion, the symbol Δ can be identified with the operator $(\partial/\partial\xi)_{P,T}$. This is one of the distinct advantages of defining ξ according to the original De Donder scheme. To prove this equivalence, let us first rewrite Eq. (8-50a) in the compact form corresponding to Eq. (8-42b):

$$\Delta_{P,T}\mathbf{J} \equiv \sum_i \nu_i \bar{J}_i. \tag{8-50b}$$

---

* Since k is a reactant, ν_k is negative.

† Occasionally written as ε, but unfortunately sometimes denoted by ξ as well. Therefore, upon encountering the symbol ξ in this context, the readers should carefully check its definition.

Since at constant P and T, **J** is only a function of (ξ and) the n_i, we also have

$$\left(\frac{\partial \mathbf{J}}{\partial \xi}\right)_{P,T} = \sum_i \left(\frac{\partial \mathbf{J}}{\partial n_i}\right)_{P,T,n_j} \frac{dn_i}{d\xi},$$

or, by Eqs. (8-13) and (8-46),

$$\left(\frac{\partial \mathbf{J}}{\partial \xi}\right)_{P,T} = \sum_i \bar{J}_i \nu_i. \tag{8-51}$$

Comparison with (8-50b) shows that

$$\Delta_{P,T}\mathbf{J} = \left(\frac{\partial \mathbf{J}}{\partial \xi}\right)_{P,T}, \tag{8-52}$$

the desired result. Once more a thermodynamic quantity can be interpreted either in terms of an infinitesimal change or in terms of a process in an infinite amount of solution (compare the two interpretations of partial molar quantities in section 8.3). The change in **J** represented by Eqs. (8-50) and (8-52) is sometimes referred to as the change due to *unit advancement* or *unit reaction.* In the opinion of the author, this terminology is not particularly meaningful. It is important to realize that this change is still dependent on our choice of stoichiometry as explained at the very beginning of this section. Once the ν_i are fixed, either by an *ad hoc* decision, or by the convention of taking the smallest possible set of integers, the change in **J**, as represented by either side of Eq. (8-52), is unambiguously determined.

It is readily seen that relations such as (8-32) and (8-34) between various $\bar{J}_i$ also hold between the corresponding $\Delta\mathbf{J}$ for chemical reactions (constant P and T implied):

$$\Delta\mathbf{G} = \Delta\mathbf{H} - T\,\Delta\mathbf{S} \tag{8-53}$$

and

$$\Delta\mathbf{C}_p = \left(\frac{\partial\,\Delta\mathbf{H}}{\partial T}\right)_P. \tag{8-54}$$

The reader should verify this.

PROBLEMS

Problem 8-1

(a) If **V** is the volume (in cubic centimeters) of a two-component solution containing 1000 g of solvent, show that

$$\bar{V}_2 = \frac{M_2 - \mathbf{V}(\partial\rho/\partial m)}{\rho},$$

in which M_2 is the weight of 1 mole of the solute, ρ is the density of the solution, and m is the molality of the solute.

(b) Show that

$$\bar{V}_1 = \frac{M_1\left(1 + \dfrac{m\mathbf{V}}{1000}\dfrac{\partial \rho}{\partial m}\right)}{\rho}.$$

These equations give $\bar{V}_1$ and $\bar{V}_2$ if, from experiments, the density ρ is known as a function of m.

Problem 8-2

(a) In 1931, W. Geffcken [*Z. Physik. Chem.* **155**, 7 (1931)] found experimentally that the apparent molar volumes of several 1-1 electrolytes in water are simple functions of the molarity, c, over a wide range of concentrations. Specifically, for NaCl at 25°C:

$$^{\phi}V_2 = 16.28 + 2.22\sqrt{c}\ \text{ml}, \tag{1}$$

in which c is in moles per liter. Obtain $\bar{V}_2$ and $\bar{V}_1$ in such a solution for $c = \frac{1}{2}$ and $c = 1$. At 25°C the density of water is 0.99707 kg ℓ^{-1}. [*Hint*: Write down (8-38) with $\mathbf{J} = \mathbf{V}$ and convert into an equation for $\bar{V}_2$ as a function of $^{\phi}V_2$ and c.]

(b) I. M. Klotz (*Chemical Thermodynamics*, 3rd ed., W. A. Benjamin, Menlo Park, Calif., 1972, p. 276) suggested the following alternative empirical equation for the same system (NaCl in water at 25°C):

$$\bar{V}_2 = 16.6253 + 2.6607m^{1/2} + 0.2388m\ \text{ml}. \tag{2}$$

From the results obtained in part (a), compute m for $c = \frac{1}{2}$. Next evaluate $\bar{V}_2$ for *this* molality, using Eq. (2). Compare with $\bar{V}_2$ obtained in part (a).

Problem 8-3

If we mix 2 moles of nitrogen and 8 moles of hydrogen and the reaction

$$N_2 + 3H_2 \rightarrow 2NH_3$$

goes to completion (in the sense that the limiting reactant has disappeared), obtain $\xi_{\max}$.

Problem 8-4

A quantity of ammonia gas is heated to some specific temperature under such conditions that the dissociation reaction

$$2NH_3 = N_2 + 3H_2$$

reaches equilibrium. The *degree of dissociation*, d, is defined by

$$d \equiv \frac{n^{\circ}_{NH_3} - n^{e}_{NH_3}}{n^{\circ}_{NH_3}}.$$

Express d in terms of $n^{\circ}_{NH_3}$ and ξ_e.

REFERENCES

Unfortunately, some of the books we can recommend in the context of this chapter are out of print at the time of this writing.

For a comprehensive discussion of partial molar quantities one may always go back to the "source":

1. G. N. Lewis and M. Randall, as revised by K. S. Pitzer and L. Brewer, *Thermodynamics*, McGraw-Hill, New York, 1961, in particular Chapter 17.

A very good discussion is also given in

2. I. M. Klotz and R. M. Rosenberg, *Chemical Thermodynamics*, 3rd ed., W. A. Benjamin, Menlo Park, Calif., 1942, in particular Chapters 13 and 17.

For the material in section 8-4, in particular the introduction and use of the degree of advancement of a reaction, see

3. I. Prigogine and R. Defay, *Chemical Thermodynamics*, D. H. Everett, trans., Longmans, Green, New York, 1954, Chapter I. These authors make maximum use of this concept, much more so than in the present book. It is an excellent introduction to equilibrium thermodynamics according to the Belgian School. Many useful references are given here as well.

Part II

The Laws of Thermodynamics; Some Basic Applications

Chapter 9

The First Law

9.1 Introductory Remarks

Unless specifically stated otherwise, we shall always consider our thermodynamic system to be at rest, and we shall disregard the presence of external fields. Consequently, we can equate the *total* energy of the system with its *internal* energy, to be denoted* by the symbol **E**. In our discussion of the *First* Law, we shall further restrict ourselves by considering *closed* systems only. As pointed out in section 1.4, this implies that we exclude any exchange of matter between these systems and their surroundings.

As mentioned in section 1.3, this book emphasizes the "traditional" approach. The corresponding formulation of the First Law is given in section 9.2. Section 9.3 contains a critical discussion of the alternative "axiomatic" treatment, which is rapidly gaining in popularity. That section should be considered optional reading.

Before proceeding, we remind readers of the sign conventions adopted for Q and W (see sections 3.4 and 4.3, respectively), and of the fact that *in our notation* we do not differentiate between exact differentials, such as $d\mathbf{E}$, and other infinitesimal increments, such as dQ and dW (see section 6.2).

9.2 The "Traditional" Approach

The First Law tells us that for all systems there exists a state function, the internal energy, **E**, and that the change of this property, in going from an arbitrary initial state, 1, to an arbitrary final state, 2, is given by the heat *absorbed by* the system (*from* the surroundings) minus the work *done by* the system (*on* the surroundings). Mathematically, this is expressed as

$$\mathbf{E}_2 - \mathbf{E}_1 \equiv \Delta\mathbf{E} = Q - W \tag{9-1}$$

---

* Some books use **U** for *internal* energy and introduce **E** for the *total* energy whenever the distinction is relevant.

for macroscopic changes, or as

$$d\mathbf{E} = dQ - dW \tag{9-2}$$

for infinitesimal changes.

Thus introduced, the First Law appears as the definition of (the change in) **E**. But how did this *definition* become a *law*? Why couldn't we, similarly, define a quantity **X** by means of

$$\Delta\mathbf{X} = 3Q - 2W \tag{9-3}$$

and call this a law as well? The crucial point is that we claim **E** to be a *state function*, while **X** is *not*. In other words, Eq. (9-1) is extremely useful because $\Delta\mathbf{E}$, thus defined, is *path independent*, while the definition (9-3) is useless, because $\Delta\mathbf{X}$ is *not*. For this vital *assertion* about **E**, just stated, to become a *law*, it is necessary to convince ourselves that *it is based on some fundamental fact of experience* (cf. section 1.2).

In elementary discussions it is often stated that (9-1) "simply" expresses the *Principle of the Conservation of Energy*. It certainly appears intuitively obvious to most of us that if in some macroscopic change, a system absorbs more (less) heat than it performs work, its energy will increase (decrease) by a corresponding amount. Unfortunately, this can never *be confirmed, or refuted, experimentally*, since we have *no way to measure* $\Delta\mathbf{E}$ *directly*. In fact, if we had to get an empirical value of $\Delta\mathbf{E}$, we would measure Q and W separately, and subsequently use Eq. (9-1).

On an atomistic level, several types of energy changes may be identified and obtained directly, e.g., by means of spectroscopic measurements. In this context as well, the "energy principle" has experienced its share of controversies. At one time, it appeared to be violated in certain nuclear experiments, but the discovery of the neutrino restored the energy balance sheet.* It is interesting that at the height of this "crisis," Niels Bohr was perfectly willing to relinguish, or at least drastically modify, the "principle" for phenomena on a nuclear scale.† This attitude was certainly unusual. For example, in order to account for Einstein's interconversion of mass and energy ($\mathbf{E} = mc^2$), very few were willing to abandon energy conservation. Instead, the "principle" was saved by declaring that *mass is a form of energy*. Some have argued that our acceptance of the "energy principle" is an act of faith; if we discover (apparent) violations, we promptly invent new forms of energy in order to maintain it. In Redlich's terminology (cf. the last part of section 3.2) the "principle" becomes a "rule of order."

---

* See, e.g., L. M. Brown, *Physics Today* **31** (9), 23 (1978). For some general comment on "extrapolating" phenomenological laws, see also section 1.3.

† Ibid.

There are two related ways to subject the First Law, as formulated in Eq. (9-1), to experimental verification:

1. We can show that for any change, the quantity $(Q - W)$ is completely determined by initial and final states, independent of the path taken between them (while, in general, Q and W *individually* are path dependent).
2. We can show that for any system, along any *closed* path, $Q = W$.

The first type of experiment demonstrates that $\Delta \mathbf{E}$ is path independent, while the second type shows that $\oint d\mathbf{E} = 0$. Either observation confirms that **E** is a state function (cf. section 6.2), and either way the crucial comparison of heat flow and work is based on our knowledge that *heat is a form of energy*. This was first shown unequivocally by the experiments of James Prescott Joule, who thus established the fundamental fact of experience from which the First Law is ultimately "derived" along the lines indicated above.

Joule carried out a lengthy series of experiments between the years 1843 and 1848. One usually elaborates upon the "paddlewheel experiments," which established the equivalence of heat and mechanical work with remarkable accuracy and precision. Far more important, for our purpose, were the much cruder experiments which illustrated that *work performed in a variety of ways could always be converted into the same amount of heat*; the two are truly equivalent to each other and interconvertible.

In the "traditional" approach, an *adiabatic process* is simply *defined* as one for which $Q = 0$. Hence, for such a change, we obtain, from Eq. (9-1),

$$\Delta \mathbf{E} = -W_{\text{adiabatic}}. \tag{9-4}$$

On the other hand, for all processes in which $W = 0$, we have

$$\Delta \mathbf{E} = Q_{W=0}. \tag{9-5}$$

There is no special name for changes of *this* type. Equations (9-4) and (9-5) show that Q and W, which are not state functions *in general*, become equal to $\Delta \mathbf{E}$, hence are path independent, *under certain restrictive circumstances.*

9.3 The "Axiomatic" Approach; Critique

Since the beginning of this century, attempts have been made to define thermal concepts entirely in mechanical terms. This led Carathéodory to formulate the *First Law* as

$$\mathbf{E}_2 - \mathbf{E}_1 \equiv \Delta \mathbf{E} = -W_{\text{adiabatic}}. \tag{9-6}$$

Subsequently, he *defined* Q for a nonadiabatic path, between the *same* states 1 and 2, through the relation

$$Q \equiv W - W_{\text{adiabatic}}. \tag{9-7}$$

Finally, eliminating $W_{\text{adiabatic}}$ between Eqs. (9-6) and (9-7), he recovered the familiar form

$$\Delta \mathbf{E} = Q - W. \tag{9-1}$$

There are *some* analogies with the "traditional" formulation of section 9.2. Once more, the First Law appears as the *definition* of the state function **E**. Again, this has to be justified experimentally, in this case by showing that any *adiabatic* path, between specific initial and final states, is always associated with the same amount of work. Last, but not least, the definition (9-7) implies that heat is a form of energy, which, once more, can be based on the results of Joule's experiments. Besides such similarities, there are substantial differences between the two approaches. Most obviously, in the "axiomatic" formulation we must define an adiabatic process *prior to* the introduction of the concept of (the flow of) heat. A critical discussion of this problem has been given in section 3.2(c). We shall now turn to two other, much more subtle issues.

As Born, one of the early and active proponents—perhaps even the "instigator"—of the "axiomatic" approach, pointed out some time ago:* "... this formulation of the First Law ... *presupposes that mechanical work is measurable* however it is applied, ... *even for the most violent reactions*" (the italics are the present author's). He admits that this may be difficult in practice, and suggests that we should *restrict* ourselves, *whenever possible, to reversible processes.* The difficulties that may arise in obtaining W for irreversible changes have already been discussed, in some detail, in the first part of this book (see section 4.4). The obvious question arises as to whether such difficulties, *if* they are real, do not affect *both* formulations of the First Law to the same extent. After all, the expression (9-1) would *also* become useless if W cannot be measured. To understand why this issue has much more of an impact on the "axiomatic" than on the "traditional" approach, we must first mention another serious drawback, associated with the former.†

Ironically, this problem concerns a very important observation made by Carathéodory himself, known as the *Carathéodory Principle* (*of Adiabatic Inaccessibility*), which serves as the basis for his "derivation" of the *Second* Law of Thermodynamics (see Chapter 13). This observation, which has

---

* Max Born, *Natural Philosophy of Cause and Chance*, based on the Waynflete Lectures, delivered in Oxford in 1948, Dover New York, 1964, p. 37.

† J. de Heer, "Some Comments on the "Axiomatic" Formulation of the First Law of Thermodynamics," *Am. J. Phys.* **45**, 1225 (1977).

attained the status of a fundamental fact of experience, tells us that *it is not always possible to reach any state*, 2, *from any other state*, 1, *by means of an adiabatic process.* For every initial state, this inaccessibility involves an infinite number of points, albeit located in a restricted region. Thus, for all transitions concerned, the definition (9-6) becomes strictly inoperative. The author has been able to locate only two books in which this unsatisfactory state of affairs is even mentioned.* In each text, the dilemma is resolved (?) by observing (correctly) that in these cases, it is always possible to execute the *reverse* of the desired process adiabatically,† so that one can obtain $\Delta\mathbf{E}$ as $W_{\text{adiabatic; } 2\rightarrow 1}$ rather than as $-W_{\text{adiabatic; } 1\rightarrow 2}$. Unfortunately, here we are prematurely using a property of a *state function*, namely $\Delta\mathbf{E}_{1\rightarrow 2} = -\Delta\mathbf{E}_{2\rightarrow 1}$ (see section 6.1.b), while it is the purpose of the First Law to *establish* this characteristic of the internal energy. Moreover, in all cases in which state 2 is adiabatically inaccessible from 1,‡ the allowed reverse processes are, *of necessity, irreversible.* This creates exactly the type of problem (in the evaluation of W) which we would like to avoid. Note that here the traditional formulation gives us an important extra degree of flexibility: Since it does not tie us down to an *adiabatic* change, we are not restricted by the Principle of Inaccessibility either, so that we can heed Born's advice (p. 81) and look for some convenient *reversible* path between the pair of states in question.

Finally, it should be stressed that the critique given in this section by no means prejudges Carathéodory's ingenious approach to the *Second* Law, which we discuss in Chapter 13.

---

* A. Münster, *Classical Thermodynamics*, E. S. Halberstadt, trans. Wiley-Interscience, New York, 1970, p. 23. In M. W. Zemansky, *Heat and Thermodynamics* (McGraw-Hill, New York), the relevant issue is raised through the 4th edition (1957, see p. 60); it is dropped entirely in the 5th edition.

† This will be made clear in section 13.1.

‡ An example is given in section 13.1, in which the Inaccessibility Principle is discussed in some detail.

Chapter 10

Some Basic Applications of the First Law

Unless stated otherwise, we shall assume throughout this chapter that the only possible work is $P,\mathbf{V}$ work. For reversible changes, the First Law [Eq. (9-2)] then becomes

$$dQ = d\mathbf{E} + P\,d\mathbf{V}. \tag{10-1}$$

10.1 Some General Formulas; Mathematical Manipulations

Due to the existence of an *equation of state* between P, $\mathbf{V}$, and T, we can use the techniques outlined in section 6.3 to obtain a number of useful relationships.

First, *let* $\mathbf{V}$ *and* T *be the independent variables*, i.e., consider $\mathbf{E} = \mathbf{E}(\mathbf{V}, T)$. Substitution of Eq. (6-20) into (10-1) then yields

$$dQ = \left(\frac{\partial \mathbf{E}}{\partial T}\right)_{\mathbf{V}} dT + \left[P + \left(\frac{\partial \mathbf{E}}{\partial \mathbf{V}}\right)_T\right] d\mathbf{V}. \tag{10-2}$$

Since

$$\mathbf{C}_v \equiv \left(\frac{\partial Q}{\partial T}\right)_{\mathbf{V}} = \left(\frac{\partial \mathbf{E}}{\partial T}\right)_{\mathbf{V}}, \tag{10-3}$$

this expression for dQ simplifies to

$$dQ = \mathbf{C}_v\,dT + \left[P + \left(\frac{\partial \mathbf{E}}{\partial \mathbf{V}}\right)_T\right] d\mathbf{V}. \tag{10-4}$$

Evidently,

$$d\mathbf{E} = \mathbf{C}_v\,dT + \left(\frac{\partial \mathbf{E}}{\partial \mathbf{V}}\right)_T d\mathbf{V}. \tag{10-5}$$

The only mysterious-looking term left is $(\partial\mathbf{E}/\partial\mathbf{V})_T$. This quantity, known as the *internal pressure*, can be evaluated from the equation of state of the system by means of the important Eq. (2-8), which is a consequence of the

Second Law. As is also done in most other textbooks, we have opted, in a few instances, to use Eq. (2-8) prematurely (see, e.g., Exercise 2-1, as well as Exercise 10-5 below). For gases, $(\partial \mathbf{E}/\partial \mathbf{V})_T$ is usually small compared to P; in the limiting case of an *ideal* gas, it becomes zero [see Chapter 2, in particular section 2.2(c)]. However, for condensed phases, where there are strong cohesive forces present, $(\partial \mathbf{E}/\partial \mathbf{V})_T$ may no longer be negligible compared to P.

Next, let us treat *P and T as independent variables.* In this case we can proceed in several ways. For example, we can go all the way back to Eq. (10-1) and consider both **E** and **V** as functions of P and T:

$$dQ = \left(\frac{\partial \mathbf{E}}{\partial T}\right)_P dT + \left(\frac{\partial \mathbf{E}}{\partial P}\right)_T dP + P\left(\frac{\partial \mathbf{V}}{\partial T}\right)_P dT + P\left(\frac{\partial \mathbf{V}}{\partial P}\right)_T dP$$

or

$$dQ = \left[\left(\frac{\partial \mathbf{E}}{\partial T}\right)_P + P\left(\frac{\partial \mathbf{V}}{\partial T}\right)_P\right] dT + \left[\left(\frac{\partial \mathbf{E}}{\partial P}\right)_T + P\left(\frac{\partial \mathbf{V}}{\partial P}\right)_T\right] dP. \qquad (10\text{-}6)$$

Since

$$\mathbf{C}_p \equiv \left(\frac{\partial Q}{\partial T}\right)_P = \left(\frac{\partial \mathbf{E}}{\partial T}\right)_P + P\left(\frac{\partial \mathbf{V}}{\partial T}\right)_P, \qquad (10\text{-}7)$$

we have obtained

$$dQ = \mathbf{C}_p\, dT + \left[\left(\frac{\partial \mathbf{E}}{\partial P}\right)_T + P\left(\frac{\partial \mathbf{V}}{\partial P}\right)_T\right] dP. \qquad (10\text{-}8)$$

Exercise 10-1

Derive Eq. (10-8), starting with Eq. (10-2). [This is a clumsy way of deriving Eq. (10-8), but it is an excellent exercise in the relevant mathematical manipulations.]

Exercise 10-2

Prove that

$$dQ = \mathbf{C}_v\left(\frac{\partial T}{\partial P}\right)_{\mathbf{V}} dP + \mathbf{C}_p\left(\frac{\partial T}{\partial \mathbf{V}}\right)_P d\mathbf{V}. \qquad (10\text{-}9)$$

[In Eq. (10-9), *P and* **V** *are the independent variables.*] For an application, see Exercise 10-8.

Upon subtracting Eq. (10-3) from (10-7), we get

$$\mathbf{C}_p - \mathbf{C}_v = \left(\frac{\partial \mathbf{E}}{\partial T}\right)_P - \left(\frac{\partial \mathbf{E}}{\partial T}\right)_{\mathbf{V}} + P\left(\frac{\partial \mathbf{V}}{\partial T}\right)_P, \qquad (10\text{-}10)$$

or, by Eq. (6-21),

$$\mathbf{C}_p - \mathbf{C}_v = \left[P + \left(\frac{\partial \mathbf{E}}{\partial \mathbf{V}}\right)_T\right]\left(\frac{\partial \mathbf{V}}{\partial T}\right)_P. \tag{10-11}$$

For 1 mole,

$$C_p - C_v = \left[P + \left(\frac{\partial E}{\partial V}\right)_T\right]\left(\frac{\partial V}{\partial T}\right)_P. \tag{10-1$\widetilde{1}$}$$

In fact, *all* equations given above can be expressed in terms of molar quantities; we simply have to replace the *total* energy, volume, and heat capacity ($\mathbf{E}$, $\mathbf{V}$, and $\mathbf{C}$) by the corresponding (mean) molar quantities (E, V, and C), as defined in section 8.2 [see Eq. (8-30)].

When P and T are the independent variables, it is often more convenient to work in terms of a new state function, the *enthalpy*,* $\mathbf{H}$, rather than in terms of $\mathbf{E}$. Define

$$\mathbf{H} \equiv \mathbf{E} + P\mathbf{V}. \tag{10-12}$$

Since Eq. (10-1) can be written as

$$dQ = d\mathbf{E} + d(P\mathbf{V}) - \mathbf{V}\,dP = d(\mathbf{E} + P\mathbf{V}) - \mathbf{V}\,dP,$$

we thus obtain a very useful alternative form of the First Law:

$$dQ = d\mathbf{H} - \mathbf{V}\,dP. \tag{10-13}$$

Exercise 10-3

Convince yourself that $\mathbf{H}$ is a state function. This is the basis for much of thermochemistry (see section 10.4).

Exercise 10-4

By analogous manipulations as used above, derive

(a) $\mathbf{C}_p = \left(\dfrac{\partial \mathbf{H}}{\partial T}\right)_P$ and (10-14)

(b) $\mathbf{C}_p - \mathbf{C}_v = \left[\mathbf{V} - \left(\dfrac{\partial \mathbf{H}}{\partial P}\right)_T\right]\left(\dfrac{\partial P}{\partial T}\right)_{\mathbf{V}}$ (10-15)

(i) "from scratch," i.e., working entirely in terms of $\mathbf{H}$, and
(ii) by transformation of the results (10-7) and (10-11), respectively.

---

* Sometimes called the *heat content*. This designation has its origin in the validity of Eqs. (10-14) and (10-41). Mathematicians call the derivation of (10-13) from (10-1) a *Legendre transformation*. For an excellent discussion, see H. B. Callen, *Thermodynamics*, Wiley, New York, 1963, pp. 90ff.

Exercise 10-5

By means of Eq. (2-8), Eq. (10-11) becomes

$$C_p - C_v = T\left(\frac{\partial P}{\partial T}\right)_V \left(\frac{\partial V}{\partial T}\right)_P.$$

If we define the coefficients of thermal expansion, α, and of isothermal compressibility, κ, by

$$\alpha \equiv \frac{1}{V}\left(\frac{\partial V}{\partial T}\right)_P \quad \text{and} \quad \kappa \equiv -\frac{1}{V}\left(\frac{\partial V}{\partial P}\right)_T,$$

respectively, prove that

$$C_p - C_v = TV\frac{\alpha^2}{\kappa}. \tag{10-16}$$

For any stable phase, κ is always positive, since $(\partial V/\partial P)_T$ is always negative. For the time being, the reader is asked to accept the latter statement as being "intuitively" reasonable (*increasing* the pressure *decreases* the volume); it will be proven in section 17.1. *Equation* (10-16) *is exact* and allows us to compute $(C_p - C_v)$ if α and κ are known experimentally (see Problem 10-3). Actually, α and κ may be easier to measure than C_p and C_v. The result of this exercise also illustrates the fact that C_p is always larger than C_v. For a rare exception, where $C_p = C_v$, see Problem 10-3(a).

It is the author's experience that manipulations such as those encountered in this section often present serious problems for students. To eliminate these difficulties, a thorough familiarity with the appropriate mathematics, as summarized in Chapter 6, is absolutely essential. In addition, attentive study of the material of the present section, *coupled with the execution of all exercises*, should slowly convince the reader that there is some logical "system" employed in these proceedings. To recognize this methodology is of special importance in carrying out derivations involving other parameter couples in lieu of pressure and volume [see, e.g., Eqs. (4-17)]. By way of illustration, let us denote a completely general pair of parameters by X and x [cf. Eq. (4-18); X may or may not carry a minus sign], so that the First Law reads initially

$$dQ = d\mathbf{E} + X\,dx. \tag{10-17}$$

If we are asked to find an expression for $\mathbf{C}_X$, we should immediately realize that Eq. (10-17) must be transformed into

$$dQ = f\,dT + g\,dX, \tag{10-17'}$$

in which f and g denote expressions analogous to those multiplying dT and dP in Eq. (10-6). If *all* we need is $\mathbf{C}_X$, there is no need whatsoever to write out g in detail, since $\mathbf{C}_X$, *defined* as $(\partial Q/\partial T)_X$, is immediately seen to be equal to f.

Exercise 10-6

For an isotropic system in a uniform magnetic field, $\mathscr{H}$, we may write: $dW = -\mathscr{H}\, d\mathscr{M}$, where $\mathscr{M}$ is the "magnetization." The relation

$$\mathscr{M} = \chi(T)\mathscr{H},$$

in which χ is the magnetic susceptibility, can be looked upon as a "magnetic equation of state." Prove that

$$\mathbf{C}_{\mathscr{H}} - \mathbf{C}_{\mathscr{M}} = \left[\left(\frac{\partial \mathbf{E}}{\partial \mathscr{M}}\right)_T - \mathscr{H}\right]\left(\frac{\partial \mathscr{M}}{\partial T}\right)_{\mathscr{H}}. \tag{10-18}$$

To what extent is it necessary to know about the existence of the "magnetic equation of state"?

10.2 Applications to Some Gaseous Model Systems

In section 10.1 we obtained, under the single assumption that only $P,\mathbf{V}$ work is possible, a large number of equations of very general validity. Most of these would soon "die from lack of usefulness" unless, to repeat Uhlenbeck's metaphor,* "we give them something to eat." As the heading of the present section indicates, we shall now "feed" them some gaseous models.

The three models we shall consider are:

1. The single-component ideal gas, consisting of noninteracting mass points. This model has been discussed in considerable detail in section 2.2.
2. The elastic hard-sphere model, in which the particles do have a non-negligible volume, leading to a *repulsive* potential. This model was mentioned briefly in exercise 2.1.
3. The van der Waals gas, which derives from the preceding one by superimposing (long-range) *attractive* forces between the atoms or molecules.

Obviously, these models become progressively more realistic, and therefore also harder to deal with mathematically, in this order. They will be introduced by means of their phenomenological definition (see Chapter 2). Our presentation will be most detailed for the ideal gas. Analogous manipulations for the other two models will, to a large extent, be assigned as exercises.

---

* See footnote, p. 9, in the context of the basic discussion in section 1.5.

(a) *First Model: The Single-Component Ideal Gas*

On the basis of the discussion in section 2.2, this model system is to be defined by

$$P\mathbf{V} = nRT \tag{10-19a}$$

and

$$\mathbf{E} = nE(T), \tag{10-19b}$$

the "thermal" and "caloric" equations of state, respectively. In this case

$$d\mathbf{E} = \mathbf{C}_v\, dT, \tag{10-20}$$

since $\mathbf{E}$ is a function of T only, and Eq. (10-4) simplifies to

$$dQ = \mathbf{C}_v\, dT + P\, d\mathbf{V}. \tag{10-21}$$

Exercise 10-7

Show that for a reversible, isothermal expansion of 1 mole of an ideal gas,

$$Q = W = RT \ln \frac{V_2}{V_1} = RT \ln \frac{P_1}{P_2}. \tag{10-22}$$

The readers should verify that for this model the general result (10-11) reduces to an equation given in many introductory chemistry courses:

$$C_p - C_v = R. \tag{10-23}$$

To obtain these important heat capacities *individually*, we would have to know the energy as a function of the temperature, i.e., the detailed form of $E(T)$. This involves the use of statistical mechanics, hence falls outside the scope of this book.

In the limiting case that *classical* mechanics is valid, E is an integer multiple of $\frac{1}{2}RT$. Then C_v is the corresponding multiple of R and C_p is obtained by adding another unit of R, so that *both are constant.* Outside the classical limit, C_p and C_v become *functions of the temperature.* At the same time, as long as we confine ourselves to the ideal gas model, Eq. (10-23) *must* retain its validity; the *difference,* $C_p - C_v$, *remains constant.* For a discussion of possible ambiguities in the *definition* of an ideal gas, as it is related to these matters, we refer readers back to Chapter 2.

Approximate values of the *individual* C_p and C_v may be obtained by combining the *exact* relation (10-23) with an *approximate* one for the *ratio* of heat capacities $(C_p/C_v) \equiv \gamma$. To this purpose we consider the reversible adiabatic change involving 1 mole of an ideal gas, for which

$$\gamma\, d \ln V + d \ln P = 0. \tag{10-24}$$

Exercise 10-8

Derive Eq. (10-24) from Eq. (10-9) of Exercise 10-2.

In general, this equation cannot be integrated analytically. Assume, however, that the changes involved in the adiabatic expansion or compression are small enough to make γ a constant. Then the integration of Eq. (10-24) immediately yields

$$P_2V_2^{\gamma} = P_1V_1^{\gamma} \quad \text{or} \quad PV^{\gamma} = \text{constant}. \tag{10-25}$$

This is one of three *Poisson Laws.** Adiabatic compression experiments were first carried out by Clément and Désormes, in 1819. While Boyle's Law, $PV = \text{constant}$, is occasionally referred to as the "*isothermal* equation of state" for an ideal gas, Poisson's Law [Eq. (10-25)] has been called its "*adiabatic* equation of state." The former is exact (for the model), the latter is approximate. We should remember that in an adiabatic change, even for an ideal gas, the temperature is, in general, not constant.

Clément–Désormes-type experiments allow for a fairly crude determination of γ. More accurate values may be obtained from measurements of the velocity of sound. Since sound waves in a gas involve local changes in pressure which are so fast that there is no time for heat to flow from one part of the gas to another, the expansions and compressions involved can be considered adiabatic. If the pressure of the gas is low enough, Poisson's Law [Eq. (10-25)] can be used in the derivation of the relevant equations, for which we must refer to appropriate textbooks in physics. Experimental data on γ, properly interpreted on the basis of molecular theories, have provided strong evidence in history (e.g., in the discovery of the noble gases) to decide whether one was dealing with a monatomic gas ($\gamma \approx \frac{5}{3}$) or a diatomic gas ($\gamma \approx \frac{9}{7}$ if classical mechanics were valid; experimentally, it is usually closer to 1.4).

Exercise 10-9

Derive the other two Poisson Laws:

$$T^{\gamma}P^{1-\gamma} = \text{constant} \tag{10-26}$$

and

$$TV^{\gamma-1} = \text{constant}, \tag{10-27}$$

---

* Poisson's Laws were first developed, in collaboration with Laplace, between 1810 and 1820, *within the framework of the caloric theory of heat.* For a brief account of the fascinating history of this "early thermodynamics" (predating the work of, among others, Carnot, Clapeyron, Rumford, and Joule), see E. Mendoza, *Physics Today* **14**, 32 (1961).

under assumptions similar to those that led to Eq. (10-25), each by the following two methods:

(a) By modifying Eq. (10-25), i.e., by eliminating the unwanted variable through the use of the (thermal) equation of state.

(b) By going back to the First Law for adiabatic changes, with P and T, or V and T, as independent variables.

Method (b) is more tedious, but more instructive.

(b) Second Model: Elastic Hard Spheres (without attractive forces)

Once more, this is a model system which requires, at this stage, a two-pronged phenomenological definition. Its thermal equation of state reads

$$P\mathbf{V} = nRT + \mathbf{B}P. \tag{2-10a}$$

For 1 mole

$$PV = RT + BP \tag{2-10b}$$

or

$$P(V - B) = RT, \tag{10-28a}$$

in which the constants $\mathbf{B}$ and B represent the *excluded volume*, which is proportional to the real volume of the spherical particles. In addition, this model must satisfy the caloric equation of state,

$$\mathbf{E} = nE(T). \tag{10-28b}$$

Just as for an ideal gas, Eq. (10-28b) implies that for any *isothermal* process, $\Delta\mathbf{E} = 0$. But while for an ideal gas, in such a change, $\mathbf{H}$ is *also* constant, we have for the present model

$$\Delta\mathbf{H} = \mathbf{B}(P_2 - P_1). \tag{10-29}$$

Readers should be able to verify other analogies and differences themselves (see Exercises 10-10 and 10-12).

Exercise 10-10

(a) Show that Eq. (10-23) is still valid for the elastic hard-sphere model.

(b) Show that for an isothermal change involving 1 mole, Eq. (10-22) has to be replaced by

$$Q = W = RT \ln \frac{V_2 - B}{V_1 - B} = RT \ln \frac{P_1}{P_2}.$$

(c) Show that, provided γ is still constant, the three Poisson equations have to be modified simply by replacing V everywhere by $(V - B)$.

(c) *Third Model: The van der Waals Gas*

This system satisfies the equation of state

$$\left(P + \frac{n^2 a}{\mathbf{V}^2}\right)(\mathbf{V} - nb) = nRT, \tag{10-30}$$

or, for 1 mole,

$$\left(P + \frac{a}{V^2}\right)(V - b) = RT. \tag{10-$\widetilde{30}$}$$

In this book we generally use capital boldface sans serif letters (such as $\mathbf{V}$ in Eq. (10-30) for extensive quantities and capital lightface italic letters (such as V in Eq. (10-$\widetilde{30}$)) for the corresponding (mean) molar properties. However, in Eqs. (10-30) and (10-$\widetilde{30}$) we conform to conventional usage and write a and b for the "van der Waals constants." Thus b represents the "excluded volume," associated with hard spheres, just as B did in Eq. (10-28a). Consequently, nb is identical with $\mathbf{B}$ of Eqs. (2-10a) and (10-29). The constant a determines the (long-range) attractive potential, characteristic for the van der Waals model, and lacking in the preceding model. For some representative numerical values, see Problem 10-2(c).

A very important difference with the preceding two models is that for a van der Waals gas $\mathbf{E}$ is no longer a function of the temperature only. In fact, we shall see in section 14.1 that the internal pressure, $(\partial E/\partial V)_T$, is just (a/V^2). For other properties, see Exercise 10-11 and Problem 10-2.

Exercise 10-11

(a) Obtain a general expression for the work done by 1 mole of a van der Waals gas if it undergoes a reversible isothermal change.

(b) Accepting $(\partial E/\partial V)_T = a/V^2$, obtain ΔE and ΔH for this process.

Verify that all these results reduce to the correct ones for the elastic hard-sphere model if we put *only* a equal to zero (and $b \equiv B$), and to the corresponding ones for the ideal gas if we put *both* a *and* b *equal to zero.*

10.3 The Joule–Kelvin Effect

(a) As early as 1806, Gay-Lussac performed experiments in order to check whether or not the adiabatic expansion of a gas into a vacuum would result in a temperature change. It was hoped that this might be a sensitive way to investigate the forces between particles in the gaseous state. Gay-Lussac thought such forces would be negligible, in opposition to Dalton's conviction that there would be interactions of considerable magnitude. In 1811, Gay-Lussac received strong support from Avogadro, but the matter would not be firmly settled for another 50 years or so.

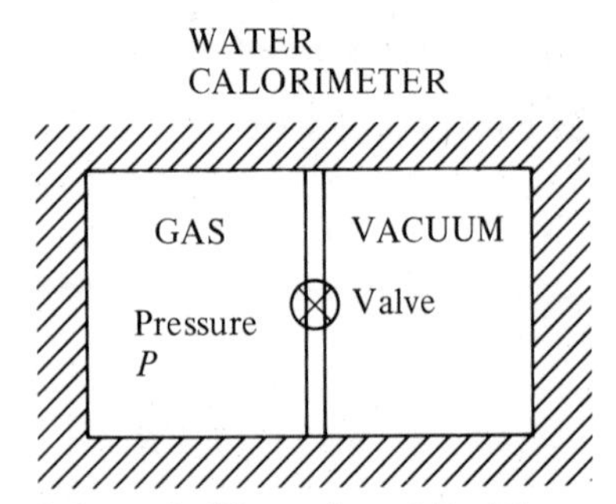

Fig. 3. Schematic illustration of Joule's experiment.

In 1843, Joule carried out a related experiment (Fig. 3). The container on the left was filled with air ($P \approx 22$ atm), the one on the right evacuated, and the entire apparatus submerged in a water calorimeter. Upon opening the valve, a "free expansion" took place. *No temperature change of the water in the calorimeter could be detected.*

Later it was realized that the effect was simply too small to have been observed within the limits of the accuracy of Joule's experiments. More refined measurements indeed show a small temperature change, which decreases as the initial pressure of the gas is lowered. This strongly suggests that the temperature change will be zero only in the limiting case of an ideal gas.

Consider the *joint* containers (with their contents) as our thermodynamic system. In the idealized Joule experiment (Fig. 3; $P \to 0$), the absence of a change in the calorimeter temperature shows that $Q = 0$. We shall *define a Joule expansion* as one that is "free" ($W = 0$) and adiabatic ($Q = 0$), irrespective of P. (In Fig. 3, replace the calorimeter by an adiabatic enclosure.) Then, according to the First Law, $\Delta\mathbf{E} = 0$, *a Joule expansion is always isoenergetic.* The quantity $(\partial T/\partial \mathbf{V})_{\mathbf{E}}$, called the (differential) *Joule coefficient*, does not vanish for all values of P. However, as $P \to 0$ (i.e., as we approach the ideal gas limit), the expansion becomes isothermal (as in the idealized Joule experiment), and $(\partial T/\partial \mathbf{V})_{\mathbf{E}} \to 0$. In this limit, evidently,

$$\left(\frac{\partial \mathbf{E}}{\partial \mathbf{V}}\right)_T = 0. \tag{10-31a}$$

Hence also

$$\left(\frac{\partial \mathbf{E}}{\partial P}\right)_T = 0, \tag{10-31b}$$

so that

$$\mathbf{E} = \mathbf{E}(T); \tag{10-32}$$

$\mathbf{E}$ is a function of the temperature only. This analysis of the results of Joule's experiments, along the lines originally suggested by W. Thomson (who later became Lord Kelvin), was anticipated in section 2.2.

(b) After Joule's failure to produce a detectable temperature change in a *single* expansion, Joule and Thomson, between 1854 and 1856, investigated the effect of a *cumulative* process. To this purpose, they designed an apparatus, schematically illustrated in Fig. 4a. The gas flows repeatedly through an adiabatically isolated porous plug (made, e.g., of cotton and asbestos), and with each passage through the cycle the pressure is lowered from P_1 to P_2. Each time, the initial pressure is restored by means of a compressor. This Joule–Thomson (or Joule–Kelvin) experiment did indeed show a measurable temperature change. We proceed to interpret this result thermodynamically.

As a starting point, the presence of the compressor immediately suggests to us that in contrast to what we saw in the Joule experiment, work *is* performed here. Consequently, the application of the First Law is somewhat more involved. Let us concentrate our analysis on that part of the experiment involving only the adiabatic expansion; hence Q is still zero. As our system we take a fixed amount of gas, initially located between the imaginary boundaries A and B (see Fig. 4b). After a certain time has elapsed, A has moved to A' and B to B'. Let the volume between A and A' be $\mathbf{V}_1$, and that between B and B', $\mathbf{V}_2$. Obviously, these two volumes will contain the same amount of gas. During the process considered, the gas situated on the left of our designated system has performed an amount of work, $P_1\mathbf{V}_1$, *on* it, while the system itself has performed an amount of work, $P_2\mathbf{V}_2$, on the gas to its right. Hence, in this expansion, the total amount of work done *by* the system is

$$W = P_2\mathbf{V}_2 - P_1\mathbf{V}_1. \tag{10-33}$$

Since $Q = 0$, the First Law tells us that $\Delta\mathbf{E} = -W$, hence

$$\Delta\mathbf{E} = -(P_2\mathbf{V}_2 - P_1\mathbf{V}_1). \tag{10-34}$$

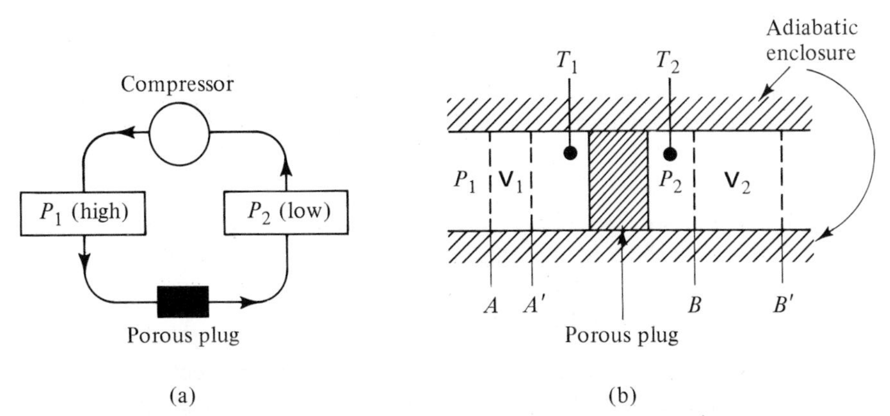

Fig. 4. Schematic illustration of the Joule-Kelvin experiment.

Since the status between the boundaries A' and B has not changed, we can disregard this region in our calculation of $\Delta\mathbf{E}$. Hence $\Delta\mathbf{E}$ is just the energy of the gas in $\mathbf{V}_2$, which we call $\mathbf{E}_2$, minus the energy of the (same amount of) gas in $\mathbf{V}_1$, $\mathbf{E}_1$. Thus Eq. (10-34) becomes

$$\mathbf{E}_2 - \mathbf{E}_1 = -P_2\mathbf{V}_2 + P_1\mathbf{V}_1$$

or

$$\mathbf{E}_2 + P_2\mathbf{V}_2 = \mathbf{E}_1 + P_1\mathbf{V}_1. \tag{10-35}$$

By the definition (10-12), Eq. (10-35) simply reads

$$\mathbf{H}_2 = \mathbf{H}_1. \tag{10-36}$$

Hence, when a certain amount of gas passes through the porous plug in the experiment, its enthalpy remains constant. *The Joule–Kelvin expansion is isenthalpic* [just as *the Joule expansion*, of subsection (a), *was isoenergetic*].

We now *define* the (differential) Joule–Kelvin coefficient, $\mu_{\text{J.T.}}$, by

$$\mu_{\text{J.T.}} \equiv \left(\frac{\partial T}{\partial P}\right)_{\mathbf{H}}. \tag{10-37}$$

By means of a cycle relationship, of the type (6-16), between $\mathbf{H}$, P, and T, Eq. (10-37) can be converted into

$$\mu_{\text{J.T.}} = \frac{-(\partial\mathbf{H}/\partial P)_T}{(\partial\mathbf{H}/\partial T)_P}, \tag{10-38}$$

or, by virtue of Eq. (10-14) (see Exercise 10-4),

$$\mu_{\text{J.T.}} = -\frac{1}{\mathbf{C}_p}\left(\frac{\partial\mathbf{H}}{\partial P}\right)_T. \tag{10-39}$$

For an ideal gas, $\mathbf{H}$ is a function of the temperature only, hence $\mu_{\text{J.T.}} = 0$; i.e., there is no temperature change, just as in the Joule expansion. For most real gases, under most circumstances, $\mu_{\text{J.T.}}$ is *not* zero. Its sign and magnitude will be discussed in section 14.2 after we have obtained a useful expression for $(\partial\mathbf{H}/\partial P)_T$. This "Joule–Kelvin effect" has great technical utility as a method of achieving the cooling necessary for the liquefaction of gases.

Exercise 10-12

For (a) the elastic hard-sphere model and (b) the van der Waals gas, check whether there can be a temperature change:

(i) in a Joule expansion, and
(ii) in a Joule–Kelvin expansion.

As stated at the beginning of this section, the experimental researches of Joule and Kelvin were initiated as a consequence of their desire to learn something about the forces between gas molecules. As we shall see in section

14.2, the results of the measurements of Joule–Kelvin coefficients will indeed give us *some* useful information about these forces, but unfortunately, the situation will turn out to be quite complex. One reason for this is already evident from a critical study of the present section, which shows clearly that *both enthalpy arguments and energy considerations are needed* in the correct treatment of the relevant problems. For processes in *condensed* phases, the difference between $\Delta \mathbf{E}$ and $\Delta \mathbf{H}$ may be very small, but for *gases* $\Delta(P\mathbf{V})$ is usually of considerable magnitude. For this reason, all conclusions drawn from a discussion based on energetic factors *exclusively* should be regarded with the utmost care.

10.4 The Basis of Thermochemistry

Since we are assuming that the only *possible* work is $P,\mathbf{V}$ work, Eq. (9-5) can be written as

$$Q_v = \Delta \mathbf{E}, \tag{10-40}$$

i.e., the heat flow *at constant volume* becomes path independent. In the language of chemistry, the heat of a reaction, when carried out under these conditions, is dependent only on the initial state of the reactants and the final state of the products, and is independent of any intermediates that may (or may not) be involved. Most of the time, chemists carry out their reactions *at constant pressure*, and the question immediately arises whether the heat of this more common type of reaction is also path independent. The affirmative answer, based on experimental evidence, was given by G. H. Hess (1840), in his famous *Law of Constant Heat Summation.** At the time this was considered to be an "independent" fundamental fact of experience. Therefore, thermodynamics gained one of its early triumphs when it could be shown that Hess's discovery followed immediately from the First Law. In fact, all we have to do is to integrate $dQ = d\mathbf{H} - \mathbf{V}\,dP$ [Eq. 10-13] at constant pressure, to yield

$$Q_p = \Delta \mathbf{H}, \tag{10-41}$$

and remind ourselves that $\mathbf{H}$ is a state function.

Today, applications are relegated to general chemistry courses and will not be reviewed in this book. They often appear to us as trivial exercises in "bookkeeping," but we should not forget that they provide us with a splendid illustration of the practical significance of thermodynamics (see section 1.2). That is, we obtain heats of reactions, which are *very hard* to measure

* A special case, the discovery that the heat of a forward reaction is the negative of the heat of the corresponding reverse reaction, is attributed to Lavoisier and Laplace (1786).

directly, from those that are *much easier* to measure and/or have already been tabulated.

In the rest of this section we shall touch briefly on some miscellaneous items in this general area. First, we should never forget that the fundamental result (10-41) *is valid only when the only possible work is of the* $P,\mathbf{V}$ *type*! If we are dealing with a chemical reaction occurring in a galvanic cell, Eq. (10-41) must be replaced by

$$Q_p = \Delta\mathbf{H} + W_{\text{electric}}. \tag{10-42}$$

Exercise 10-13

Derive Eq. (10-42).

Let us return to the situation where only $P,\mathbf{V}$ work is possible. As already hinted above, Q_p is often *more interesting* to chemists than Q_v. But for gas reactions, Q_v is *easiest to obtain experimentally*, namely, by carrying out the reaction in a closed calorimeter. Fortunately, thermodynamics gives us a way to relate the two *at a given temperature*. It is instructive to give first a "bogus derivation," which readers may have encountered in their general chemistry courses. Write Eq. (10-41) as

$$Q_p = \Delta\mathbf{E} + \Delta(P\mathbf{V}), \tag{10-43}$$

and subtract Eq. (10-40) to yield

$$Q_p - Q_v = \Delta(P\mathbf{V}). \tag{10-44}$$

In general, this is *incorrect* since Eqs. (10-40) and (10-43) refer to different processes, so that $\Delta\mathbf{E}$ in these two equations is not the same. The correct replacement for Eq. (10-44) is

$$Q_p - Q_v = (\Delta\mathbf{E}_p - \Delta\mathbf{E}_v) + P\,\Delta\mathbf{V}_p. \tag{10-45}$$

This is not a very useful result, unless certain justifiable simplifications can be made. Let us consider briefly some special cases.

(a) *For a reaction in an ideal mixture of ideal gases*, a model system tacitly assumed in most general chemistry books and discussed in detail in Chapter 23, we know that:

1. $\mathbf{E}$ is a function of T only, hence $\Delta\mathbf{E}_p = \Delta\mathbf{E}_v$, and
2. $P\,\Delta\mathbf{V}_p = RT\,\Delta\nu$, if $\Delta\nu$ is defined by Eqs. (8-43).

In this case Eq. (10-45) becomes

$$Q_p - Q_v = RT\,\Delta\nu. \tag{10-46}$$

At "ordinary" temperatures this difference rarely exceeds a few kilocalories per mole, and is therefore small compared to the typical heat of a reaction. As long as the model is accepted, Eq. (10-46) is strictly correct.

(b) *For a reaction involving exclusively condensed phases*, the "conventional wisdom" tells us that $\Delta\mathbf{E}_p \approx \Delta\mathbf{E}_v$ *and*, unless we have an exceedingly high P and/or an exceptionally high $\Delta\mathbf{V}_p$, $P\,\Delta\mathbf{V}_p \approx 0$, so that $Q_p = Q_v$. For a reaction in which *some components* are present *in condensed phases*, while the *remaining ones* form an *ideal mixture of ideal gases*, the preceding arguments are readily combined with the conclusions drawn in (a), to yield

$$Q_p - Q_v = RT\,\Delta\nu^g, \tag{10-47}$$

in which $\Delta\nu^g$ has the meaning discussed on p. 70. Unfortunately, there are few, if any, experimental data on $(\Delta\mathbf{E}_p - \Delta\mathbf{E}_v)$ and $P\,\Delta\mathbf{V}_p$ for reactions in condensed phases. In this context corpuscular theories are not of much help either. In the absence of such information, we should make allowances for the possibility that $(\Delta\mathbf{E}_p - \Delta\mathbf{E}_v)$ and $P\,\Delta\mathbf{V}_p$, although small with respect to Q_p and Q_v individually, may *also* have an order of magnitude of several times RT.* Thus Q_p does not necessarily equal Q_v in condensed phases, and Eq. (10-47) may require corrections.

Finally, there is the problem of obtaining a relation between the Q_p's of a given reaction (at the same pressure and) at two different temperatures. Since

$$(Q_p)_{T_2} - (Q_p)_{T_1} = (\Delta\mathbf{H})_{T_2} - (\Delta\mathbf{H})_{T_1},$$

we have, by Eq. (8-54),

$$(Q_p)_{T_2} - (Q_p)_{T_1} = \int_{T_1}^{T_2} \Delta\mathbf{C}_p\,dT, \tag{10-48}$$

in which, in accordance with Eq. (8-50b),

$$\Delta\mathbf{C}_p \equiv \sum_i \nu_i \bar{C}_{pi}. \tag{10-49}$$

To integrate Eq. (10-48), we would have to know each $\bar{C}_{pi}$† as a function of T, at the pressure of interest. Since such data are rarely available, one usually equates the $\bar{C}_{pi}$ with the corresponding heat capacity for the pure component, i.e., with $C^\bullet_{pi}$. This can be justified strictly for certain model systems, among which is the aforementioned ideal mixture of ideal gases. Subsequently, the $C^\bullet_{pi}$ are taken from published data, which frequently take empirical forms such as

$$C^\bullet_{pi} = a_i + b_i T + c_i T^2 + \ldots \tag{10-50a}$$

---

* A somewhat *related* issue emerges in Problem 10-3. We shall see that for a *single substance* in the liquid state, $(C_p - C_v)$ may reach a value of many times R. [For solids, $(C_p - C_v)$ is usually much smaller.]

† For definitions and notation, see section 8.2.

or

$$C^{\bullet}_{pi} = a'_i + b'_i T + \frac{c'_i}{T^2}. \qquad (10\text{-}50b)$$

For each component one of these expressions is substituted into Eq. (10-49). Subsequently, the resulting $\Delta \mathbf{C}_p$ is entered into Eq. (10-48), which can now be integrated. As is implied in Eqs. (10-50a) and (10-50b), the various $C^{\bullet}_{pi}$ are essentially functions of the temperature only (see also the discussion in section 14.1, as well as Exercises 14-6 and 14-8b).

If the temperature interval $T_1 \to T_2$ is not too large, one sometimes *approximates* $\Delta \mathbf{C}_p$ by a constant [see Problem 10-4(b) for an example], so that Eq. (10-48) can be integrated to yield

$$(Q_p)_{T_2} - (Q_p)_{T_1} = \Delta \mathbf{C}_p(T_2 - T_1), \qquad (10\text{-}51)$$

*which is Kirchhoff's Law** (*1858*). *This equation is of some historical* significance, and is useful if we just want to have an estimate of the *order of magnitude* of $(Q_p)_{T_2} - (Q_p)_{T_1}$.

PROBLEMS

(Some problems, frequently given in a chapter dealing with the First Law, have been presented in Chapters 3 and 4. For applications of Hess's Law, readers are referred to current general chemistry textbooks, written for students majoring in chemistry or chemical engineering.)

Problem 10-1

A mole of helium gas ($C_v = \frac{3}{2}R$), with an initial temperature of 300 K, is expanded reversibly from a volume of 10 ℓ to a volume of 20 ℓ.

(a) *Assume that the gas is ideal.* Obtain W, Q, ΔE, and ΔH if the expansion is carried out:

(i) isothermally,

(ii) adiabatically.

(b) *Assume that the gas behaves as an elastic hard-sphere model.* With $B = 0.024$ ℓ mole^{-1}, repeat the calculations assigned in part (a).

Problem 10-2

The coefficients of thermal expansion, α, and of isothermal compressibility, κ, have been defined in Exercise 10-5. For a van der Waals gas:

(a) obtain explicit expressions for α and κ, and verify that

$$\frac{\kappa}{\alpha} = \frac{V - b}{R}.$$

---

* G. R. Kirchhoff, *Ann. Physik* **103**, 454 (1858).

(b) Derive from Eq. (10-16)

$$C_p - C_v = R(1 - \varepsilon)^{-1},$$

in which

$$\varepsilon \equiv \frac{2a}{RTV}\left(\frac{V-b}{V}\right)^2.$$

(c) Given the values of a and b in the following table:

	a (in atm ℓ^2 mole^{-2})	b (in ℓ mole^{-1})
He	0.034	0.0237
H_2	0.244	0.0266
NO_2	5.28	0.0442

compute ε for 1 mole of each of these three gases, at 300 K, 100 atm.

Problem 10-3

(a) Use Eq. (10-16), the definition of α (Exercise 10-5), and your knowledge of $\rho(T)$ for water, the density of this substance as a function of the temperature, to conclude that at 1 atm pressure, there is *one* temperature (which one?) at which $(C_p)_{H_2O} = (C_v)_{H_2O}$ *exactly*.

(b) For the same substance, at 25°C, 1 atm,

$$\alpha = 2.57 \times 10^{-4}\ \mathrm{K}^{-1},$$

$$\kappa = 45.66 \times 10^{-6}\ \mathrm{atm}^{-1},$$

$$\rho = 0.9971\ \mathrm{g\ ml}^{-1}.$$

Compute $(C_p - C_v)$ for water under *these* conditions.

(c) For CCl_4, at 25°C, 1 atm,

$$\alpha = 12.4 \times 10^{-4}\ \mathrm{K}^{-1},$$

$$\kappa = 107 \times 10^{-6}\ \mathrm{atm}^{-1},$$

$$\rho = 1.5940\ \mathrm{g\ ml}^{-1}.$$

Compute $(C_p - C_v)$.

Experimental values for C_p, easiest to measure *for condensed phases*, are about 75 J (18 cal) mole^{-1} K^{-1} for H_2O and 134 J (32 cal) mole^{-1} K^{-1} for CCl_4. Your answers to parts (b) and (c) should indicate that for H_2O, $C_p \approx C_v$, but for CCl_4 the difference is substantial. This shows that contrary to what your intuition may have suggested, $(C_p - C_v)$ for a condensed phase can be of considerable magnitude. This point has already been referred to in the text (see footnote, p. 97).

Problem 10-4

The heat of the reaction

$$\mathrm{C(graphite)} + \mathrm{O_2(g)} \rightarrow \mathrm{CO_2(g)}$$

at 25°C, 1 atm, is $-393 \cdot 509$ kJ. [This is the "*standard heat of formation* of CO_2 (from the elements.)"] Compute the heat of this reaction, at 500°C, 1 atm,

(a) by integration of Eq. (10-48) if, in the temperature interval concerned, the relevant $\bar{C}_{pi}$ are given (in units J K^{-1} $mole^{-1}$) by

$$\bar{C}_p(\text{graphite}) \approx C_p^{\bullet}(\text{graphite})$$

$$= -5.293 + 5.861 \times 10^{-2}T - 4.323 \times 10^{-5}T^2, \qquad (1a)$$

$$\bar{C}_p(O_2) \approx C_p^{\bullet}(O_2) = 25.723 + 1.298 \times 10^{-2}T - 0.386 \times 10^{-5}T^2, \qquad (1b)$$

and

$$\bar{C}_p(CO_2) \approx C_p^{\bullet}(CO_2) = 26.000 + 4.350 \times 10^{-2}T - 1.483 \times 10^{-5}T^2. \qquad (1c)$$

(b) using Kirchhoff's Law [Eq. (10-51)] by evaluating an *approximate* $\Delta \mathbf{C}_p$ from the three average $C_p^{\bullet}$'s between 25 and 500°C. Compute these average values by the use of Eqs. (1a) through (1c) at 25°C and at 500°C.

Problem 10-5

For the reaction

$$CH_4(g) = C(g) + 4H(g)$$

at 25°C, 1 atm, $\Delta \mathbf{H} = +1652$ kJ $mole^{-1}$. The quantity q_{C-H}, computed as $1652/4 = 413$ kJ $mole^{-1}$, is known as the *average bond energy* (of the C—H bond) in methane. It is obvious that q_{C-H} is really the bond *enthalpy* instead. Assuming ideal behavior of the gases involved, compute the true average bond *energy*. Note that the difference is small (possibly within the experimental error).

Problem 10-6

When the process

$$Pb + Hg_2Cl_2 \rightarrow PbCl_2 + 2Hg$$

is carried out in a calorimeter at 25°C, 1 atm, the heat evolved is 95.1 kJ per mole of lead reacting. When the same reaction is carried out in a reversibly operating galvanic cell [see section 5.4(b)], at the same pressure and temperature, 103.4 kJ $mole^{-1}$ is made available as electrical energy. What is the heat evolved or absorbed during the cell reaction? (Being able to give the correct thermodynamic *justification* for your computational procedure is more important here than merely obtaining the right numerical answer.)

Problem 10-7

One liter of a monatomic ideal gas, at a pressure of 10 atm and a temperature of 300 K, is expanded isothermally and reversibly to a volume of 2 ℓ. The work freed, in this process, is stored in a mechanical device. Subsequently, *all* this stored work is used to compress the gas adiabatically and reversibly to a final volume, **V**. Try to *predict* whether **V** will be larger than, smaller than, or equal to 1 ℓ.

(a) Compute **V** (take $C_v = \frac{3}{2}R$).

(b) Compute $\Delta \mathbf{E}$ for the overall process.

Chapter 11

Introduction to the "Derivation" of the Second Law

11.1 Limitation of the First Law; the Need for—and Scope of—the Second Law; Preliminary Discussion*

Perhaps it is best to start our discussion with a specific example. Let us heat a quantity of a gas, enclosed in a cylinder with a movable piston. We can write the First Law [Eq. (10-4)] as

$$dQ = \underbrace{\mathbf{C}_v\, dT}_{\text{(i)}} + \underbrace{\left(\frac{\partial \mathbf{E}}{\partial \mathbf{V}}\right)_T d\mathbf{V}}_{\text{(ii)}} + \underbrace{P\, d\mathbf{V}}_{\text{(iii)}}. \tag{11-1}$$

This expresses the fact that the heat flowing into the system is used:

(i) To raise the temperature of the gas (associated with the translational, rotational, and vibrational energy of the molecules).
(ii) To raise the internal potential energy of the gas (associated with the mutual interaction of the molecules).
(iii) To perform work on the surroundings (associated with the expansion of the gas).

Thus the First Law sets up an "energy balance," but it does not tell us how much of the heat is channeled into each of the three categories. To appreciate that this is not just an exercise in irrelevant abstraction, consider such very important practical issues as: Can an amount of heat, under certain circumstances, be *entirely* converted into work? This is precisely the type of question which the Second Law addresses.

---

* This section owes much to *De beide hoofdwetten der thermodynamica* by G. L. de Haas-Lorentz (M. Nijhoff, The Hague, 1938), the textbook in the course given by E. C. Wiersma which confronted the present author with thermodynamics for the very first time. In turn, Mrs. de Haas acknowledged having been strongly influenced by the views of T. Ehrenfest-Afanassjewa.

Let us next put this in a more general framework. The First Law fails to reveal the *direction* of a process; it does not tell us whether a system will undergo a spontaneous (irreversible) change or whether it is at equilibrium. *The task of the Second Law is to provide us with the appropriate criteria for spontaneity* (and, ipso facto, for impossibility), *as well as for equilibrium.* Within this context, however, two qualifications should be made immediately. In the first place, equilibrium is to be taken only in a macroscopic sense; fluctuations are disregarded. Second, *thermodynamics will provide us only with necessary, but not sufficient conditions.* That is, if thermodynamics tells us that a process is impossible, *it is*! But if thermodynamic criteria give a process the green light, it may still go infinitely slowly, which in practical terms means that it does not occur at all. *Thermodynamics does not deal with dynamics*, so to speak, and some have even suggested the designation *thermostatics* instead!

In 1878, Berthelot,* in his famous "Essai de Mécanique Chimique," proposed a very simple criterion: "All processes that occur spontaneously in nature, are exothermic." This was also adhered to by Julius Thomsen. Today we know that this "Berthelot Principle" is incorrect: *Many* endothermic processes do occur spontaneously, and *many* exothermic processes do not. At the same time we cannot ignore the fact that Berthelot is right *in most cases.* We shall see that all this becomes clear when we properly apply the Second Law, utilizing a *new state function, the entropy*, **S**. Actually, this is the task of what we shall designate as the "*second part* of the Second Law"; *the first part "proves" the existence of this new state function* (and deals with some other consequences). All this will be the subject of the next four chapters.

At this stage it is instructive to introduce the entropy concept in a qualitative, "intuitive" way. This *does not prove anything*, but it *makes* the existence of such a state function *more plausible.* In section 9.2 we stressed that ultimately the First Law is based on the fundamental realization that *heat is a form of energy.* Thus energy passes from system A to system B:

1. When A performs work on B.
2. When heat flows from A to B.

Let us *compare* these *two modes of energy transfer* for the specific case that A and B are gases, separated by a movable diathermal wall [section 3.2(c)]. Let the pressure and temperature of A and B differ by (at most) only infinitesimal amounts, so that the *processes* involved (if they take place) are *reversible.*

---

* Pierre Eugène Marcellin Berthelot (1827–1907). He should not be confused with Claude Louis Berthollet (1748–1822), who played an important part in the development of the atomic-molecular theory (stoichiometry and the Law of Constant Proportions).

We observe:

A performs work on B	Heat flows from A to B
if	
the pressure in A is higher than that in B	the temperature of A is higher than that of B
in which case	
the volume of A increases at the cost of that in B.	an as yet unknown quantity **X** of A increases at the cost of that in B.

We know that for the *reversible* work done by A, we can write the product of the original pressure of A and the *change* in the volume of A. The analogy, we are trying to establish, would be complete if for the heat *reversibly* given off by A, we also could write the product of the original temperature of A and the *change* in the quantity **X** of A. The new state function **S**, introduced by the (first part of) the Second Law, actually turns out to be the negative of this quantity **X**. Unfortunately, it cannot be *visualized* as easily as the volume, **V**.

"Analogy reasoning" is both enlightening and dangerous; again, nothing is *proven* this way. We do not wish to attach too much importance to the analogy introduced here, since there are vital differences between mechanical work being performed and heat being transferred. Let us just point out two very striking ones. First, *mechanical equilibrium requires an equality of forces.* Only when we have the situation carefully chosen as in the example above, in which contact between the two gases is effectuated by means of a single movable partition, can this requirement be replaced by one of *equality of pressures.* In a more complicated setup, making proper use of a set of levers, it is possible to have equilibrium between gases of different pressure and to have gas A perform work on gas B even though B has the higher pressure. But when we look at heat transfer, equilibrium will *always* be associated with equality of temperatures and, without the occurrence of auxiliary processes (see the more precise discussion in section 11.2), heat will *always* flow from the higher to the lower temperature, *never* the other way: *There exist no temperature levers!* Second, the absence of P, **V** work *always* implies **V** to be constant, but $Q = 0$ implies $\Delta \mathbf{X} = 0$ only for *reversible* processes.

Some final comments are in order. In the first place we scrupulously abstain, at this level of discussion, from talking about entropy on a corpuscular, statistical, level. We shall treat the (*first part of*) *the Second Law* just as the First Law, namely, *as the definition of a state function.* (The reader may wish to review the first few paragraphs of section 9.2 before proceeding to the next chapters.) Accordingly, the *change in entropy* will be *defined by*

$$d\mathbf{S} \equiv \frac{dQ_{\text{rev}}}{T} \tag{11-2}$$

for infinitesimal changes, or by

$$\Delta \mathbf{S} \equiv \int_1^2 \frac{dQ_{\text{rev}}}{T} \tag{11-3}$$

for macroscopic processes. We have to prove our claim that **S**, thus defined, is a *state function*, and that this ultimately reflects some fundamental fact of experience in a powerful, i.e., mathematical, language. In later chapters (in particular, Chapters 15 and 16), we shall discuss the entropy briefly in a corpuscular way. It should be noted at this point that *the absolute temperature T has not been defined as yet* (see section 3.3); hence, strictly speaking, Eqs. (11-2) and (11-3) are presently meaningless. In the next two chapters we shall therefore reintroduce the empirical temperature t, until the appropriate time arrives to define T and to prove the latter's equivalence to the ideal gas temperature of section 3.3.

11.2 The Fundamental Facts of Experience Underlying the Second Law; the Clausius and Kelvin–Planck Principles

Let us immediately state the two "principles" involved (of course a slightly different wording is possible):

> It is impossible to devise a machine which, *operating in a cycle*, shall produce no effect other than the transfer of heat from a colder to a hotter body (without at the same time introducing some work into the system, which is converted into heat).
>
> —*Clausius*

> It is impossible to devise a machine which, *operating in a cycle*, shall produce no effect other than the extraction of heat from a reservoir and the performance of an equal amount of work (without at the same time transferring a certain amount of heat from a hotter to a colder body).
>
> —*Kelvin–Planck*

The two parts in parentheses *need not* be there, but in the author's opinion they have a clarifying effect. In what follows we shall always refer to a hypothetical experimental setup that would *violate* these two "principles" as an "anti-Clausius machine" (ACM) and an "anti-Kelvin machine" (AKM), respectively. Such machines would create what Wilhelm Ostwald has called a *perpetuum mobile of the second kind. They would in no way violate the First*

*Law** or, in Ostwald's terminology, they would *not* constitute a perpetuum mobile of the *first* kind. The qualification *operating in a cycle* should not be forgotten; its absence would allow "auxiliary" changes to take place, which, in turn, could make the main process possible.

Let us now put these two assertions into a practical framework. The Clausius Principle tells us, in so many words, that it is impossible to operate a refrigerator (which, after all, is designed to transfer heat from a colder to a hotter body) without cost (i.e., without putting work into it via an electrical outlet). The Kelvin–Planck Principle tells us that we cannot operate a power plant by extracting heat from "the air," and cannot run a ship across the ocean, and back, by extracting heat from said ocean ("the air" and the oceans to be considered reservoirs with unlimited amounts of heat available). Unfortunately, the ship needs to burn fuel!

At first sight the two principles appear to be totally unrelated, but it is possible to show that *if* one of them could be violated, so could the other. The easiest way to do this involves the concept of a Carnot cycle as a model for a heat engine or a refrigerator (section 12.1). Consequently, we postpone the analysis involved.†

By the time we shall have formulated the Second Law mathematically, and manipulated it to obtain the more interesting consequences, the Clausius and Kelvin–Planck Principles will be far behind us. But we should never forget that if *any* prediction on the basis of the Second Law would turn out to be false (that is, not in accordance with experiment), it should be possible to go "all the way back" and construct an AKM and/or an ACM, i.e., a *perpetuum mobile* of the second kind.

11.3 The Traditional—and the Axiomatic—Approach

We return briefly to the distinction already made in section 1.3, also encountered in conjunction with our discussion of the *First* Law (see, in particular, section 9.3). It should be reiterated, however, that the choice of method in presenting the *Second Law* is more or less independent of how we elected to formulate the *first*. In fact, every possible combination has probably found its way into *some* textbook(s). That is, some books give the traditional approach throughout, others consistently adhere to the axiomatic point of view,

---

* The energy balance is perfectly in order. In fact, the *reverse* of both processes *may occur* (*irreversibly*). Hence the Clausius and Kelvin–Planck Principles immediately introduce a *directional* feature, which was lacking within the framework of the First Law (see section 11.1).

† See the end of section 12.2.

while still others give the First Law axiomatically and follow this up by a traditional treatment of the Second Law (very few authors do just the reverse). As mentioned before, we shall give the traditional approach in great detail, followed by an outline of the axiomatic method. Those who don't want to study *both* alternatives can make their choice at this stage.

The traditional approach is associated with the names of Carnot, Clapeyron, Clausius, Thomson (Kelvin), Planck, and Poincaré. Our presentation will not attempt to follow any single person's work consistently; it is a "hybrid" in which an attempt has been made to incorporate the best features of several historical papers and existing textbooks. The axiomatic method* is usually named in one breath with (Born and) Carathéodory; elaborations and clarifications have been published by T. Ehrenfest-Afanassjewa, H. A. Buchdahl, and L. A. Turner, among others. Again, our presentation has profited from several sources.

REFERENCES

Most of these will be given in subsequent chapters. At this stage we draw attention to *The Second Law of Thermodynamics*, edited by J. Kestin (Benchmark Papers on Energy, Vol. 5, Dowden, Hutchinson & Ross, Stroudsburg, Pa., 1976). This is a series of annotated reprints (translated where necessary) of papers of historical interest. As far as the present author is aware, it is the *only* source for an English translation of C. Carathéodory's famous paper in *Math. Ann.* (*Berlin*) **67**, 355 (1909).

---

* As we shall see in Chapter 13, there is nothing really "axiomatic" about this elegant part of Carathéodory's treatment, provided that we are willing to interpret what he calls his *Axiom II* as (an immediate consequence of) a *Fundamental Fact of Experience.*

Chapter 12

The Traditional "Derivation" of the Second Law

12.1 The Carnot Cycle and the Carnot Theorem

In 1824, Sadi Carnot* devised an *idealized* cyclic process, which can represent the operation of either a heat engine or a refrigerator. This *Carnot cycle* consists of two reversible isotherms, at empirical temperatures t_1 and t_2, respectively ($t_1 > t_2$), linked by two isentropes (reversible adiabatic paths), as illustrated in Fig. 5(a). Note that, in our terminology, which is shared by many but not all other authors, a Carnot cycle is *reversible, by definition.* In discussing the heat flow along the two isotherms, and the work done by (or on) the cycle as a whole, it is convenient to deal with the absolute values of these quantities, to be denoted by $|Q|$ and $|W|$, respectively. If the Carnot cycle is operated in a *clockwise* direction [Fig. 5(b)], it represents a *heat engine*; $|Q_1|$ is absorbed from the reservoir at t_1, $|Q_2|$ (smaller than $|Q_1|$) is given off to the reservoir at t_2, and the difference, $|W|$, is *freed* in the form of work. In order to make the cycle run *counterclockwise* [Fig. 5(c)], we have to *put in* this amount of work. The cycle now could represent a *refrigerator*; it takes $|Q_2|$ from the reservoir at t_2, and gives off $|Q_1|$ at t_1. The equality

$$|W| = |Q_1| - |Q_2| \tag{12-1}$$

is an immediate consequence of the First Law, since $\Delta \mathbf{E} = Q - W = 0$ for *any* cyclic process.

Next, we define the *efficiency* of the cycle, ξ, by

$$\xi \equiv \frac{|W|}{|Q_1|} = 1 - \frac{|Q_2|}{|Q_1|}. \tag{12-2}$$

The name *efficiency* can be rationalized by considering the clockwise operation of the cycle. The goal of a heat engine is to convert a maximum fraction of the amount of heat, $|Q_1|$, into work. An engine with $\xi = 1$ ($|W| = |Q_1|$, $|Q_2| = 0$) would constitute an

---

* *"Reflections on the Motive Power of Fire," and Other Papers*, E. Mendoza, ed., Dover, New York, 1960.

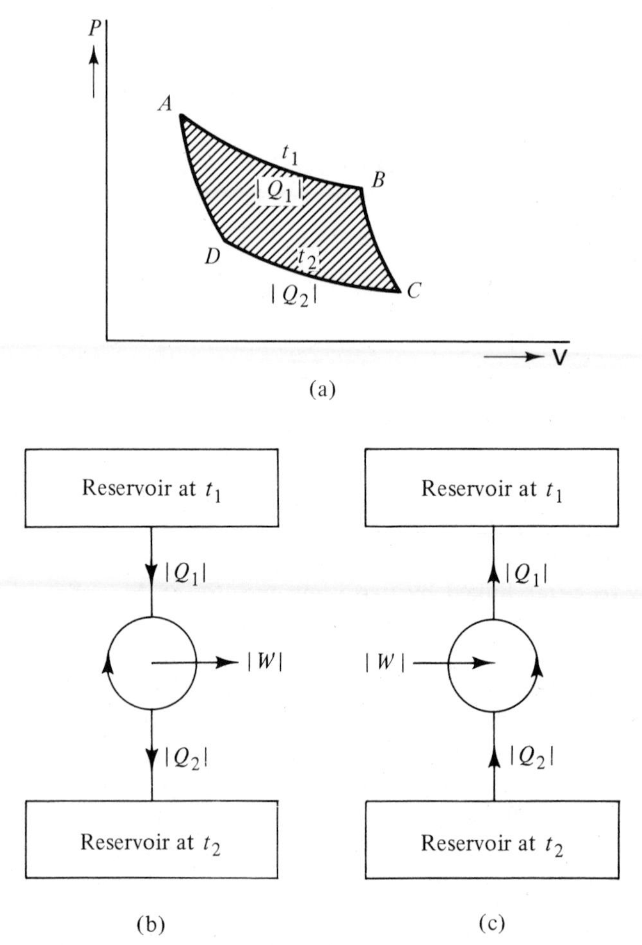

Fig. 5. The Carnot cycle. (a) Indicating reversible isotherms (*AB* and *CD*) and "isentropes," reversible adiabatic paths (*BC* and *AD*). The shaded area is the work, $|W|$, associated with the cycle. Here $t_1 > t_2$. The detailed shape of the cycle depends on the nature of the working substance. (b) Schematic representation of a Carnot cycle running clockwise (when it acts as a heat engine). (c) Schematic representation of a Carnot cycle running counterclockwise (when it acts as a refrigerator; for alternative uses see Problem 12-2).

AKM, since it would, operating in a cycle, extract an amount of heat from the reservoir at t_1 and convert this entirely into work, in violation of the Kelvin–Planck Principle. Hence ξ *must be* less than unity. Denbigh,* who gives an excellent discussion of these matters, prefers to call ξ the *conversion factor*. He points out that owing to

* K. G. Denbigh, *The Principles of Chemical Equilibrium*, 3rd ed., Cambridge University Press, Cambridge, 1971, p. 32.

the Carnot Theorem (see below), ξ is completely determined by t_1 and t_2, and has nothing to do with how efficiently or inefficiently we operate our engine. In other words, for the *idealized* reversible machine, $|W|$, as given by Eqs. (12-1) and (12-2), already represents the *maximum* amount of *work* [cf. section 5.3(a)] that can possibly be obtained from $|Q_1|$, once t_1 and t_2 are fixed. Any *real* engine operating between these two temperatures would certainly be associated with an even *smaller* $|W|/|Q_1|$. For this reason, some authors call ξ the *ideal* (thermodynamic) efficiency. Although this analysis is very much to the point, we shall conform to the more generally accepted terminology, and continue to refer to ξ, as defined by Eq. (12-2), as "the efficiency" of our Carnot cycle, which is reversible *by definition.* When such a cycle is utilized as a refrigerator, $|Q_2|/|W|$ is sometimes referred to as its "coefficient of performance."

The *Carnot Theorem* tells us:

> All Carnot cycles operating between the same temperatures have the same efficiency.

To prove the theorem for *all* cycles, it is necessary and sufficient to show its validity for *any two arbitrary* cycles operating between the reservoirs at t_1 and t_2. Since these two cycles will be characterized by primed and unprimed quantities (ξ' and ξ, $|W'|$ and $|W|$, etc.), we shall refer to them, colloquially, as the "primed cycle" and the "unprimed cycle," respectively.

It is most convenient, as well as illustrative, to give the proof in several stages.

(*a*) *First Stage*

We impose the restriction

$$|W| = |W'|; \tag{12-3}$$

the two cycles are associated with the same amount of work. Since, in a $P,\mathbf{V}$ diagram, this work is represented by the area inside the cycle (the shaded area in Fig. 5a), one expresses the restriction (12-3) by stipulating that *the two cycles must have the same "size"* (Fig. 6a). We shall now prove that

$$\xi = \xi', \tag{12-4}$$

by showing that if ξ were *larger* than ξ', we would be able to construct an ACM, i.e., violate the Clausius Principle.* To this purpose, we couple the

* In Exercise 12-1 the reader will be asked to rule out the case that ξ is *smaller* than ξ'. Please note that the acronyms ACM and AKM, introduced in the last paragraph of section 11.2, will be used throughout this chapter.

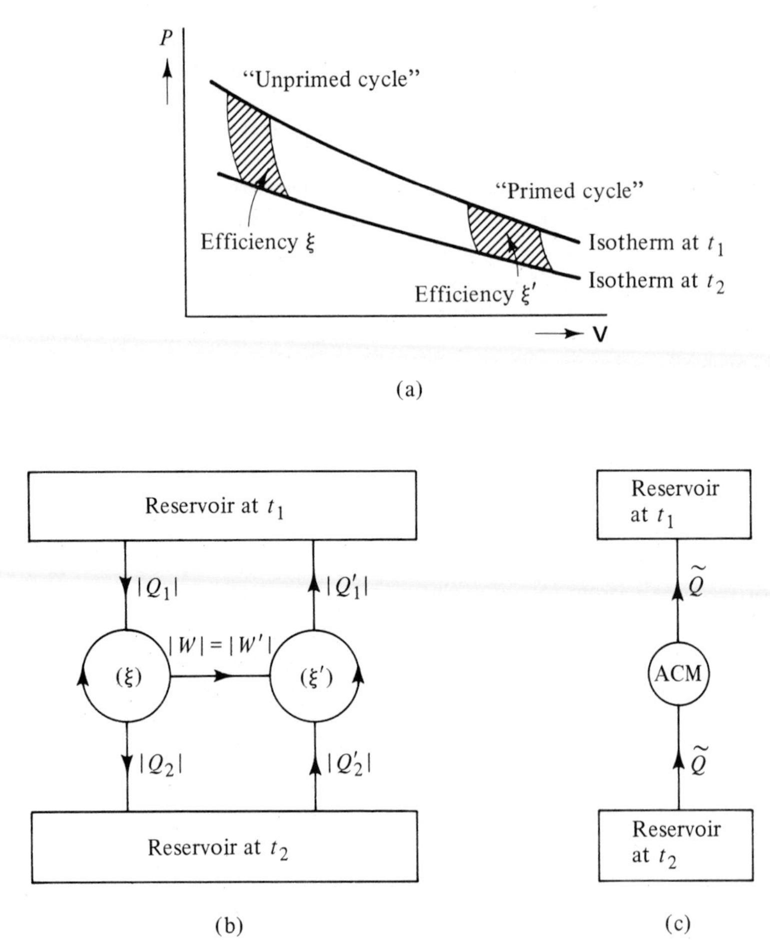

Fig. 6. (a) Two Carnot cycles of the same "size" (shaded portions have the same area), operating between the same two isotherms. (b) Under the assumption that $\xi > \xi'$, these two cycles are coupled in the sense that the unprimed one "drives" the primed one. (c) The resulting composite cycle is an ACM. ($\tilde{Q} \equiv |Q'_1| - |Q_1| = |Q'_2| - |Q_2|$ is positive.)

two cycles in such a manner that the more efficient one "drives" the less efficient one (Fig. 6b). In other words, the work produced by the unprimed cycle, running clockwise, is used for the counterclockwise operation of the primed one. The resulting composite cycle (Fig. 6c), for which the net work vanishes, removes a net amount of heat, $|Q'_2| - |Q_2|$, from the reservoir at the lower temperature, t_2, and discharges the amount $|Q'_1| - |Q_1|$ into the reservoir at the higher temperature, t_1. Since $\mathbf{\Delta E}$ for the composite cycle (as for *any* cycle) is zero, these two amounts of heat must be equal, in order to satisfy the *First* Law. In fact, by Eqs. (12-1) and (12-3),

$$|Q_1| - |Q_2| = |Q_1'| - |Q_2'|, \tag{12-5}$$

and hence

$$|Q_1'| - |Q_1| = |Q_2'| - |Q_2| \equiv \tilde{Q}, \text{ say.} \tag{12-6}$$

In order for us to conclude that the composite cycle is indeed an ACM, we still have to show that $\tilde{Q}$ is positive. (If $\tilde{Q}$ were negative, a net amount of heat would, in effect, flow from a higher to a lower temperature, and this situation would be entirely acceptable.) Therefore, to complete the proof of the first stage, we use our assumption of unequal efficiencies, $\xi > \xi'$, which by virtue of the definition (12-2) can be written

$$\frac{|W|}{|Q_1|} > \frac{|W'|}{|Q_1'|},$$

or by (12-3),

$$\frac{1}{|Q_1|} > \frac{1}{|Q_1'|};$$

hence

$$|Q_1'| > |Q_1|. \tag{12-7a}$$

From Eqs. (12-5) and (12-7a), we also have

$$|Q_2'| > |Q_2|. \tag{12-7b}$$

The equations (12-7a) and (12-7b) show that $\tilde{Q}$ of Eq. (12-6) is a positive quantity, so that our composite cycle would indeed violate the Clausius Principle.

Exercise 12-1

By similar reasoning, show that $\xi < \xi'$ is also impossible. *Note*: To prove this we must let the primed cycle "drive" the unprimed one (as in the text above, we must still let the *more* efficient cycle drive the *less* efficient one); the reverse arrangement would *not* yield an ACM. Readers should convince themselves that this does not introduce any logical inconsistencies. On the contrary, this simply confirms the *directional* features, associated with the foundations of the *Second Law.*

(*b*) *Second Stage*

Let us next impose the restriction

$$|W| = 2|W'|; \tag{12-8}$$

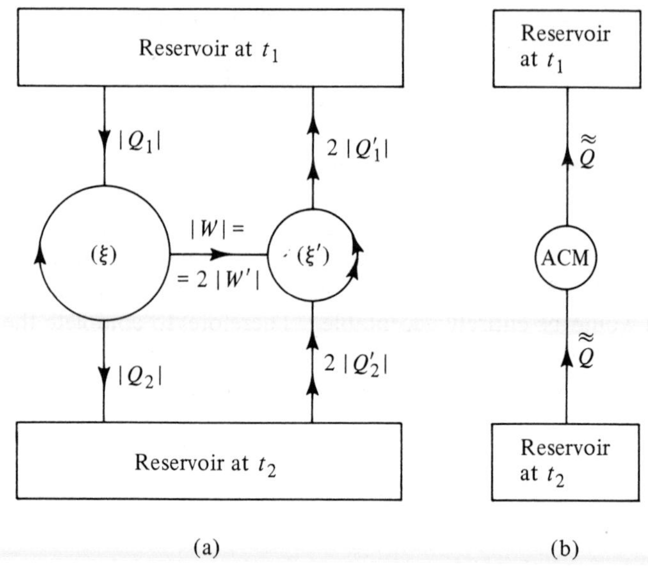

Fig. 7. (a) Coupling of two Carnot cycles as in Fig. 6(b), but with $|W| = 2|W'|$, again with the assumption that $\xi > \xi'$. (b) The resulting composite cycle is an ACM. ($\tilde{\tilde{Q}} \equiv 2|Q'_1| - |Q_1| = 2|Q'_2| - |Q_2|$ is a positive quantity.)

the unprimed cycle has twice the size of the primed one. As in the first stage, we shall prove that unless $\xi = \xi'$, we can construct an ACM. If ξ were larger than ξ', we would utilize the work obtained from *one* clockwise cycle of the unprimed system to "drive" the primed cycle counterclockwise *twice* (Fig. 7). The rest of the proof is left as an exercise for the readers.

Exercise 12-2

Prove the Carnot theorem for two cycles, satisfying Eq. (12-8). Refer to Fig. 7.

(c) *Third Stage*

Generalizations are now obvious. If

$$q|W| = p|W'|, \tag{12-9}$$

with $\xi > \xi'$, we would run the unprimed cycle clockwise q times, which provides us with just the right amount of work to "drive" the primed cycle counterclockwise p times, and so on. This completes the proof of the theorem.

Occasionally, at this point in the author's lectures, a student will claim that the proof of the third stage cannot be carried out whenever p/q is an irrational number. This should be countered by stressing that $|W|$ and $|W'|$ are not mathematical entities but *macroscopic observables.* The ratio $|W|/|W'|$ can *always* be represented by a (positive) rational number, *within the accuracy of the relevant observations.* To conclude that $|W|/|W'|$ is an irrational number, we would have to be able to measure W and W' with unlimited accuracy. Of course this is impossible, for a number of reasons. For example, it is important to realize that long before we reach the limits of our observational skills, *fluctuations* would interfere, thus reducing the accuracy of our measurements. Such fluctuations are disregarded in conventional phenomenological thermodynamics (see also the introduction to the Second Law; section 11.1).

12.2 The Absolute Temperature and Its Relation to the Ideal Gas Temperature

(a) In the preceding section we derived the Carnot Theorem *for an arbitrary system.* We found that the efficiency of a Carnot cycle, ξ, is some universal function of the two temperatures, t_1 and t_2, between which the cycle operates. By the definition (12-2), this implies that the quantity $|Q_1|/|Q_2|$ is *also* completely determined by these temperatures. Mathematically we express this as

$$\frac{|Q_1|}{|Q_2|} = f(t_1, t_2). \tag{12-10}$$

In Fig. 8 we show three isotherms, at t_1, t_2, and t_3, respectively ($t_1 > t_2 > t_3$), intersecting two isentropes *for an arbitrary system.* Let the heat connected with the three isothermal sections be $|Q_1|$, $|Q_2|$, and $|Q_3|$, respectively ($|Q_1| > |Q_2| > |Q_3|$). We have the mathematical identity

$$\frac{|Q_1|}{|Q_3|} = \frac{|Q_1|}{|Q_2|}\frac{|Q_2|}{|Q_3|}. \tag{12-11}$$

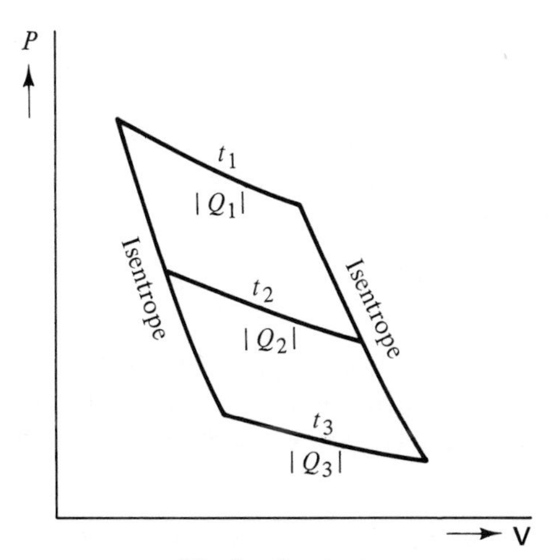

Fig. 8. See text.

But each of the three quotients in this equation, by Eq. (12-2), determines the efficiency of a Carnot cycle; the left-hand side refers to the "big" cycle between t_1 and t_3, and the right-hand side refers to those involving t_1, t_2 and t_2, t_3, respectively. Therefore, by Eq. (12-10), Eq. (12-11) becomes

$$f(t_1, t_3) = f(t_1, t_2) f(t_2, t_3). \tag{12-12}$$

This "cycle relation" implies that

$$f(t_i, t_j) = \frac{T(t_i)}{T(t_j)} \quad \text{for all } t_i, t_j, \tag{12-13}$$

in which T is a universal function of the empirical temperature t.

It is trivial to show that Eq. (12-13) is *a possible* solution of Eq. (12-12), but it is not so easy to prove that it is *the only* solution, i.e., that Eq. (12-12) *requires* f to be of the form (12-13). The interested reader will find a proof in Denbigh's monograph.*

From Eqs. (12-10) and (12-13) we have the three identities

$$\frac{|Q_1|}{|Q_2|} = \frac{T(t_1)}{T(t_2)}, \qquad \frac{|Q_2|}{|Q_3|} = \frac{T(t_2)}{T(t_3)}, \qquad \frac{|Q_1|}{|Q_3|} = \frac{T(t_1)}{T(t_3)}. \tag{12-14}$$

But from our study of Carnot cycles we know that, given the condition

$$t_1 > t_2 > t_3, \tag{12-15}$$

it follows that

$$|Q_1| > |Q_2| > |Q_3|, \tag{12-16}$$

so that the three equations (12-14) give us the inequality

$$T(t_1) > T(t_2) > T(t_3). \tag{12-17}$$

Therefore, we have not only shown that *T is a universal function of t*, but also that *T increases monotonically with t*.

The combined results (12-14) and (12-17) suggested to Lord Kelvin that one should, in effect, use the Carnot cycle as an *absolute thermometer*, and call T the *absolute temperature*:

$$\frac{|Q_i|}{|Q_j|} = \frac{T_i}{T_j} \quad \text{for all } t_i, t_j. \tag{12-18}$$

The designation *absolute* seemed appropriate to Kelvin, because this thermometer is *independent of* (the) *substance* (with which one chooses to execute the cycle). Of course, Eq. (12-18), by itself, does not completely specify T;

* See the footnote on page 30 of the reference given in our footnote on p. 108.

to determine unambiguously this *absolute, thermodynamic,* or *Kelvin* scale,* we require, in addition,

$$T_{\substack{\text{normal boiling}\\ \text{point of water}}} - T_{\substack{\text{normal melting}\\ \text{point of ice}}} = 100. \tag{12-19}$$

This ensures that the *size* of the Kelvin degree is the same as that of the centigrade degree, which, in turn, equals that of the ideal gas scale (section 3.3). Equation (12-18) suggests that all T_i have the same sign; we shall deal with positive absolute temperatures exclusively. As has been pointed out by Pippard,† negative absolute temperatures never occur for entire systems in equilibrium, only for some very special subsystems that can occasionally be identified. These matters fall beyond the scope of this book.‡

(b) We shall now prove that the Kelvin temperature and the ideal gas temperature are identical, as we have tentatively assumed ever since the end of section 3.3. To this purpose we take 1 mole of an ideal gas through the Carnot cycle, drawn in Fig. 5a, in the clockwise direction. Along the two isotherms, AB and CD, ΔE is zero by the definition of our model [see Eq. (10-19b)]; hence $W_{A\to B} = |Q_1|$ and $-W_{C\to D} = W_{D\to C} = |Q_2|$. Therefore, by (3-5),

$$|Q_1| = \int_A^B P\,dV = R\theta_1(t)\ln\frac{V_B}{V_A}$$

and

$$|Q_2| = \int_D^C P\,dV = R\theta_2(t)\ln\frac{V_C}{V_D}.$$

Hence

$$\frac{|Q_1|}{|Q_2|} = \frac{\theta_1(t)}{\theta_2(t)}\left[\frac{\ln(V_B/V_A)}{\ln(V_C/V_D)}\right]. \tag{12-20}$$

In order to evaluate the expression in brackets, we have to make use of the fact that the line segments BC and AD (Fig. 5a) represent isentropes, i.e., reversible adiabatic processes. For these paths we adapt the First Law, in

---

* See also footnote, p. 22.

† A. B. Pippard, *The Elements of Classical Thermodynamics*, Cambridge University Press, Cambridge, 1947, p. 52.

‡ The interested reader may consult:

(i) N. F. Ramsey, "Thermodynamics and Statistical Mechanics at Negative Absolute Temperatures," *Phys. Rev.* **103**, 20 (1956).

(ii) L. P. Bazarou, *Thermodynamics*, F. Immirzi, trans., and A. E. J. Hayes, ed., Pergamon Press, Oxford, 1964, Chapter X.

the form (10-21), to give for 1 mole of an ideal gas,

$$0 = C_v(t)\,dt + \frac{R\theta(t)}{V}\,dV.$$

In going from B to C,

$$R \ln \frac{V_C}{V_B} = -\int_{t_1}^{t_2} \frac{C_v(t)}{\theta(t)}\,dt, \tag{12-21a}$$

and in going from A to D,

$$R \ln \frac{V_D}{V_A} = -\int_{t_1}^{t_2} \frac{C_v(t)}{\theta(t)}\,dt. \tag{12-21b}$$

Since the right-hand sides of the two Eqs. (12-21) are identical,* the left-hand sides must also be equal:

$$\ln \frac{V_C}{V_B} = \ln \frac{V_D}{V_A}, \quad \text{hence} \quad \frac{V_C}{V_B} = \frac{V_D}{V_A} \quad \text{or} \quad \frac{V_B}{V_A} = \frac{V_C}{V_D};$$

therefore,

$$\ln \frac{V_B}{V_A} = \ln \frac{V_C}{V_D},$$

so that (12-20) reduces to

$$\frac{|Q_1|}{|Q_2|} = \frac{\theta_1(t)}{\theta_2(t)};$$

hence, by Eq. (12-18),

$$\frac{T_1}{T_2} = \frac{\theta_1}{\theta_2}. \tag{12-22}$$

This result, *by itself*, would only allow us to conclude that T is some constant times θ, but *by convention* we have made the size of the respective degrees equal [cf. Eqs. (3-6) and (12-19)]; hence we may conclude that

$$T_i = \theta_i \qquad \text{for all } i, \tag{12-23}$$

the desired result.

As an appendix to this section we shall demonstrate the equivalence of the Clausius and Kelvin–Planck Principles (see section 11.2) by showing that, *given* an AKM, *we can construct* an ACM, as illustrated in Fig. 9. (By now the reader should be convinced that drawing the appropriate diagram accounts for ninety percent of the proof.) The AKM takes an amount of

---

* Note that in order to conclude this, we have to know only that C_v is a function of t exclusively (cf. section 10.2), not the detailed form of this functional dependence.

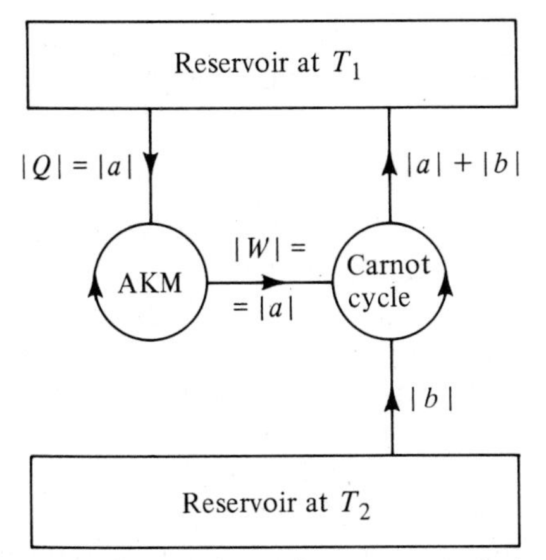

Fig. 9. An AKM drives a Carnot cycle counterclockwise. The resulting composite machine is an ACM (see text).

heat, called $|a|$, out of the reservoir at T_1 and converts this completely into work. The latter is used to "drive" the specified Carnot cycle, between T_1 and T_2, counterclockwise. [T_2 and $|b|$ are not entirely arbitrary but, through Eq. (12-18), related by

$$\frac{T_1}{T_2} = \frac{|a| + |b|}{|b|}.]$$

The resulting *composite* machine is a cycle, hence $\Delta \mathbf{E}$ is zero. But since W vanishes as well, the *net* effect is to remove an amount of heat, $|b|$, from the reservoir at T_2 and deposit it into the reservoir at the higher temperature T_1. This constitutes an ACM.

Exercise 12-3

Show that given an ACM, we can construct an AKM. See also Problem 12-1.

12.3 Proof that the Entropy Is a State Function; the Clausius Inequality

We have shown that for any Carnot cycle

$$\frac{Q_1}{T_1} + \frac{Q_2}{T_2} = 0, \tag{12-24}$$

in which Q_1 and Q_2 have opposite signs. Since, by definition, a Carnot cycle is reversible, this equation can be considered a special case of

$$\oint \frac{dQ_{\text{rev}}}{T} \equiv \oint d\mathbf{S} = 0. \tag{12-25}$$

In this section, we shall prove that (12-25) holds for an *arbitrary* system taken through an *arbitrary reversible* cycle. As we pointed out in Chapter 6, the validity of (12-25), under these circumstances, is a necessary and sufficient condition for $d\mathbf{S}$, as defined by (11-2), to be an exact differential, and $\mathbf{S}$ to be a state function.

Consider an arbitrary cycle (I), Fig. 10, without stipulating as yet that it is reversible, though we shall require that T is well defined at every stage (compare the discussion in section 3.2). We imagine (I) to be subdivided into very small parts, so that along each of these the temperature may be considered constant. Next, we concentrate on one such small segment, i, going from a to b. Along this segment, which may be reversible or irreversible, the system will absorb an amount of heat, δQ_i (which can be positive or negative), at the temperature T_i. We shall assume that this amount of heat is obtained from (or given to) a Carnot cycle, which operates (reversibly) between T_i and a heat reservoir, R, at T_o. In Fig. 10 this cycle is indicated *in a purely schematic way*; its heat flows, at T_i and T_o, are denoted by δQ_{i1} and δQ_{i2}, respectively. Let δQ_{io} be the heat absorbed by the reservoir, R, in order to drive

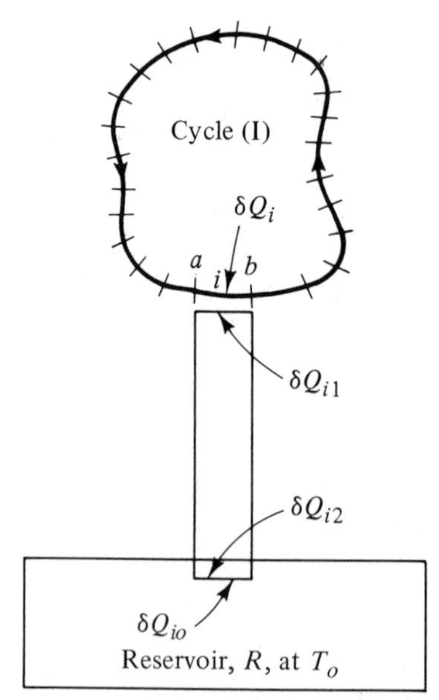

Fig. 10. Illustrating the proof given in section 12.3.

this particular Carnot cycle. For this cycle, by (12-24),

$$\frac{\delta Q_{i1}}{T_i} + \frac{\delta Q_{i2}}{T_o} = 0. \tag{12-26}$$

But since $\delta Q_{i1} = -\delta Q_i$, and $\delta Q_{i2} = -\delta Q_{io}$, Eq. (12-26) is equivalent to

$$\frac{\delta Q_i}{T_i} + \frac{\delta Q_{io}}{T_o} = 0. \tag{12-27}$$

Let us now traverse the entire cycle (I) (until we return to our initial position), with the aid of an infinite number of such "narrow" Carnot cycles, all linked to the *same* reservoir, R. Summing Eq. (12-27) over all small segments, i, we get

$$\sum_i \frac{\delta Q_i}{T_i} = -\frac{1}{T_o}\sum_i \delta Q_{io}. \tag{12-28}$$

In the limit of infinitesimally small segments, the left-hand side can be replaced by the integral $\oint_{(I)} dQ/T$; hence we can write

$$\oint_{(I)} \frac{dQ}{T} = \frac{-Q_o}{T_o}, \tag{12-29}$$

in which $-Q_o\,(\equiv -\sum_i \delta Q_{io})$ *is the total amount of heat taken from* R, in order to complete cycle (I).

Next we consider the effects of a composite machine, consisting of cycle (I) plus all the necessary Carnot cycles. When (I) has returned to its initial position, so have all these auxiliary cycles, and the composite machine has returned to its original state as well. Therefore, for this composite machine, $\Delta \mathbf{E} = 0$, and the net effect of its operation is that a certain amount of heat, $|Q_o|$, is given to, or taken from, reservoir R and, by the First Law, an equivalent amount of work is obtained from, or given to, the environment.

Finally, we must consider the vital *directional* features of the overall process under consideration; hence we must take into account the possible *signs* of the net effects involved. It should be reiterated that *we have not yet assumed that cycle (I) is reversible** (although all auxiliary Carnot cycles *are*). This allows us to consider several cases:

Case I

$$-Q_o > 0, \qquad \oint_{(I)} \frac{dQ}{T} > 0.$$

* The fact that we have chosen a counterclockwise direction for its operation, in Fig. 10, is entirely immaterial.

In this case the net effect the composite machine produces, in a cyclic operation, would be to extract heat from reservoir R and completely transform it into work. This would violate the Kelvin–Planck Principle; hence we can conclude that for our arbitrary cycle

$$\oint_{(I)} \frac{dQ}{T} > 0 \quad \textit{is impossible under all conditions, reversible or irreversible.} \tag{A}$$

Case II

$$-Q_o < 0 \qquad \oint_{(I)} \frac{dQ}{T} < 0.$$

In this case we further distinguish between the two remaining possibilities:

Case IIa. Cycle (I) is irreversible.

This situation is perfectly legitimate. With our earlier stipulation that T be well defined at every stage, we can conclude that

$$\oint_{(I)} \frac{dQ}{T} < 0 \quad \textit{for any irreversible cycle.} \tag{12-30a}$$

This is the famous Clausius inequality.

Case IIb. Cycle (I) is reversible.

Now the entire operation of the composite machine can be reversed, which would lead us right back to Case I, in violation of the Kelvin–Planck Principle. Conclusion:

$$\oint_{(I)} \frac{dQ}{T} < 0 \quad \textit{is impossible for all reversible cycles.} \tag{B}$$

Taken together, statements (A) and (B) tell us that for a reversible process Q_o can be neither positive nor negative; hence it must be identically zero. We have therefore obtained the important result [cf. Eq. (12-29)]

$$\oint_{(I)} \frac{dQ}{T} = 0 \quad \textit{for any reversible cycle.} \tag{12-30b}$$

Referring back to Eq. (12-25) and the discussion in the ensuing paragraph, we see that in obtaining Eq. (12-30b), we have indeed completed the proof that **S** is a state function. The Clausius inequality, Eq. (12-30a) appears as a "bonus," which will form the basis for what we already called (see section 11.1) the *second part of* the Second Law. Its important implications will be taken up in Chapter 15. Frequently, one combines two equations, such as

Eqs. (12-30a) and (12-30b) into one, and writes [henceforth discarding the designation (I)]

$$\oint \frac{dQ}{T} \leq 0. \tag{12-31}$$

In equations of this type, it is always implied:

1. That the *inequality* sign, *as written*, pertains to *irreversible* (spontaneous, "natural") processes.
2. That the *equality* sign refers to the limiting case of reversible (equilibrium) processes.
3. That the same equation with the *reverse inequality*, in this case $\oint dQ/T > 0$, reflects an *impossible** process.

A few concluding comments. The proof in this section utilized the Kelvin–Planck Principle. Many alternative procedures are possible. Some of these employ the Clausius Principle, which we already know to be entirely equivalent to the Kelvin–Planck formulation (see the end of section 12.2 and Exercise 12-3). It would be instructive for the reader to skim through this *entire* chapter again and note that we have used nothing but the ("nondirectional") First Law, and the ("directional") fundamental facts of experience, introduced in section 11.2.

PROBLEMS

Problem 12-1

An anti-Clausius machine (ACM) and a Carnot cycle (CC) can operate between the same two reservoirs, at 1500 K and at 500 K, respectively. The ACM transports 300 kJ of heat from the low-temperature to the high-temperature reservoir, and the CC produces 200 kJ of work every time it executes one cycle (clockwise). Combine the operation of the ACM with that of the CC in such a way that the resulting machine violates the Kelvin–Planck Principle (i.e., is an AKM),

(a) by letting this AKM extract its heat from the reservoir at 500 K, and
(b) by letting this AKM extract its heat from the reservoir at 1500 K.

Problem 12-2

Most problems dealing directly with the subject matter of this chapter belong to the realm of engineering thermodynamics and fall outside the scope of this book. In the present problem we give an example of practical importance.

A Carnot cycle, running counterclockwise, can be considered as a model for a *heat pump*, which can be used to heat a home during the winter and can act as an air conditioner in the summer.

---

* Some prefer *nonspontaneous* or *unnatural* (Planck).

(a) Let the house temperature be T_1 and the outside temperature be T_2 ($T_1 > T_2$). Obtain the ratio $|Q_1|/|W|$ ($|Q_1|$ is the heat *pumped into* the house) in terms of T_1 and T_2. Compute this ratio for a winter day when the outside temperature is 10°C, while the house temperature is to be maintained at 20°C. Of course, in practice a heat pump does not operate reversibly, and its performance characteristics will be considerably less than that suggested by the value you just computed. Nevertheless, $|Q_1|/|W|$ will still be considerably larger than unity (which is the maximum it could be for an electric heater).

(b) In summer we have to reverse the heat flow. Now the outside temperature is T_1 and the house temperature is T_2 (still $T_1 > T_2$). Express the ratio $|Q_2|/|W|$ ($|Q_2|$ is now the heat *extracted* from the house) in terms of T_1 and T_2. Compute this ratio for a summer day when the outside temperature is 30°C, while the house temperature is still to be maintained at 20°C.

Note that either mode of operation of the *heat pump* is the more effective, the smaller ($T_1 - T_2$). This means that the outside should not be *too* cold in winter nor *too* hot in summer. Of course, this makes eminent sense intuitively.

Problem 12-3

Two students live in an apartment with a volume of 300 m^3. On a typical summer day, its temperature reaches 27°C. The apartment came with a refrigerator, which operates with an efficiency as if it were a Carnot cycle; its cooling (expansion) coils are kept at 0°C. After the students have ascertained that under these conditions, the refrigerator is capable of freezing 1 kg of ice *per hour of continuous operation*, they decide to cool their apartment by opening the refrigerator door.

(a) Will this idea work?

(b) What is the final temperature, t_f, of the apartment after 10 hours of continuous refrigerator operation? (Assume that C_v for air is constant and equal to $\approx 2.5R$ per mole; the heat of fusion of ice is 335 kJ/kg.)

(c) If the refrigerator did not operate with its ideal thermodynamic efficiency, would t_f be higher or lower?

REFERENCES

There are obviously many sources for this "traditional derivation." We abstain from recommending any particular one, but instead point to various references and acknowledgments in footnotes. Of historical interest are (see also the end of Chapter 11):

1. R. Clausius, *The Mechanical Theory of Heat, etc.* T. Archer Hirst, ed., Taylor & Francis, London, 1867.
2. M. Planck, *Treatise on Thermodynamics*, 3rd English edition, translated from the 7th German edition by A. Ogg, Dover, New York, 1926. The first German edition was published in 1897.
3. H. Poincaré, *Thermodynamique*, Carré, Paris, 1892.

Chapter 13

Outline of the Carathéodory "Derivation" of the Second Law

13.1 The (Restricted) Carathéodory Principle and Its Relation to the Clausius and Kelvin–Planck Principles

The various cycles and machines used in the traditional derivation of the Second Law (as presented in Chapter 12) appear to be, in the words of Max Born,* "strange conceptions, obviously borrowed from engineering." Although he expressed admiration for the pioneers of this approach, Born confessed that even as a student he already objected to its deviation from the "ordinary methods of physics." Consequently, he discussed the problem with his friend, the mathematician Carathéodory, which resulted in the development of an entirely new methodology in deriving the Second Law, much more satisfactory to Born and many of his contemporaries. It is this derivation which is the subject of the present chapter.

Carathéodory's analysis[†] utilizes the mathematical concepts and techniques discussed in Chapter 7. As we shall see in section 13.2, it combines the *purely mathematical* Carathéodory *Theorem* of section 7.3 with a newly discovered *fundamental fact of experience*,[‡] the Carathéodory *Principle*,[§] also known as the *Principle of Adiabatic Inaccessibility*:

> In the vicinity of *any* given state of *any* thermodynamic system, there exist *some* states that cannot be reached from it by *any* adiabatic process.

---

* Max Born, reference 2 of Chapter 7, p. 38.

[†] C. Carathéodory; see the references of Chapter 11.

[‡] In Carathéodory's paper it is presented as his second *Axiom*.

[§] Unfortunately, the *Theorem* and the *Principle* are sometimes confused.

If for some reason, one wishes to consider *reversible** processes *only*, one may, of course, replace the words "*any* adiabatic process" in the statement above by "*any reversible* adiabatic process." With this stipulation one obtains what is frequently referred to as the *Restricted Carathéodory Principle.*

To bring the Carathéodory Principle "down to earth," let us give a simple example. As our system we take 1 mole of a gas, with its initial state specified by the variables V and t.† We can easily see that the states $(V + dV)$, t are generally inaccessible from V, t by a reversible adiabatic process. For any gas undergoing such a change, the First Law gives us

$$0 = C_v\,dt + \left[P + \left(\frac{\partial E}{\partial V}\right)_t\right] dV, \tag{13-1}$$

i.e., an adiabatic volume increase is normally associated with a temperature change. Carathéodory himself gave this example in 1909, and it has been repeated in nearly all monographs that present his methodology. A more general analysis of this principle was not published until more than fifty years later, when Crawford and Oppenheim‡ proved its equivalence to the "old" Clausius and Kelvin (–Planck) Principles of section 11.2. The rest of this section is devoted to some aspects of this proof.

Figure 11 represents a system characterized by two state variables, the empirical temperature, t, and a single generalized coordinate (see section 4.5),

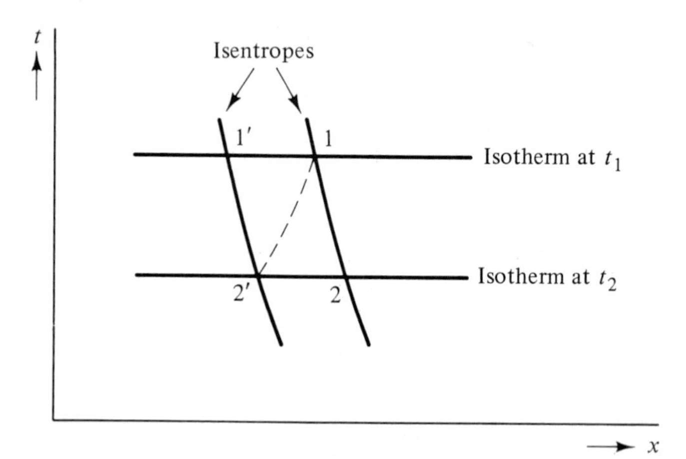

Fig. 11. Illustrating the relation between the Kelvin-Planck Principle and the Carathéodory Principle.

---

* *Reversible* to be taken as *retraceable* or *quasi-static* in the original Carathéodory sense. See the discussion of these concepts in Chapter 5.

† As in Chapter 12, we shall have to work with the empirical temperature, t, until the absolute temperature, T, is introduced.

‡ Bryce Crawford, Jr., and I. Oppenheim, *J. Chem. Phys.* **34**, 1621 (1961).

x, e.g., the volume **V**. For simplicity (the restriction is *not* essential), we assume that in the relevant range of t,x values no phase transitions occur, so that we may assume continuity of heat absorption along the isotherms. To be specific, we shall stipulate that Q is positive (the system absorbs heat from the environment) as x increases along an isotherm. This would indeed be the case if x were **V**. (With alternative choices of generalized coordinates, an increase in x might be associated with negative Q. The entire reasoning below would still be valid, provided that we take the isentrope $1'$-$2'$ to the *right* of the isentrope 1-2.)

Let point 1 in Fig. 11 represent an entirely arbitrary initial state. We claim that any point $2'$ to the left of the isentrope through 1 is inaccessible from 1 by *any* adiabatic process, reversible or irreversible. (In Fig. 11 we have chosen $t_2 < t_1$. The case $t_2 = t_1$ is trivial, and readers should have no difficulty in adapting the ensuing arguments for the case $t_2 > t_1$.) To prove this let us assume that we *could* reach $2'$ from 1 by *some* adiabatic path, say, the dashed line in Fig. 11. Then we could complete the cycle $1 \rightarrow 2' \rightarrow 1' \rightarrow 1$ (the direction is vital!). Along this path, the only heat exchanged between the system and its surroundings would be $Q_{1' \rightarrow 1}$, which is *positive*. Since for *any* cycle $\Delta \mathbf{E}$ vanishes, W $(= Q_{1' \rightarrow 1})$ is also positive. We thus violate the Kelvin (–Planck) Principle; our cycle would constitute an AKM in the terminology introduced in section 11.2.

Since t_2 is arbitrary, the entire region *to the left of* the isentrope through 1 is adiabatically inaccessible from 1. This is a *finite* region containing an *infinite* number of points, some of which lie arbitrarily close to 1 or, colloquially speaking, "in the vicinity (neighborhood) of 1." (Compare the formulation of the Principle of Adiabatic Inaccessibility at the beginning of this section.) In passing we note that states corresponding to *points on the isentrope* through 1 are adiabatically *accessible* from 1 by a *reversible* process. It is left to the reader to verify that points to the right of this isentrope *can be* reached adiabatically from 1, but by means of irreversible changes only. Further implications of these results are discussed in section 15.1.

Exercise 13-1

Prove that two isentropes can never cross. *Hint*: If they did, we could construct an AKM as in the proof given above.

So far we have proven the equivalence of the Carathéodory and the Kelvin Principles only for a system that can be characterized by *two* state variables, but generalizations are obvious. If a system has to be specified by t, and r generalized coordinates x_i $(i = 1, 2, \ldots, r)$, a similar analysis can be given for any t,x_i plane in the $(r + 1)$-dimensional space involved. The adiabatically *inaccessible region* becomes *a volume*, which includes points arbitrarily close to the state 1. The allowed *reversible* adiabatic processes originating from 1 are now confined to an isentropic *surface*, which separates

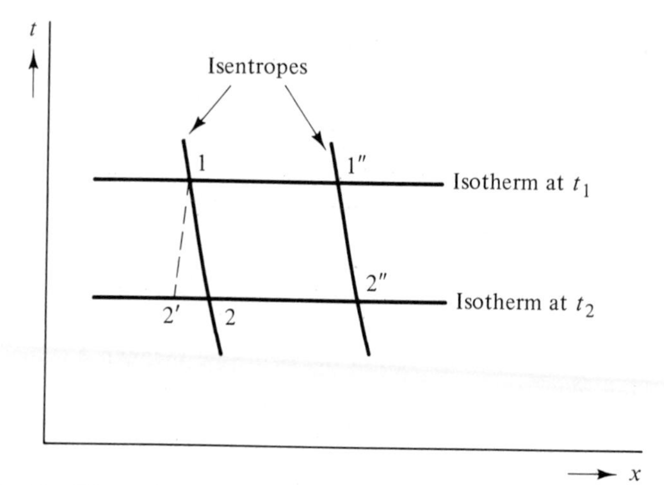

Fig. 12. Illustrating the relation between the Clausius Principle and the Carathéodory Principle.

the inaccessible volume from a volume accessible by *irreversible* changes only.

Since we have already established the equivalence of the Kelvin–Planck Principle and the Clausius Principle (see the end of section 12.2), it is strictly unnecessary to link the latter with the Carathéodory Principle directly. However, since the procedure is instructive, it is desirable that readers work it out as an exercise. As an aid, the author provides Fig. 12, which pertains to a system similar to that considered in Fig. 11. In particular, we associate the same continuity property and directional sign with the heat flow along the isotherms. If we, once more, focus our attention on processes originating from an arbitrary point 1, at t_1, the final result, of course, has to be the same as before: No point to the left of the isentrope through 1 should be adiabatically accessible from 1 by *any* process, reversible or irreversible. In Fig. 12, as in Fig. 11, 2′ is taken to be such a point, at an arbitrary temperature t_2 [$t_2 < t_1$; see also Exercise 13-2(c)]. The diagram is completed by drawing the isentrope 1″-2″ in such a way that $Q_{1 \to 1''} = Q_{2' \to 2''}$ (positive quantities).

Exercise 13-2

(See Fig. 12, and the preceding paragraph.)

(a) Show that if we could reach 2′ from 1 by any adiabatic process, the cycle $1 \to 2' \to 2'' \to 1'' \to 1$ would constitute an ACM.

(b) Reflect on the nature of the cycle $1 \to 2 \to 2'' \to 1'' \to 1$, with $x \equiv \mathbf{V}$.

(c) How would you modify Fig. 12, and the analysis given in part (a) above, if the point 2′ (still to the left of the isentrope through 1) were located at a temperature *above* t_1? (Watch out for directional features: the reverse of an ACM is perfectly legitimate!)

As a result of the analysis of this section, we may consider the *three principles* as occupying *equivalent* corners of a triangle. Some authors prefer to consider the *Carathéodory* Principle as a *consequence* of the two others. This is a matter of personal opinion, based on historical precedent and "philosophical" preference.

13.2 The Absolute Temperature as an Integrating Denominator

(For a proper understanding of this section, mastery of the contents of Chapter 7 is essential.)

For a reversible* process, we previously wrote (see Chapter 4; the X_i and x_i are generalized forces and generalized coordinates, respectively)

$$dW = \sum_{i=1}^{r} X_i \, dx_i. \tag{4-18}$$

Since $\mathbf{E}$ is a function of t and the x_i, the First Law gives

$$dQ = \mathbf{C}_{x_1, \ldots, x_r} \, dt + \sum_{i=1}^{r} \left(\frac{\partial \mathbf{E}}{\partial x_i} + X_i \right) dx_i. \tag{13-2}$$

Formally, Eq. (13-2) can be written

$$dQ = \sum_{i=1}^{r+1} Y_i \, dy_i, \tag{13-3}$$

and thus represents a Pfaffian expression in $(r + 1)$ variables for the inexact differential dQ. The (restricted) Carathéodory *Principle*, discussed in the preceding section, tells us that in the vicinity of any point in this $(r + 1)$-dimensional space there exist adiabatically inaccessible points, i.e., states that cannot be reached (reversibly) along surfaces satisfying $dQ = 0$. Hence the Carathéodory *Theorem* of section 12.3 tells us that dQ must have an (infinite number of) integrating factor(s) [or denominators]. Therefore, we can write

$$dQ = T(t, x_1, \ldots, x_r) \, d\mathbf{S}(t, x_1, \ldots, x_r), \tag{13-4}$$

in which T is such a denominator and $d\mathbf{S}$ *is an exact differential* ($\mathbf{S}$ *is a state function*). Although we admit that this derivation required a lot of groundwork (parts of Chapters 4 and 5, the entire Chapter 7, and section 13.1), this section so far has been very brief, simple, and straightforward. But we have by no means finished! In fact, the T and $\mathbf{S}$ of Eq. (13-4) are, as yet, fairly arbitrary quantities. Our next task is to show that we may take T to be a

* We repeat: "Reversible" is to be taken in the sense of *retraceable*, hence *quasi-static* in the Carathéodory sense (Chapter 5). This reversibility is *implied* throughout the rest of this section.

function of the empirical temperature, t, only. This can be done in a variety of ways;* some authors give a lengthy very mathematical proof, but we shall join those who are satisfied with a more "intuitive" approach, which makes no claim to complete rigor.

Let us impose the restriction that $\mathbf{S}$, the entropy, is extensive. This is not implausible since for any reversible process, by Eq. (13-4), $d\mathbf{S}$ is proportional to dQ, and the heat absorbed (from the environment) by the system as a whole is the sum of those heats for any arbitrary set of subsystems into which this system can be divided. If there are N such subsystems, the restriction requires that

$$\mathbf{S} = \sum_{k=1}^{N} \mathbf{S}_k, \tag{13-5}$$

so that

$$T\,d\mathbf{S} = T \sum_{k=1}^{N} d\mathbf{S}_k. \tag{13-6}$$

Equation (13-4) holds for all systems; hence it must also be true for each subsystem k:

$$dQ_k = T_k\,d\mathbf{S}_k \qquad (k = 1, \ldots, r). \tag{13-7}$$

But since

$$dQ = \sum_{k=1}^{N} dQ_k, \tag{13-8}$$

it follows from Eq. (13-7) that

$$dQ = \sum_{k=1}^{N} T_k\,d\mathbf{S}_k. \tag{13-9}$$

Finally, we equate the right-hand sides of Eqs. (13-4) and (13-9) to yield

$$T\,d\mathbf{S} = \sum_{k=1}^{N} T_k\,d\mathbf{S}_k. \tag{13-10}$$

Since Eqs. (13-6) *and* (13-10) *must both be true for any arbitrary subdivision,* it necessarily implies that all T_k's are the same and equal to T: The integrating denominator has the same value for all subsystems as well as for the system as a whole. This, in turn, means that T *cannot* be a function of any of the coordinates $x_1, \ldots, x_r$, since these can vary widely as we *arbitrarily* divide our system into subsystems. Neither can T be a constant, since then dQ and all the dQ_k's would also be exact differentials, which, in general, they are not.

* See the references at the end of the chapter.

Thus the only remaining possibility *forces* T to be *a function of the empirical temperature*, t, *exclusively*; t is obviously the same for all subsystems. Since our original system was entirely arbitrary, T has to be a *universal* function of t.

To complete this "derivation," we still have to show, as in the traditional approach (section 12.2), that $T(t)$ and the ideal gas temperature $\theta(t)$ of Chapter 3 are identical. We start with the First Law for a single-component gas:

$$dQ = \left(\frac{\partial \mathbf{E}}{\partial t}\right)_{\mathbf{V}} dt + \left[P + \left(\frac{\partial \mathbf{E}}{\partial \mathbf{V}}\right)_t\right] d\mathbf{V}. \tag{13-11}$$

By (13-4) this tells us that

$$d\mathbf{S} = \frac{1}{T}\left(\frac{\partial \mathbf{E}}{\partial t}\right)_{\mathbf{V}} dt + \frac{1}{T}\left[P + \left(\frac{\partial \mathbf{E}}{\partial \mathbf{V}}\right)_t\right] d\mathbf{V}. \tag{13-12}$$

Since $d\mathbf{S}$ is now known to be an exact differential, the Euler reciprocal relation (section 6.2) yields

$$\frac{1}{T}\frac{\partial^2 \mathbf{E}}{\partial \mathbf{V}\,\partial t} = -\frac{1}{T^2}\frac{dT}{dt}\left[P + \left(\frac{\partial \mathbf{E}}{\partial \mathbf{V}}\right)_t\right] + \frac{1}{T}\frac{\partial^2 \mathbf{E}}{\partial t\,\partial \mathbf{V}} + \frac{1}{T}\left(\frac{\partial P}{\partial t}\right)_{\mathbf{V}}.$$

The second differentials of $\mathbf{E}$ cancel, and the remaining terms can be rearranged to give

$$T\left(\frac{\partial P}{\partial t}\right)_{\mathbf{V}} = \frac{dT}{dt}\left[P + \left(\frac{\partial \mathbf{E}}{\partial \mathbf{V}}\right)_t\right]. \tag{13-13}$$

At this stage we introduce the ideal gas model (see sections 2.2 and 10.2) by the dual requirements

$$P = \frac{nR\theta(t)}{\mathbf{V}} \equiv \frac{nR\theta}{\mathbf{V}} \tag{13-14}$$

and

$$\left(\frac{\partial \mathbf{E}}{\partial \mathbf{V}}\right)_t = 0. \tag{13-15}$$

We can write

$$\left(\frac{\partial P}{\partial t}\right)_{\mathbf{V}} = \left(\frac{\partial P}{\partial \theta}\right)_{\mathbf{V}}\frac{d\theta}{dt} = \frac{nR}{\mathbf{V}}\frac{d\theta}{dt}, \tag{13-16}$$

and obviously

$$\frac{dT}{dt} = \frac{dT}{d\theta}\frac{d\theta}{dt}. \tag{13-17}$$

Upon substitution of Eqs. (13-14) through (13-17) into (13-13), we obtain

$$\frac{nRT}{\mathbf{V}}\frac{d\theta}{dt} = \frac{dT}{d\theta}\frac{d\theta}{dt}\frac{nR}{\mathbf{V}};$$

hence

$$\frac{dT}{T} = \frac{d\theta}{\theta} \quad \text{or} \quad d\ln T = d\ln\theta;$$

therefore,

$$T = (\text{const}\cdot)\,\theta. \tag{13-18}$$

By making the *size* of the degree on both scales equal, i.e., by requiring Eqs. (3-6) and (12-19) to hold, the integration constant is fixed at unity, and we obtain the desired result:

$$T = \theta. \tag{13-19}$$

This completes our "derivation" of the (first part of) the Second Law according to the method of Carathéodory.

REFERENCES

1. For Carathéodory's original paper, see the references in Chapter 11.
2. H. A. Buchdahl, reference 3 of Chapter 1.
3. M. Born, reference 2 of Chapter 7.
4. S. Chandrasekhar, reference 4 of Chapter 7.
5. Several papers by L. A. Turner, e.g., *Am. J. Phys.* **28**, 781 (1960); **30**, 506 (1962); and *J. Chem. Phys.* **38**, 1163 (1963).

Chapter 14

Some Basic Applications of the First Part of the Second Law

In this chapter we summarize the most important implications of the first part of the Second Law, i.e., of our knowledge that

$$d\mathsf{S} = \frac{dQ_{\text{rev}}}{T} \tag{14-1}$$

is an exact differential. The first and last sections are of *general* importance; a good grasp of the subject matter involved is essential to all students of thermodynamics. Many details will be assigned as exercises. If these cause mathematical difficulties, a rereading of section 6-2 may be in order.

Sections 14.2 and 14.3 are concerned with two famous *specific* applications, of both historical and practical significance. They should be considered optional for the uninitiated, and for those primarily interested in "*chemical* thermodynamics." Of course, further applications will be given throughout Parts III and IV.

14.1 Applications of the Euler Reciprocal Relations and Related Features

In accordance with the author's preference, we shall give a general formulation, followed by an analysis of the most important special case. The reader has the option to reverse the order and go through the simpler special case first. In more elementary discussions, part (a) may be dispensable altogether.

(a) General Formulation

For a system described by the generalized coordinates $x_1, \ldots, x_r$, the combination of the First Law and the (first part of the) Second Law, for a reversible process, can be put in the form

$$d\mathbf{S} = \frac{1}{T}\left(\frac{\partial \mathbf{E}}{\partial T}\right)_{x_1,\ldots,x_r} dT + \frac{1}{T}\sum_{i=1}^{r}\left[X_i + \left(\frac{\partial \mathbf{E}}{\partial x_i}\right)_{T,x_j \neq i}\right] dx_i, \quad (14\text{-}2)$$

where we have used the general expression (4-18) for dW.

Our knowledge that $d\mathbf{S}$ is an exact differential leads to an important set of equations, which can be summarized systematically as follows:

Zero-order relations

$$\left(\frac{\partial \mathbf{S}}{\partial T}\right)_{x_1,\ldots,x_r} = \frac{1}{T}\left(\frac{\partial \mathbf{E}}{\partial T}\right)_{x_1,\ldots,x_r} \quad (14\text{-}3)$$

and

$$\left(\frac{\partial \mathbf{S}}{\partial x_i}\right)_{T,x_j \neq i} = \frac{1}{T}\left[X_i + \left(\frac{\partial \mathbf{E}}{\partial x_i}\right)_{T,x_j \neq i}\right] \quad \text{for all } i = 1, \ldots, r. \quad (14\text{-}4)$$

These are immediately verified "by inspection."

First-order relations

$$X_i + \left(\frac{\partial \mathbf{E}}{\partial x_i}\right)_{T,x_j \neq i} = \mathbf{T}\left(\frac{\partial X_i}{\partial T}\right)_{x_1,\ldots,x_r} \quad \text{for all } i = 1, \ldots, r \quad (14\text{-}5)$$

and

$$\left(\frac{\partial X_i}{\partial x_k}\right)_{T,x_j \neq k} = \left(\frac{\partial X_k}{\partial x_i}\right)_{T,x_j \neq i} \quad \text{for all } i \neq k = 1, \ldots, r. \quad (14\text{-}6)$$

These follow from a straightforward application of the Euler reciprocal relations. If there is only one X,x parameter couple, as in the case discussed in section 14.1(b), no relations of the type (14-6) will occur.

Exercise 14-1

Show that the first-order equations can be obtained from the Euler reciprocal relations as applied to the exact differential (14-2). [For the underlying theory, see section 6.2. Compare also the details leading up to Eq. (13-13).]

Exercise 14-2

Show that

$$\left(\frac{\partial Q}{\partial x_i}\right)_{T,x_j \neq i} = \mathbf{T}\left(\frac{\partial X_i}{\partial T}\right)_{x_1,\ldots,x_r} \quad \text{for all } i = 1, \ldots, r. \quad (14\text{-}7)$$

Second-order relations

These result after differentiating any of the first-order relations with respect to T or one of the x_i. To give just one example, differentiation of both sides of Eq. (14-5) with respect to T, at constant $x_1, \ldots, x_r$, yields

$$\left(\frac{\partial \mathbf{C}_x}{\partial x_i}\right)_{T, x_{j \neq i}} = T\left(\frac{\partial^2 X_i}{\partial T^2}\right)_{x_1, \ldots, x_r} \qquad \text{for all } i = 1, \ldots, r, \tag{14-8}$$

in which $\mathbf{C}_x$ is the heat capacity at constant $x_1, \ldots, x_r$.

Exercise 14-3

Derive Eq. (14-8).

Higher-order relations

These can be obtained in similar fashion, but they are rarely needed.

(*b*) *Special Case of a Single Coordinate*, **V**

$$d\mathbf{S} = \frac{1}{T}\left(\frac{\partial \mathbf{E}}{\partial T}\right)_{\mathbf{V}} dT + \frac{1}{T}\left[P + \left(\frac{\partial \mathbf{E}}{\partial \mathbf{V}}\right)_T\right] d\mathbf{V}. \tag{14-2$'$}$$

[Primed equations are a special case of the corresponding unprimed ones in the general case of subsection (a) above.] Obviously, since **S** is an exact differential, we have

$$\left(\frac{\partial \mathbf{S}}{\partial T}\right)_{\mathbf{V}} = \frac{1}{T}\left(\frac{\partial \mathbf{E}}{\partial T}\right)_{\mathbf{V}} = \frac{\mathbf{C}_v}{T} \tag{14-3$'$}$$

and

$$\left(\frac{\partial \mathbf{S}}{\partial \mathbf{V}}\right)_T = \frac{1}{T}\left[P + \left(\frac{\partial \mathbf{E}}{\partial \mathbf{V}}\right)_T\right]. \tag{14-4$'$}$$

Proceeding as in the derivation of Eq. (13-13) from Eq. (13-12), the application of the Euler reciprocal relation yields the important result

$$P + \left(\frac{\partial \mathbf{E}}{\partial \mathbf{V}}\right)_T = T\left(\frac{\partial P}{\partial T}\right)_{\mathbf{V}}, \tag{14-5$'$}$$

anticipated as Eq. (2-8) and used in earlier chapters. Further implications are discussed in subsection (c).

Exercise 14-4

Obtain Eq. (14-5′) from Eq. (14-2′) as a consequence of the Euler reciprocal relation. In an entirely equivalent procedure, we may first obtain $\partial^2\mathbf{S}/\partial\mathbf{V}\,\partial T$ by partial differentiation of Eq. (14-3′) with respect to **V** and $\partial^2\mathbf{S}/\partial T\,\partial\mathbf{V}$ by partial differentiation of Eq. (14-4′) with respect to T, after which the resulting expressions are equated.

From Eqs. (14-4′) and (14-5′), we obtain

$$T\left(\frac{\partial \mathbf{S}}{\partial \mathbf{V}}\right)_T = T\left(\frac{\partial P}{\partial T}\right)_{\mathbf{V}}. \tag{14-7′}$$

Differentiation of Eq. (14-7′) with respect to T, at constant $\mathbf{V}$, yields

$$\left(\frac{\partial \mathbf{C}_v}{\partial \mathbf{V}}\right)_T = T\left(\frac{\partial^2 P}{\partial T^2}\right)_{\mathbf{V}}. \tag{14-8′}$$

Exercise 14-5

Verify Eqs. (14-7′) and (14-8′). Those readers who studied subsection (a) should convince themselves that the primed equations are indeed special cases of the corresponding unprimed ones.

In this special case of a single coordinate, $\mathbf{V}$, we can also start with the combination of Eqs. (14-1) and (10-13):

$$T\,d\mathbf{S} = d\mathbf{H} - \mathbf{V}\,dP. \tag{14-9}$$

This is expedient if we wish to work with P and T (rather than $\mathbf{V}$ and T) as independent variables. We expand $d\mathbf{H}$ in terms of dP and dT and proceed along similar lines as given above. The most important result is the first-order equation

$$\left(\frac{\partial \mathbf{H}}{\partial P}\right)_T = \mathbf{V} - T\left(\frac{\partial \mathbf{V}}{\partial T}\right)_P. \tag{14-10}$$

Exercise 14-6

Derive Eq. (14-10).

Exercise 14-7

Derive:

$$\left(\frac{\partial \mathbf{C}_P}{\partial P}\right)_T = -T\left(\frac{\partial^2 \mathbf{V}}{\partial T^2}\right)_P. \tag{14-11}$$

(c) *Application of the Results of Subsection (b) to Some Gaseous Model Systems*

The important result (14-5′) is sometimes called the *thermodynamic equation of state*. Referring back to the discussion in section 2.2, this equation was already anticipated in the form (2-8). We showed that with this result the *thermal equation of state*

$$p\mathbf{V} = nRT \tag{2-4}$$

could be used as the *sole* phenomenological definition of a single-component ideal gas. It is recommended that readers recheck the important Exercise 2-1.

For 1 mole of a van der Waals gas, the combination of the (thermal) equation of state [Eq. (10-3$\tilde{0}$)] with the fundamental result (14-5′) yields

$$\left(\frac{\partial E}{\partial V}\right)_T = \frac{a}{V^2} \tag{14-12}$$

for the *internal pressure* of the gas.

Exercise 14-8

Derive Eq. (14-12).

In conjunction with Eq. (10-3$\tilde{0}$), Eq. (14-8′) tells us that for a van der Waals gas, as for an ideal gas, $(\partial \mathbf{C}_v/\partial \mathbf{V})_T$ is zero and $\mathbf{C}_v$ is a function of T only, even though for the former model $(\partial \mathbf{E}/\partial \mathbf{V})_T$ does *not* vanish. In fact, this situation pertains to a large class of model systems, as we shall illustrate in Exercise 14-9.

Exercise 14-9

(a) Show that for any gas satisfying an equation state

$$p = f(\mathbf{V})T + g(\mathbf{V}), \tag{14-13}$$

in which $f(\mathbf{V})$ and $g(\mathbf{V})$ are entirely arbitrary,

$$\left(\frac{\partial \mathbf{E}}{\partial \mathbf{V}}\right)_T = -g(\mathbf{V}) \tag{14-14}$$

and

$$\left(\frac{\partial \mathbf{C}_v}{\partial \mathbf{V}}\right)_T = 0. \tag{14-15}$$

[For an ideal gas, $f(\mathbf{V}) = nR/\mathbf{V}$; $g(\mathbf{V}) = 0$. For *1 mole* of an elastic hard-sphere model (see section 10.2),

$$f(V) = \frac{R}{V - B} \qquad g(V) = 0.$$

For *1 mole* of a van der Waals gas: $f(V) = R/(V - b)$; $g(V) = -a/V^2$.]

(b) In similar fashion, if $\mathbf{V} = f'(P)T + g'(P)$, analyze $(\partial \mathbf{C}_p/\partial P)_T$.

To account for the experimental fact that at high pressures $\mathbf{C}_p$ and $\mathbf{C}_v$ depend on $\mathbf{V}$ (as well as T), we have to use more realistic, and therefore more complicated equations of state than will be used in this book.

14.2 The Joule–Kelvin Effect Revisited

The Joule–Kelvin effect of a gas was defined in section 10.3 in its *differential* form through the Joule–Kelvin *coefficient*:

$$\mu_{\text{J.T.}} \equiv \left(\frac{\partial T}{\partial P}\right)_{\mathrm{H}}. \tag{10-38}$$

Some authors define, in addition, an *integral* Joule–Kelvin effect:

$$\frac{T_2 - T_1}{P_2 - P_1} = \frac{1}{P_2 - P_1} \int_{P_1 \,(\mathrm{H})}^{P_2} \mu_{\text{J.T.}} \, dP. \tag{10-38$'$}$$

This isenthalpic expansion is used technically in the cooling (and, ultimately, liquefaction) of gases. For this method to be effective, $\mu_{\text{J.T.}}$ should be positive and large. Generally, $\mu_{\text{J.T.}}$ decreases with increasing temperature. To give an example, expansion of nitrogen from 200 atm to 5 atm causes a temperature lowering of about twelve degrees at 200 K, but less than half of this amount at 500 K. At the so-called *inversion temperature*, T_i, $\mu_{\text{J.T.}}$ changes sign. For most gases this temperature is quite high. For example, for N_2 it is generally above 600 K; *its precise value depends on the pressure* (see below)! By way of contrast, for helium the inversion temperature is about 50 K. Therefore, we must "precool" helium gas below this temperature by other means before we can utilize the Joule–Kelvin effect for further cooling toward the condensation point.

In Chapter 10 we applied the First Law to obtain

$$\mu_{\text{J.T.}} = -\frac{1}{C_p}\left(\frac{\partial H}{\partial P}\right)_T. \tag{10-39}$$

By virtue of Eq. (14-10), the Second Law allows us to write

$$\mu_{\text{J.T.}} = \frac{1}{C_p}\left[T\left(\frac{\partial V}{\partial T}\right)_P - V\right], \tag{14-16}$$

which makes $\mu_{\text{J.T.}}$ much more accessible. For example, if we are dealing with an ideal gas, Eq. (14-16) immediately confirms the earlier conclusion (section 10.3) that $\mu_{\text{J.T.}} = 0$. For a real gas, $[T(\partial V/\partial T)_P - V]$ is not zero *in general* and can be evaluated provided that we know its (thermal) equation of state. Unfortunately, in most cases, the simpler (and therefore the more manageable) the latter, the less accurately it reproduces the experimental results. Thus, in practice, for the best agreement, one may prefer to use tabulations and/or graphical representations of empirical equation of state data.

One such procedure utilizes the compressibility factor

$$Z \equiv \frac{PV}{RT} \tag{14-17}$$

as an "intermediate." In terms of Z, $\mu_{\text{J.T.}}$ becomes

$$\mu_{\text{J.T.}} = \frac{1}{C_p}\frac{RT^2}{P}\left(\frac{\partial Z}{\partial T}\right)_P. \tag{14-18}$$

Exercise 14-10

Derive Eq. (14-18).

It is interesting to analyze how, in a qualitative sense, the experimental data are correctly predicted by the van der Waals equation of state:

$$\left(P + \frac{a}{V^2}\right)(V - b) = RT. \tag{10-$\widetilde{30}$}$$

Upon evaluation of $(\partial V/\partial T)_p$, substitution in Eq. (14-16) yields

$$\mu_{\text{J.T.}} = \frac{1}{C_p}\left[\frac{RT}{\dfrac{RT}{V-b} - \dfrac{2a(V-b)}{V^3}} - V\right], \tag{14-19}$$

certainly not a simple expression. Unfortunately, this has led to the common practice of converting it into an approximate form by some appropriate series expansion and the neglect of all but those terms that are linear in a and b (both positive):

$$\mu_{\text{J.T.}} \approx \frac{1}{C_p}\left(\frac{2a}{RT} - b\right). \tag{14-20}$$

Exercise 14-11

Verify Eq. (14-19), and derive Eq. (14-20), along the lines indicated in the text.

This confirms that at low temperatures $\mu_{\text{J.T.}}$ is positive (the technically important cooling effect) and at high temperatures it is negative (heating as a result of the isenthalpic expansion). At the inversion temperature T_i, $\mu_{\text{J.T.}} = 0$; from Eq. (14-20),

$$T_i \approx \frac{2a}{bR}, \tag{14-21}$$

which may be compared with the familiar van der Waals expression for the *critical* temperature:*

$$T_{\text{Cr}} = \frac{8a}{27bR},$$

* This is derived in most textbooks on physical chemistry. Cf. Exercise 27-1.

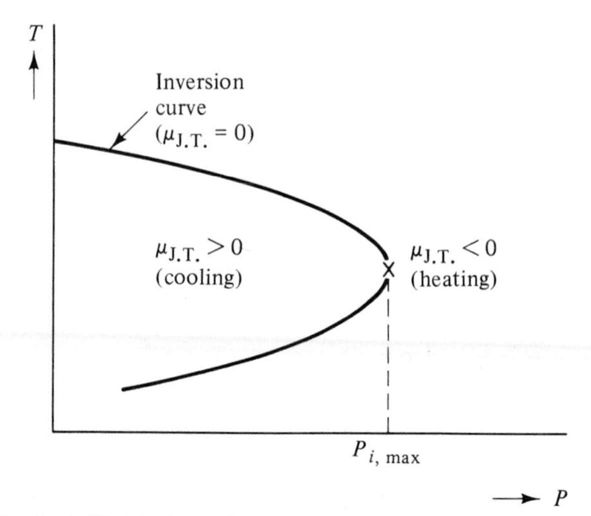

Fig. 13 Joule-Kelvin inversion curve (P_i–T_i curve). This curve separates cooling and heating regions ($\mu_{J.T.} > 0$ and $\mu_{J.T.} < 0$ respectively). For most gases, room temperature is inside the cooling region; important exceptions are H_2, He and Ne, for which room temperature lies above the entire P_i–T_i curve. These three gases would have to be "pre-cooled" if we wish to utilize the Joule-Kelvin effect for their liquefaction. (The low temperature portion of the inversion curve should end at its intersection with the coexistence line.)

suggesting a ratio T_i/T_{Cr} of 6.75. Experimental values average out to about 6. For comparison with numerical T_i values given above, T_{Cr} for He is 5.21 K, for N_2 126.3 K.

Note that the replacement of Eq. (14-19) by the *approximate* Eq. (14-20) has given us a single *inversion point, independent of pressure.* However, experimentally we find an *inversion curve*, looking somewhat like a skew-oriented parabola (Fig. 13); T_i is a function of P! We shall show that the existence of such a curve is embodied in Eq. (14-19), hence in the van der Waals theory, provided that we go through the necessary algebra *without any approximations.* We immediately go after the inversion curve as such, initially in terms of T_i and V_i, by putting the right-hand side of Eq. (14-19) equal to zero. This readily leads to the result

$$\frac{V_i - b}{V_i} = \sqrt{\frac{bRT_i}{2a}}. \tag{14-22}$$

We have to eliminate V_i between this equation and

$$\left(P_i + \frac{a}{V_i^2}\right)(V_i - b) = RT_i. \tag{10-$\widetilde{30}$}$$

The result is*

$$(b^2P_i + \tfrac{3}{2}bRT_i + a)^2 = 8abRT_i, \qquad (14\text{-}23)$$

which is of the correct form (Fig. 13).

Exercise 14-12

Verify that for $P_{i,\max} = a/3b^2$, Eq. (14-23) has only a single solution, $T_i = 8a/9bR$, and that the approximate result (14-21) is obtained formally as the upper inversion temperature at $P_i = 0$. For $P > P_{i,\max}$, Eq. (14-23) no longer has real solutions, and we are definitely in the region $\mu_{J.T.} < 0$. Below $P_{i,\max}$ there is an upper- and a lower-inversion temperature; Eq. (14-23) has two different real solutions. For some numerical values, see Problem 14-5.

Exercise 14-13

For those who are familiar with, and interested in, theories involving reduced equations of state, we suggest the conversion of Eq. (14-23) to

$$(\pi_i + 12\tau_i + 27)^2 = 1728\tau_i,$$

in which $\pi \equiv P/P_{Cr}$ and $\tau \equiv T/T_{Cr}$, the reduced pressure and temperature, respectively. We need the expressions for a and b in terms of the critical constants:

$$a = \frac{27R^2T_{Cr}^2}{64P_{Cr}} \quad \text{and} \quad b = \frac{RT_{Cr}}{8P_{Cr}}.$$

Plot the universal (van der Waals) τ_i–π_i inversion curve.

The existence of an "inversion pressure" at fixed temperature has been interpreted, on a corpuscular level, as follows:

1. At high pressure, *repulsive* intermolecular forces dominate; compressing the gas *increases* the potential energy of the molecules, $(\partial E/\partial P)_T > 0$, hence $(\partial H/\partial P)_T > 0$, and by Eq. (10-39), $\mu_{J.T.} < 0$.
2. At low pressures, *attractive* forces are the most important ones; compression *decreases* the potential energy of the molecules, reversing the inequalities above; hence $\mu_{J.T.} > 0$.

The "double inversion" if we go from high temperature to low temperature, at a fixed $P < P_{i,\max}$, can be "interpreted" by making the following observations:†

---

* The courageous readers may wish to work through the gruesome details. See also S. M. Blinder, reference 3 of Chapter 6, pp. 352–353. Blinder gives an excellent discussion of this subject.

† The author is indebted to Russell T. Pack for suggesting these "pictures."

3. At a very high temperature, the molecules fly through the "attractive wall" and hit only the "repulsive wall." We get a negative μ as explained under item 1 above.
4. At "medium" temperatures, the average encounter is more strongly influenced by the attractive potential than by the repulsive one. The situation is as in item 2 above.
5. At very low temperatures, we are effectively dealing with a very dense fluid, where the total volume becomes of the order of the "excluded volume," $4b$; any compression causes an increase in potential energy, etc.

It is striking that the simple van der Waals equation of state apparently balances the two factors (attractive and repulsive interactions) with sufficient accuracy to give at least a qualitatively correct account of what we already acknowledged to be (see the end of section 10.3) a very complex situation.*

14.3 Adiabatic Demagnetization

Ever since the beginning of this century, physicists have attempted to extend their experimental researches closer and closer to absolute zero. To this purpose a number of new refrigeration techniques have been developed,† among which *adiabatic demagnetization* is probably the best known. In this method, proposed by P. Debye and W. F. Giauque in 1926, a crystalline substance with a high paramagnetic susceptibility, such as the salt gadolinium sulfate,‡ is first precooled (e.g., by means of liquid helium) to as low a temperature as is feasible. Then the crystal is magnetized, adiabatically isolated, and finally demagnetized, which results in a further temperature reduction of the sample, as we shall explain below. In 1950, C. J. Gorter and co-workers, at Leiden, reached a temperature of 0.0014 K in this fashion. In the next decade, N. Kurti and collaborators, at Oxford, applied the same principle to *nuclear* spin systems, and reached temperatures as low as 10^{-6} to 10^{-7} K.

When the author mentions these achievements to his physical chemistry classes, he usually encounters the "who cares" attitude. The typical chemistry students justifiably do not care whether a reaction is carried out at 300.001 K or at 300.000001 K. They should realize that near absolute zero we are dealing with a very different scale, so to speak; when we reduce the temperature from 10^{-3} to 10^{-6} K, entirely new phenomena may be discovered! (Compare the discovery of superconductivity at slightly higher temperatures.) The quantity $|\Delta T|/T$ is vital!

---

* For a treatment on the basis of the virial equation of state, see Thomas R. Rybolt, *J. Chem. Educ.* **58**, 620 (1981).

† For a summary, see Olli V. Lounasmaa, *Physics Today* **32**, 32 (1979).

‡ Gadolinium, Gd, one of the lanthanides, has the electron configuration $[\mathrm{Xe}]6s^2 4f^7 5d^1$. In the crystal concerned it is present as Gd^{+3} with configuration $[\mathrm{Xe}]4f^7$, possessing *seven* unpaired electron spins.

A thorough theoretical analysis of these matters, which would include the use of statistical mechanics, falls beyond the scope of this book. We merely wish to show how a relatively straightforward application of the (first part of the) Second Law leads to an understanding of adiabatic demagnetization, at least on a phenomenological level. It also gives us a relatively rare opportunity to illustrate some of the techniques presented in section 14.1 in a situation *not* governed by the parameters P and $\mathbf{V}$.

Let us consider an isotropic crystal in a uniform magnetic field $\mathscr{H}$. The magnetization per unit volume is given by

$$\mathscr{M} = \chi(T)\mathscr{H}. \tag{14-24}$$

in which χ is the magnetic susceptibility. For our purposes this relation can be looked upon as a "magnetic equation of state." The combination of the First Law and (first part of) the Second Law gives us initially

$$T\,d\mathbf{S} = dQ = d\mathbf{E} - \mathscr{H}\,d\mathscr{M}. \tag{14-25}$$

The validity of Eq. (14-25) apparently implies that

$$dW = -\mathscr{H}\,d\mathscr{M}. \tag{α}$$

In many sources, the reader may encounter a different expression:

$$dW = -\frac{\mathbf{V}}{8\pi}\,d\mathscr{H}^2 - \mathscr{H}\,d\mathscr{M}. \tag{β}$$

The additional term corresponds to the energy of the vacuum field, and is of no real concern to us in the ensuing discussion. Whether this term occurs or not depends on the manner in which we define the internal energy of our magnetized body or, to put it another way, whether this vacuum field energy is incorporated in $\mathbf{E}$ or not. For an excellent treatment of these intricate matters, see, e.g., the books by Pippard and Becker, references 7 and 8 of Chapter 4.

Combining the techniques presented in sections 10.1 and 14.1, we obtain

$$dQ = \mathbf{C}_{\mathscr{H}}\,dT + T\left(\frac{\partial \mathscr{M}}{\partial T}\right)_{\mathscr{H}} d\mathscr{H}. \tag{14-26}$$

Exercise 14-14

Derive Eq. (14-26).
Hints:

1. Considering T and $\mathscr{H}$ as independent variables, put dQ in the form: $dQ = f(T, \mathscr{H})\,dT + g(T, \mathscr{H})\,d\mathscr{H}$. This duplicates part of Exercise 10.6; $f(T, \mathscr{H})$ can, of course, be identified with $\mathbf{C}_{\mathscr{H}}$ [see Eq. (10-17′) and the paragraph below it].

2. To convert $g(T, \mathscr{H})$ into the desired form, apply the reciprocal relation

$$\frac{\partial}{\partial \mathscr{H}}\left(\frac{f(T, \mathscr{H})}{T}\right)_T = \frac{\partial}{\partial T}\left(\frac{g(T, \mathscr{H})}{T}\right)_{\mathscr{H}}.$$

Use Eq. (14-24) to replace $(\partial \mathscr{M}/\partial T)_{\mathscr{H}}$ by $\mathscr{H}(\partial \chi/\partial T)_{\mathscr{H}}$ and put $dQ = 0$ (adiabatic process!), so that Eq. (14-26) becomes

$$0 = \mathbf{C}_{\mathscr{H}}\, dT + \mathscr{H} T\left(\frac{\partial \chi}{\partial T}\right)_{\mathscr{H}} d\mathscr{H},$$

from which we get the desired result:

$$dT = -\frac{\mathscr{H} T}{\mathbf{C}_{\mathscr{H}}}\left(\frac{\partial \chi}{\partial T}\right)_{\mathscr{H}} d\mathscr{H}. \tag{14-27}$$

But $\mathscr{H}$, T, and $\mathbf{C}_{\mathscr{H}}$ are all positive, while $d\mathscr{H}$ is negative (*de*magnetization!). Hence our analysis shows us that *dT has the sign of* $(\partial \chi/\partial T)_{\mathscr{H}}$. This is as far as phenomenological thermodynamics can go; the sign concerned has to be obtained experimentally or justified on the basis of corpuscular theories. The latter show that *for a paramagnetic substance* T enters χ in the form $\mu_M^2/3kT$ (μ_M is the magnetic dipole moment)*; hence $(\partial \chi/\partial T)_{\mathscr{H}}$ is negative.

14.4 Entropy Calculations

(a) In this section we confine ourselves to the situation that the only possible work is $P,\mathbf{V}$ work, since this brings out most of the essential features, and generalizations are normally straightforward. For *reversible processes* the combination of the First Law and (the first part of) the Second Law gives us

$$d\mathbf{S} = \frac{\mathbf{C}_v}{T}\, dT + \frac{1}{T}\left[P + \left(\frac{\partial \mathbf{E}}{\partial \mathbf{V}}\right)_T\right] d\mathbf{V},$$

which, by (14-5′), simplifies to

$$d\mathbf{S} = \frac{\mathbf{C}_v}{T}\, dT + \left(\frac{\partial P}{\partial T}\right)_{\mathbf{V}} d\mathbf{V},$$

or

$$d\mathbf{S} = \mathbf{C}_v\, d \ln T + \left(\frac{\partial P}{\partial T}\right)_{\mathbf{V}} d\mathbf{V}. \tag{14-28}$$

Within the limits we set ourselves above, this equation is exact and forms the basis of most "entropy calculations." However, this designation is strictly a

---

* To give a very qualitative picture: The higher T, the harder it is for the magnetic field to orient the magnetic dipoles, since this is opposed by their heat movement.

misnomer, since at best Eq. (14-28) can give us entropy *changes*. Expressions for **S** as such, which *apparently* can be obtained by *indefinite* integration, contain constants the value of which cannot be ascertained on the basis of phenomenological thermodynamics alone. On the other hand, the *definite* integral of $d\mathbf{S}$ can be obtained if we know both $\mathbf{C}_v(T)$ and $(\partial P/\partial T)_{\mathbf{V}}$, which, as usual, requires the availability of empirical data, or the introduction of models.

Let us consider the simplest such model, 1 mole of a (single-component) ideal gas, for which (14-28) becomes

$$dS = C_v \, d \ln T + R \, d \ln V. \tag{14-29}$$

Even this simple equation cannot be integrated unless we know $C_v(T)$. The crudest approximation is to take this heat capacity to be *constant* (see section 2.2 for a critical discussion). In this case the *definite* integration of (14-29) gives

$$S_2 - S_1 \equiv \Delta S = C_v \ln \frac{T_2}{T_1} + R \ln \frac{V_2}{V_1}, \tag{14-30}$$

and the *indefinite* integration yields

$$S = \tilde{S}_o + C_v \ln T + R \ln V. \tag{14-31}$$

The last relation is usually referred to as the Sackur–Tetrode equation, because (in 1911–1912) these two scientists were the first to evaluate $\tilde{S}_o$ for monatomic ideal gases, by the methods of statistical mechanics. However, *as a phenomonological equation*, (14-31) had already been given by Gibbs, in 1873!*

In Eq. (14-31), we encounter for the first time a feature the importance of which far exceeds the relevance of this equation as such. Since this has led to much misunderstanding by students in several areas of thermodynamics, we wish to consider it in some detail. *Apparently*, Eq. (14-31) involves the logarithms of volume and temperature. But this cannot be the intention, since only the logarithms of dimensionless numbers have meaning. In the present relatively simple situation it is readily seen how this difficulty arises. Consider the state 1 in Eq. (14-30) as some convenient fixed reference state, in this context denoted by a subscript zero, and let 2 denote the (variable) state of interest, for which all subscripts are dropped. Then

$$S - S_o = C_v \ln \frac{T}{T_o} + R \ln \frac{V}{V_o}$$

* J. W. Gibbs, *Graphical Methods in the Thermodynamics of Fluids, Trans. Connecticut Acad.*, II, pp. 309–342 (1873). See *The Scientific Papers of J. W. Gibbs*, Dover, New York, 1961. Vol. I, p. 13, Eq. (E).

or

$$S = (S_o - C_v \ln T_o - R \ln V_o) + C_v \ln T + R \ln V. \qquad (14\text{-}32)$$

Obviously this S_o is *not* identical to the $\tilde{S}_o$ of Eq. (14-31); the latter *incorporates* the term $-C_v \ln T_o - R \ln V_o$, which appears as a constant for computational purposes, but at the same time resolves the dimensional dilemma under consideration. In Eqs. (14-31) and (14-32), T and V stand for $|T|$ and $|V|$, dimensionless quantities. From a practical point of view, the very important consequence is that if we wish to use Eq. (14-31) numerically, *the units in which V is expressed* (T always refers to the Kelvin scale) *should match the value of the constant* $\tilde{S}_o$. If we are allowed a slight digression on this topic, a similar feature will occur frequently in other facets of thermodynamic formalism. For example, if in some equation we find a term $R \ln P$ (the intention is $R \ln |P|$), somewhere else in the same equation a term $-R \ln P_o$ is "hidden" in a constant. In computing $R \ln |P|$, the units in which P is expressed should match those of P_o. (Alternatively, we can say that P_o represents a conveniently chosen reference or "*standard*" state.) Again, equations containing the logarithm of those equilibrium constants that *appear* to have a dimension should be treated with similar care [cf. section 19.2(b)].

(b) For *irreversible processes* $d\mathbf{S} \neq dQ/T$, Eq. (14-28) and equivalents are no longer valid; along an irreversible path $d\mathbf{S}$ is not even *defined*! Nevertheless, we can usually get at the entropy change for such processes by making optimal use of our knowledge that $\mathbf{S}$ is a state function. This suggests that all we have to do is to find *one* convenient reversible path between the same initial and final states, and the problem is solved in principle (see Fig. 14).

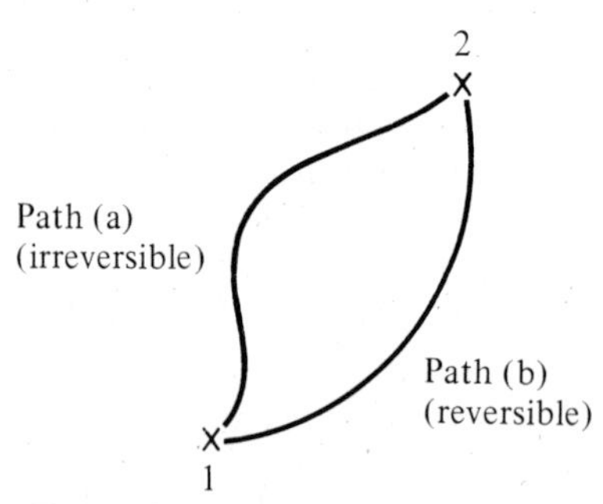

Fig. 14 Determining $\Delta\mathbf{S}$ for an Irreversible Process.

$$\underset{(a)}{\int_1^2 \frac{dQ}{T}} \neq \Delta\mathbf{S}_{(a)} = \mathbf{S}_2 - \mathbf{S}_1 = \Delta\mathbf{S}_{(b)} = \underset{(b)}{\int_1^2 \frac{dQ}{T}}$$

As an example, consider the irreversible phase transition

$$1 \text{ mole (supercooled) water} \xrightarrow[1 \text{ atm}]{-10^\circ\text{C}} 1 \text{ mole ice}, \qquad (1)$$

discussed in section 5.4. A convenient reversible path is the following:

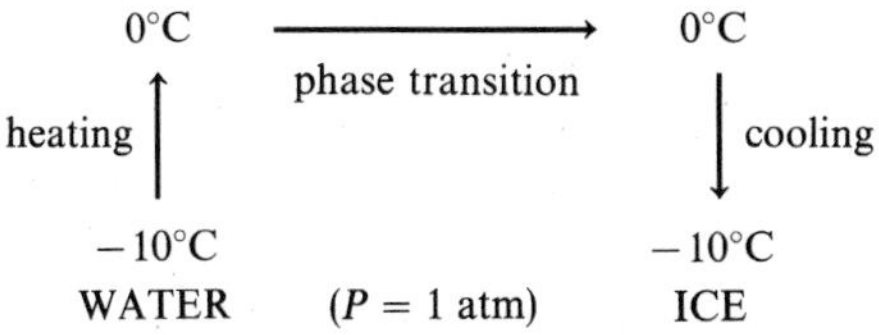

0°C	⟶ phase transition	0°C
heating ↑		↓ cooling
−10°C WATER	(P = 1 atm)	−10°C ICE

For the heating and cooling portions we can integrate:

$$dS = C_p d \ln T, \qquad (14\text{-}33)$$

making the reasonable assumption that C_p is a constant over the small temperature range involved. For the *reversible* phase transition at 0°C, 1 atm, we have

$$\Delta S = \frac{Q_p}{T} = \frac{\Delta H}{T}; \qquad (14\text{-}34)$$

the entropy change is the latent heat of the transformation divided by T. For later references we give the result (see Problem 14-3)

$$\Delta S_{(1)} = -4.92 \text{ cal mole}^{-1} \text{ K}^{-1} = -20.6 \text{ J mole}^{-1} \text{ K}^{-1}.$$

The general procedure outlined in the preceding paragraph, as illustrated in Fig. 14, immediately raises the question how to proceed if *no reversible path* (*b*) *exists.* In this situation we shall refer to the change along path (a) as *intrinsically irreversible.* Most biological processes (e.g., the aging of living organisms) fall into this class. Within the framework of "ordinary" phenomenological thermodynamics, there is no way to get at the entropy changes concerned, and we ought not even talk about them.* Attempts to resolve this dilemma have been made in the so-called *thermodynamics of irreversible processes,* referred to in Chapter 1. In addition, statistical mechanics can, *at least in principle,* lead us to "absolute entropies"[†] of initial and final states, hence to **ΔS** as well. These areas fall outside the scope of the present book. Although the author does not wish to surround biology with an undue mystique, he advises the readers to consider all statements

* P. W. Bridgman (reference 1 of Chapter 1, p. 41) has called such processes and the systems undergoing them "completely isolated by irreversibility from the universe of thermodynamics."

[†] For a brief critical discussion, see section 16.3.

regarding entropy changes of biological—and other intrinsically irreversible—processes with the utmost care. This is still a wide-open field.

PROBLEMS

Problem 14-1

Use $C_p^\bullet(O_2)$ as given in Eq. (lb) of Problem 10-4 to compute $(S_{500^\circ C} - S_{100^\circ C})$ for 1 mole of oxygen gas at constant P.

Problem 14-2

Consider the transformation of 1 mole of an ideal monatomic gas ($C_v = \frac{3}{2}R$) from an initial state at 1 atm, 300 K, to a final state of 0.5 atm, 600 K. Compute ΔS:

(a) By the use of Eq. (14-30).

(b) By splitting this transformation into an adiabatic expansion, followed by an isobaric* temperature rise, both carried out reversibly.

Problem 14-3

(a) Obtain the entropy change, $\Delta S_{(1)}$, for the phase transition

$$\text{1 mole (supercooled) water} \xrightarrow[\text{1 atm}]{-10^\circ C} \text{1 mole ice} \tag{1}$$

along the lines indicated in section 14-4. Use the following data:

1. $(\Delta H_f)_{273.15}$, the heat of fusion of ice at 0°C, 1 atm, is 6008 J mole^{-1}.
2. C_p for water between 0 and -10°C is 75.3 J K^{-1} mole^{-1}.
3. C_p for ice between 0 and -10°C is 37.65 J K^{-1} mole^{-1}.

(b) The heat of crystallization of (supercooled) water at -10°C, 1 atm, is -5619 J mole^{-1}. Verify that for reaction (1):

$$\Delta S_{(1)} \neq -\frac{(\Delta H_f)_{263.15}}{263.15}.$$

(c) An alternative pathway, in which reaction (1) is carried out *reversibly*, consists of the following three steps, all at -10°C:

1. Evaporate 1 mole of (supercooled) water at its equilibrium pressure, P_ℓ,
2. Expand the vapor from P_ℓ to P_s†, the equilibrium vapor pressure of ice, and
3. Condense the vapor to ice at P_s.

Derive the equation

$$\Delta S_{(1)} = -\frac{(\Delta H_f)_{\text{at } 263.15}}{263.15} + R \ln \frac{P_\ell}{P_s} \tag{2}$$

and note carefully under what *assumptions* this result is obtained.

---

* The term *isobaric* refers to a process at constant pressure. *Isochoric* refers to constant volume.

† The fact that this is indeed an *expansion* ($P_\ell > P_s$) is rationalized in section 18.3. Your calculations should confirm this.

(d) Use the value of $\Delta S_{(1)}$, obtained in part (a), to compute the ratio P_ℓ/P_s from Eq. (2). Note how we have obtained a ratio of equilibrium vapor pressures from caloric data exclusively.

(e) Suppose that reaction (1) takes place in a container which is placed in a large thermostat at $-10°C$. Assume that the thermostat can absorb the heat of crystallization reversibly. What is the entropy change, associated with reaction (1) for the system consisting of the container *plus* the thermostat?

Problem 14-4

A rubber band is placed in a thermostat at 300 K. First it is stretched reversibly from its equilibrium length, l_e, to $5l_e$. In the process, 2 J of heat is absorbed by the rubber band from the thermostat. Next the band is suddenly released, causing it to "snap back" spontaneously to its original length, l_e. During the latter process the band gives off 4 J of heat to the thermostat. Compute:

(a) $\Delta \mathbf{S}$ for each of the two steps, and

(b) W for the reversible stretching process.

Problem 14-5

The van der Waals constants for N_2 are

$$a = 1.390 \, \ell^2 \text{ atm mole}^{-2},$$

$$b = 0.03913 \, \ell \text{ mole}^{-1}.$$

(a) Compute the upper and lower inversion temperatures, at a pressure of 100 atm, using Eq. (14-23).

(b) Repeat the calculations of part (a), using the equation in terms of the reduced variables given in Exercise 14-12. Use the *experimental* critical constants, $T_{Cr} =$ 126.3 K and $P_{Cr} = 33.5$ atm, in relating T to τ and P to π.

(c) Compare the results obtained in parts (a) and (b) with:

(i) the experimental values of T_i, 550.4 K and 106.2 K, respectively [J.R. Roebuck and H. Osterberg, *Phys. Rev.* **48**, 450 (1935)]; and

(ii) the single result, obtained with the equation, $T_i = 2a/bR$. Note that this approximates the *upper* inversion temperature from above, in accordance with the results of Exercise 14-11.

(d) Compute $P_{i,\text{max}}$

(i) as $a/3b^2$ (cf. Exercise 14-12), and

(ii) by first obtaining $\pi_{i,\text{max}}$, and using the experimental critical pressure given above.

Chapter 15

The Second Part of the Second Law

15.1 Proof of the Second Part of the Second Law; the Maximum Theorem of the Entropy; Equilibrium and Stability

While "deriving" the first part of the Second Law along traditional lines in Chapter 12 we obtained the *Clausius inequality* as a "bonus." It tells us that provided that the temperature, T, is well defined at every stage, we have for any irreversible cycle:

$$\oint \frac{dQ}{T} < 0. \qquad \text{(12-30a)}$$

In subsection (a) this will be the starting point of our current derivation. On the other hand, we realize that those who opted to "derive" the first part of the Second Law *only* along the lines suggested by Carathéodory (see Chapter 13) never came across Eq. (12-30a). Therefore, in subsection (b) we shall show that we can also obtain the second part of the Second Law by a methodology based on the results of Chapter 13. Both procedures are instructive, each in its own way. Implications, misinterpretations, and an alternative formulation are discussed in the remainder of this section.

(*a*) *Proof from the Clausius Inequality*

Since an overall change is considered to be irreversible if any portion is, we can apply Eq. (12-30a) to the cycle:

$$\mathrm{A} \xrightarrow{\text{irrev}} \mathrm{B} \xrightarrow{\text{rev}} \mathrm{A} \qquad \text{[see Fig. 15(a)].}$$

This leads to

$$\underset{\text{(irrev)}}{\int_{\mathrm{A}}^{\mathrm{B}} \frac{dQ}{T}} + \underset{\text{(rev)}}{\int_{\mathrm{B}}^{\mathrm{A}} \frac{dQ}{T}} < 0,$$

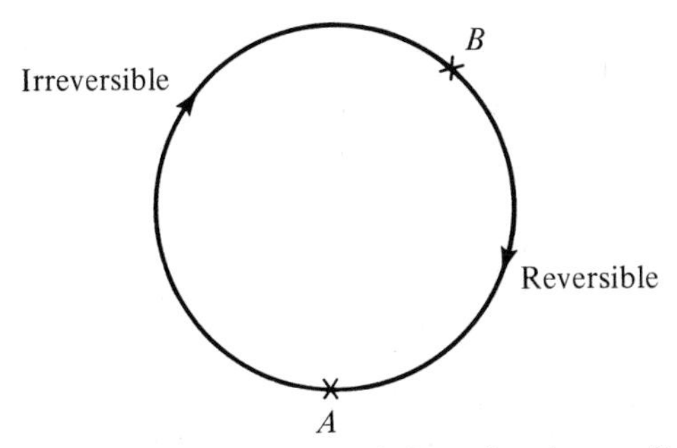

Fig. 15(a) An irreversible cycle, consisting of an irreversible path, followed by a reversible portion.

or since

$$-\int_{B\,(\text{rev})}^{A} \frac{dQ}{T} = \mathbf{S}_B - \mathbf{S}_A \equiv \Delta\mathbf{S},$$

we have

$$\int_{A\,(\text{irrev})}^{B} \frac{dQ}{T} < \Delta\mathbf{S}. \tag{15-1}$$

This fundamental inequality, and its differential analogue,

$$\frac{dQ}{T} < d\mathbf{S} \qquad \text{(for an irreversible change),} \tag{15-1a}$$

will be looked upon as the *second part of the Second Law*, since they give us the desired information regarding the *direction* of a spontaneous process (cf. section 11.1). Alternative inequalities, in terms of the so-called "free energy functions," will be derived from Eq. (15-1) in section 15.3. We often combine the two parts of the Second Law into one equation:

$$\int \frac{dQ}{T} \leqq \Delta\mathbf{S} \tag{15-2}$$

and

$$\frac{dQ}{T} \leqq d\mathbf{S}, \tag{15-2a}$$

with the integration limits in Eq. (15-2) omitted. In such formulations *it is always implied* that the equality sign pertains only to the limiting case of a *reversible* process, while the inequality covers all *irreversible* (spontaneous) paths (for which T is well defined at every stage). In thermodynamics the opposite of "spontaneous" is "impossible."* Hence Eq. (15-2) also *implies* that processes for which

$$\int \frac{dQ}{T} > \Delta \mathbf{S}$$

are *impossible.* Finally, in this context, we must remind readers (compare section 11.1) that *thermodynamics* generally *gives only necessary, not sufficient conditions.*

An important special case arises when we restrict ourselves to *adiabatic* changes. Now Eq. (15-2) becomes

$$\Delta_{\text{ad}}\mathbf{S} \geqq 0, \tag{15-3}$$

the famous *Maximum Theorem of the Entropy.* Some of its implications will be discussed in subsection (c) below. It is unfortunate that this theorem is often stated to apply to *isolated* systems only. This is too restrictive; *adiabatically enclosed* is all that is needed.

(*b*) *Proof from the T,S Diagram*†

In this alternative method we obtain the maximum theorem [Eq. (15-3)], the second part of the Second Law [Eq. (15-1)], and the Clausius inequality [Eq. (12-30a)] in reverse order. Figure 15(b) is reminiscent of Fig. 11. By

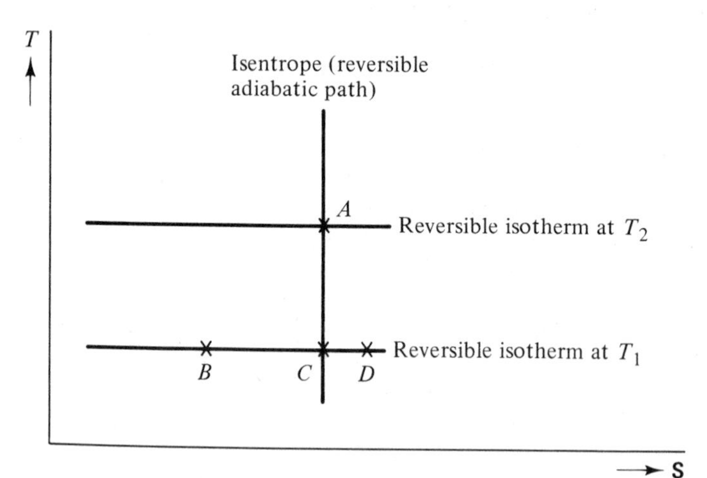

Fig. 15(b) See text, subsection (b).

---

* See footnote, p. 121.

† Bryce Crawford, Jr., and I. Oppenheim, *J. Chem. Phys.* **34**, 1621 (1961). See also John G. Kirkwood and Irwin Oppenheim, *Chemical Thermodynamics*, McGraw-Hill, New York, 1961, Chapter 4.

reasoning entirely analogous to that given in section 13.1, readers should ascertain that all points *to the left of* the isentrope AC are adiabatically inaccessible from A, by *any* process, reversible or irreversible.

Exercise 15-1

Show that if B could be reached from A by *any* adiabatic process, we could construct an anti-Kelvin machine.

Since all *reversible* adiabatic changes originating from A are represented by motion *along the* isentrope AC, all *irreversible* adiabatic processes, with A as their starting point, can only go *to the right* of this isentrope. Hence such spontaneous processes are associated with an entropy increase, and the result (15-3) is thereby confirmed.

Perhaps it is desirable at this stage to emphasize once more that *thermodynamic conditions are necessary, not sufficient*: An adiabatic process from A to D, say, *need not* proceed with a finite rate, but *if* an adiabatic process, originating from A, actually occurs spontaneously, it *must* follow a path which goes to the right of the line AC.

Next consider an adiabatically enclosed composite system, consisting of an inner container, I, in which an irreversible process occurs, surrounded by an outer system, II, which can exchange heat with I while undergoing reversible changes only (see Fig. 16). For the composite system, by Eq. (15-3), we have

$$\Delta \mathbf{S} = \Delta \mathbf{S}^{(\mathrm{I})} + \Delta \mathbf{S}^{(\mathrm{II})} > 0.$$

But since

$$\Delta \mathbf{S}^{(\mathrm{II})} = \int \frac{dQ^{(\mathrm{II})}}{T^{(\mathrm{II})}} = -\int \frac{dQ^{(\mathrm{I})}}{T^{(\mathrm{II})}},$$

this inequality becomes

$$\Delta \mathbf{S}^{(\mathrm{I})} > \int \frac{dQ^{(\mathrm{I})}}{T^{(\mathrm{II})}}. \tag{15-$\tilde{1}$}$$

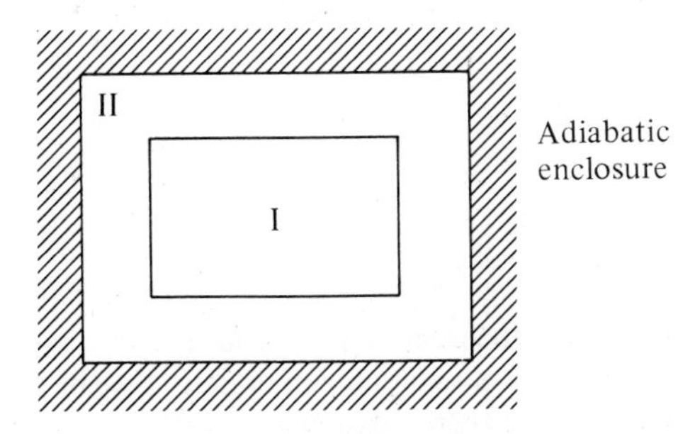

Fig. 16 See text above.

This result is neither very useful nor very revealing unless a thermal equilibrium between I and II is maintained at all times, so that $T^{(\mathrm{I})}$ equals $T^{(\mathrm{II})}$, hence is well defined. In this case Eq. (15-$\tilde{1}$) becomes

$$\Delta \mathbf{S}^{(\mathrm{I})} > \int \frac{dQ^{(\mathrm{I})}}{T^{(\mathrm{I})}},$$

which is, of course, identical to Eq. (15-1). (Note that in our earlier derivation we also stipulated that T had to be well defined.)

Exercise 15-2

From the result Eq. (15-2) or (15-2a) derive the "original" Clausius inequality, $\oint dQ/T < 0$ [Eq. (12-30a)] for any irreversible cycle.

(c) *Some Implications*

The maximum theorem of the entropy can be represented diagrammatically by a plot of **S** versus time [Fig. 17(a)] or of **S** versus some thermodynamic coordinate, x [Fig. 17(b)]. The latter could be the progress variable, also known as the degree of advancement, ξ, of a chemical reaction (see section 8.4). Figure 17(b) clearly indicates that the equilibrium position, at the maximum, may be reached from either the right or the left. Apparently, all second-order variations about this equilibrium state must *decrease* the entropy. Mathematically, we can state this as follows:

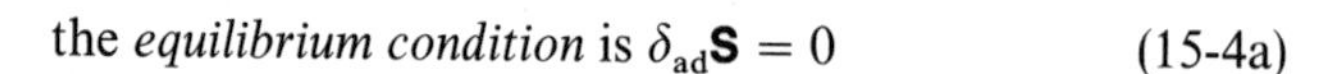

the *equilibrium condition* is $\delta_{\mathrm{ad}}\mathbf{S} = 0$ (15-4a)

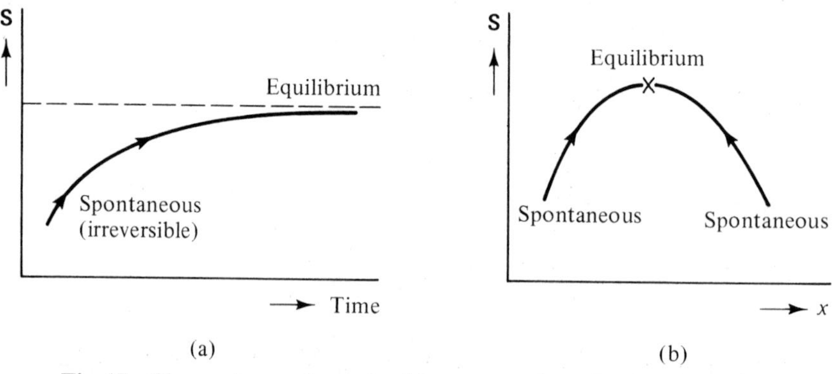

Fig. 17 The maximum theorem of the entropy (for adiabatic processes). (a) The trend to equilibrium as a function of time. (b) The trend to equilibrium as a function of a thermodynamic variable x.

and

$$\text{the } \textit{stability condition} \text{ is } \delta^2_{\text{ad}}\mathbf{S} < 0. \qquad (15\text{-}4b)$$

When no other work but P,$\mathbf{V}$ work is possible, $dQ = d\mathbf{E} + P\,d\mathbf{V}$ and the adiabatic nature of the process is *guaranteed* by constant $\mathbf{E}$ and $\mathbf{V}$. In this case Eqs. (15-4) necessarily imply:

$$\delta_{\mathbf{E},\mathbf{V}}\mathbf{S} = 0 \text{ (equilibrium condition)} \qquad (15\text{-}4a')$$

and

$$\delta^2_{\mathbf{E},\mathbf{V}}\mathbf{S} < 0 \text{ (stability condition).} \qquad (15\text{-}4b')$$

(*d*) *Some Comments and Misinterpretations*

We should never forget that the maximum theorem of the entropy is valid only for adiabatic changes; there are nonadiabatic spontaneous processes associated with a *decrease* in entropy. The example

$$1 \text{ mole (supercooled) water} \xrightarrow[1\text{ atm}]{-10^\circ\text{C}} 1 \text{ mole ice} \qquad (1)$$

has already been discussed in a different context (see section 14-4); $\Delta S_{(1)} = -20.6\text{ J K}^{-1}\text{ mole}^{-1}$. It is obvious that this process cannot possibly occur adiabatically (at fixed P and T) since the (molar) heat of crystallization is liberated. When this heat is (reversibly) absorbed by a thermostat at -10°C, 1 atm, the entropy change of the resulting composite system becomes

$$\Delta\mathbf{S} = \Delta\mathbf{S}_{(1)} + \Delta\mathbf{S}_{\text{thermostat}} = 0.81\text{ J K}^{-1}\text{ mole}^{-1}, \qquad (15\text{-}5)$$

as readers should have verified in Problem 14-3(e). The overall entropy change is indeed positive, as it *must* be, since the *composite* system can be adiabatically enclosed (and kept at -10°C, 1 atm) during the crystallization.

The most serious misinterpretation of the maximum theorem of the entropy transplants thermodynamics into cosmology. It is a consequence of Clausius's famous, but unfortunate, conclusion (1867) that *the entropy of the world (universe) strives to attain a maximum.* Philosophers and theologians have jumped on this statement, in combination with a vague corpuscular interpretation of entropy as an index of the degree of "chaos" (or "disorder") of a system, to predict confidently the "heat death of the universe"! A general critique of such extrapolations of thermodynamics has already been given in section 1.3. More specifically, in the present example, the question arises as to how one could ever properly parametrize the universe as a thermodynamic system and adiabatically separate it from its "surroundings" (?).

(e) An Alternative Formulation; "Entropy Production"

The members of the "Belgian School," founded by Th. de Donder, P. van Rysselberghe, and J. E. Verschaffelt, split all entropy changes into two parts:

$$d\mathbf{S} = d_e\mathbf{S} + d_i\mathbf{S} \tag{15-6}$$

in which $d_e\mathbf{S}$ is the entropy *flow* due to interactions with the surroundings, and $d_i\mathbf{S}$ is the "entropy *production*" inside the system. For *all* processes, reversible or irreversible, they relate the entropy flow to the heat flow as follows:

$$d_e\mathbf{S} = \frac{dQ}{T}, \tag{15-7}$$

provided that T is well defined, as usual. Comparison with the results of subsections (a) and (b) above shows that

$$d_i\mathbf{S} = 0 \qquad \text{for } \textit{reversible} \text{ processes} \tag{15-8a}$$

and

$$d_i\mathbf{S} > 0 \qquad \text{for } \textit{irreversible} \text{ processes.} \tag{15-8b}$$

As in Eqs. (15-2) and (15-3), the two Eqs. (15-8) can be combined conveniently into one, $d_i\mathbf{S} \geqq 0$, or in integrated form,

$$\Delta_i\mathbf{S} \geqq 0. \tag{15-9}$$

With the stipulations embodied in Eqs. (15-6) through (15-8), the statements (15-3) and (15-9) are equivalent for adiabatic processes; Eq. (15-9) expresses the maximum theorem of the entropy in this alternative, more general formulation. Some authors refer to $T\,d_i\mathbf{S}(\equiv dQ')$ as the "uncompensated heat," a notion already introduced by Clausius.

Exercise 15-3

Convince yourselves of the equivalence of Eqs. (15-3) and (15-9) for adiabatic processes.

All this is just a different way of expressing the Second Law; we introduce some new formal concepts, but no new science. Only a small number of authors of "equilibrium thermodynamics" have adopted this framework. However, "entropy production" has proven to be a powerful concept in the formulation of the so-called "thermodynamics of irreversible processes," already referred to several times in this book.*

---

* For example, see reference 4 of Chapter 1.

15.2 The Free Energy Functions

(a) Define:*

1. The *Helmholtz free energy*, also known as *Helmholtz function* or *work function*:

$$\mathbf{A} \equiv \mathbf{E} - T\mathbf{S}. \tag{15-10}$$

2. The *Gibbs free energy*, also known as the *Gibbs function*, or simply as "the" *free energy*:

$$\mathbf{G} \equiv \mathbf{E} - T\mathbf{S} + P\mathbf{V} = \mathbf{H} - T\mathbf{S} = \mathbf{A} + P\mathbf{V}. \tag{15-11}$$

From these definitions it is obvious that **A** *and* **G** *are state functions.* We remind readers of the definition of another state function, the enthalpy, **H**:

$$\mathbf{H} \equiv \mathbf{E} + P\mathbf{V}. \tag{10-12}$$

When only $P,\mathbf{V}$ work is permitted, a combination of the First and Second Laws gives us

$$d\mathbf{E} = T\,d\mathbf{S} - P\,d\mathbf{V}. \tag{15-12}$$

If we substitute this equation into the total differentials for $d\mathbf{A}$, $d\mathbf{G}$ and $d\mathbf{H}$, we readily obtain

$$d\mathbf{A} = -\mathbf{S}\,dT - P\,d\mathbf{V}, \tag{15-13a}$$

$$d\mathbf{G} = -\mathbf{S}\,dT + \mathbf{V}\,dP, \tag{15-13b}$$

and

$$d\mathbf{H} = T\,d\mathbf{S} + \mathbf{V}\,d\mathbf{P}, \tag{15-13c}$$

respectively. In their present form, Eqs. (15-12) and (15-13) pertain to closed systems only, and their validity is apparently restricted to reversible changes. Their applicability actually covers an important class of irreversible processes as well. This will be discussed critically in subsection (e) below. Generalizations to open systems are given in section 19.1.

Exercise 15-4

Derive Eqs. (15-13).

---

* Readers should be aware that, particularly in the older literature, there is no consensus as to nomenclature and notation. They should be especially alerted to the use of the term "free energy" (without qualification) and the symbol **F**, since these could refer to either our **A** or our **G**.

Rearrangement of Eq. (15-12) gives

$$d\mathbf{S} = \frac{1}{T}\,d\mathbf{E} + \frac{P}{T}\,d\mathbf{V}, \tag{15-14}$$

which is also a very useful relation. For example, it was taken as a starting point for the entire analysis given in section 14.1(b).

It is of some interest to apply the general mathematical techniques introduced in section 14.1(a) to the various total differentials given in Eqs. (15-12) through (15-14). Some of the results obtained in this fashion are:

Zero-order relations

$$T = \left(\frac{\partial \mathbf{E}}{\partial \mathbf{S}}\right)_{\mathbf{V}} = \left(\frac{\partial \mathbf{H}}{\partial \mathbf{S}}\right)_{P}, \tag{15-15a}$$

$$P = -\left(\frac{\partial \mathbf{E}}{\partial \mathbf{V}}\right)_{\mathbf{S}} = -\left(\frac{\partial \mathbf{A}}{\partial \mathbf{V}}\right)_{T}, \tag{15-15b}$$

$$\mathbf{V} = \left(\frac{\partial \mathbf{H}}{\partial P}\right)_{\mathbf{S}} = \left(\frac{\partial \mathbf{G}}{\partial P}\right)_{T}, \tag{15-15c}$$

$$\mathbf{S} = -\left(\frac{\partial \mathbf{A}}{\partial T}\right)_{\mathbf{V}} = -\left(\frac{\partial \mathbf{G}}{\partial T}\right)_{P}, \tag{15-15d}$$

$$\frac{1}{T} = \left(\frac{\partial \mathbf{S}}{\partial \mathbf{E}}\right)_{\mathbf{V}}, \tag{15-15e}$$

and

$$\frac{P}{T} = \left(\frac{\partial \mathbf{S}}{\partial \mathbf{V}}\right)_{\mathbf{E}}. \tag{15-15f}$$

First-order relations

These are also known as *Maxwell's relations*:*

$$\left(\frac{\partial T}{\partial \mathbf{V}}\right)_{\mathbf{S}} = -\left(\frac{\partial P}{\partial \mathbf{S}}\right)_{\mathbf{V}}, \tag{15-16a}$$

$$\left(\frac{\partial \mathbf{S}}{\partial \mathbf{V}}\right)_{T} = \left(\frac{\partial P}{\partial T}\right)_{\mathbf{V}}, \tag{15-16b}$$

$$\left(\frac{\partial \mathbf{S}}{\partial P}\right)_{T} = -\left(\frac{\partial \mathbf{V}}{\partial T}\right)_{P}, \tag{15-16c}$$

---

* They are, of course, special cases of Euler's reciprocal relations.

and

$$\left(\frac{\partial T}{\partial P}\right)_{\mathbf{S}} = \left(\frac{\partial \mathbf{V}}{\partial \mathbf{S}}\right)_P. \tag{15-16d}$$

Applications of Eqs. (15-15) and (15-16) will be given in many subsequent chapters.

(b) We digress a moment to derive those properties of **A** and **G** upon which their names are founded. We start with the basic form of the First Law,

$$d\mathbf{E} = dQ - dW, \tag{9-2}$$

in which no restrictions are imposed as yet on the type(s) of work that are permitted. For a reversible process, W is a maximum (see section 4.3), and dQ can be replaced by $T\,d\mathbf{S}$, so that we get

$$d\mathbf{E} = T\,d\mathbf{S} - dW_{\text{max}}.$$

At constant temperature* this becomes

$$d_T(\mathbf{E} - T\mathbf{S}) = -dW_{\text{max}}$$

or, by (15-10),

$$\Delta_T \mathbf{A} = -W_{\text{max}}. \tag{15-17}$$

In words:

> The maximum amount of total work obtainable from an *isothermal* change is equal to the *loss* in Helmholtz free energy (equal to the *decrease* in the value of the work function).

It is customary at this stage to consider two types of work:

1. $P,\mathbf{V}$ work, associated with a possible expansion of the system, which *in this context* is considered useless.
2. "*Useful* work" (also called "*net* work" or "*available* work"), which *in this context* can often be equated with electrical work (e.g., produced by a galvanic cell; see section 15.4). For a generalization of this concept, see also section 15.5.

Accepting this distinction and the associated nomenclature, we can write for a reversible change,

$$dW_{\text{max}} = P\,d\mathbf{V} + dW_{\text{useful, max}}. \tag{15-18}$$

It is left as an exercise to the readers to obtain

$$\Delta_{P,T}\mathbf{G} = -W_{\text{useful, max}}. \tag{15-19}$$

---

* It may suffice if the initial and final temperatures are the same, but such a process is usually carried out in a container, which is placed in—and can exchange heat with—a thermostat at temperature T.

Exercise 15-5

Derive Eq. (15-19) along the same lines as those we obtained Eq. (15-17).

In words:

> The maximum amount of "useful" (e.g., electrical) work obtainable from a change at constant P and T is equal to the *loss* in (Gibbs) free energy (equal to the decrease in the value of the Gibbs function).

Equation (15-19) is also used in evaluating the maximum amount of "other than $P,\mathbf{V}$ work" that can be obtained from certain reactions in biological systems.* (See also the remarks on biological processes in the last paragraph of section 14.4.)

Note that while *in general* W is path dependent, W_{max} and $W_{\text{useful, max}}$, *under the restriction* of constant T, and constant P *and* T, respectively, become equal to changes in state functions, hence are completely determined by the specification of the initial and final states.

(c) The quantities $\mathbf{A}$, $\mathbf{G}$, $\mathbf{E}$, $\mathbf{H}$, and $\mathbf{S}$ are sometimes referred to as the *characteristic functions* associated with the *independent sets of variables* $(\mathbf{V},T)$, (P,T), $(\mathbf{S},\mathbf{V})$, $(\mathbf{S},P)$ and $(\mathbf{E},\mathbf{V})$, respectively.† The reason is that, given $\mathbf{A}(\mathbf{V},T)$ or $\mathbf{G}(P,T)$, etc., all other relevant thermodynamic properties can readily be expressed in terms of this characteristic function and its differentials with respect to the independent variables.

As an illustration, we take P and T as independent variables and assume that $\mathbf{G}(P,T)$ is known. Readers should have no difficulty in obtaining

$$\mathbf{V} = \left(\frac{\partial \mathbf{G}}{\partial P}\right)_T \tag{15-20}$$

[which gives us the (thermal) equation of state] as well as

$$\mathbf{S} = -\left(\frac{\partial \mathbf{G}}{\partial T}\right)_P, \tag{15-21}$$

$$\mathbf{E} = \mathbf{G} - T\left(\frac{\partial \mathbf{G}}{\partial T}\right)_P - P\left(\frac{\partial \mathbf{G}}{\partial P}\right)_T, \tag{15-22}$$

$$\mathbf{A} = \mathbf{G} - P\left(\frac{\partial \mathbf{G}}{\partial P}\right)_T, \tag{15-23}$$

---

* See, e.g., A. L. Lehninger, *Bioenergetics*, W. A. Benjamin, New York, 1965, and I. M. Klotz, *Energy Changes in Biochemical Reactions*, 2nd ed., Academic Press, New York, 1967. For a brief introduction, see also I. M. Klotz and R. M. Rosenberg, *Chemical Thermodynamics*, 3rd ed., W. A. Benjamin, Menlo Park, Calif., 1972, pp. 150–153.

† Later, as various generalizations to open systems are introduced (see Chapter 19), we shall have to add the mole numbers, n_i $(i = 1, \ldots, r)$, as independent variables.

and

$$\mathbf{H} = \mathbf{G} - T\left(\frac{\partial \mathbf{G}}{\partial T}\right)_P = -T^2\left(\frac{\partial(\mathbf{G}/T)}{\partial T}\right)_P = \left(\frac{\partial(\mathbf{G}/T)}{\partial(1/T)}\right)_P. \qquad (15\text{-}24)$$

Exercise 15-6

Verify Eqs. (15-22) through (15-24). Equations (15-20) and (15-21) were already obtained as parts of Eqs. (15-15c) and (15-15d).

To drive home the essential point, assume that instead of P and T, $\mathbf{V}$ and T were the independent variables, and that we were given $\mathbf{G}$ as a function of $\mathbf{V}$ and T. From (15-13b) we obtain

$$d\mathbf{G} = \left[-S + \mathbf{V}\left(\frac{\partial P}{\partial T}\right)_{\mathbf{V}}\right] dT + \mathbf{V}\left(\frac{\partial P}{\partial \mathbf{V}}\right)_T d\mathbf{V}.$$

We see that $\mathbf{V}(\partial P/\partial \mathbf{V})_T = (\partial \mathbf{G}/\partial \mathbf{V})_T \equiv g(\mathbf{V},T)$, say. It follows that

$$P = \int_{(\text{const. T})} \frac{g(\mathbf{V},T)}{\mathbf{V}}\, d\mathbf{V}$$

or, upon integration,

$$P = f(\mathbf{V},T) + \phi(T),$$

in which $f(\mathbf{V},T)$ is a *specific* function of $\mathbf{V}$ and T, but $\phi(T)$ is an *unknown* integration "constant," so that we cannot obtain an explicit expression for $P(\mathbf{V},T)$, i.e., the (thermal) equation of state. It is crucial to realize that while working with $\mathbf{G}(P,T)$, there was no need for *integration*; we could express our results exclusively in terms of $\mathbf{G}$ and its partial *differentials*.*

Exercise 15-7

Let $\mathbf{V}$ and T be the independent variables. Express P, $\mathbf{E}$, $\mathbf{S}$, $\mathbf{H}$, and $\mathbf{G}$ in terms of $\mathbf{A}(\mathbf{V},T)$ and its partial derivatives.

(d) In many instances the states of a system have to be specified either by parameter couples *other than* P and $\mathbf{V}$, or by *additional* parameter couples *besides* P and $\mathbf{V}$. An example of the former situation was encountered in the discussion of "adiabatic demagnetization" (section 14.3), and an important illustration of the latter case will be provided by "galvanic cells" (section 15.4). There is no need to modify

$$\mathbf{A} \equiv \mathbf{E} - T\mathbf{S} \qquad (15\text{-}10)$$

---

* If $\mathbf{G}(P,T)$ were determined up to an arbitrary *constant*, this would carry over into $\mathbf{E}$, $\mathbf{A}$, and $\mathbf{H}$ [see Eqs. (15-22) through (15-24)], but Eq. (15-20) still gives us the unambiguous equation of state.

since the new parameters do not explicitly affect this definition. On the other hand, one may wish to replace $\mathbf{G}$, as defined in Eq. (15-11) by $(\mathbf{E} - T\mathbf{S} + P\mathbf{V})$, by a more relevant quantity. For example, in the thermodynamics of surfaces, some authors introduce the *Gibbs free energy of a Surface*, $\mathbf{G}^\sigma$, by

$$\mathbf{G}^\sigma \equiv \mathbf{E}^\sigma - T\mathbf{S}^\sigma - \gamma\mathscr{A}, \tag{15-25}$$

in which $\mathscr{A}$ is the surface area and γ the surface tension.

If we consider systems in external fields, *generalized free energy functions* may take over the role that $\mathbf{A}$ and $\mathbf{G}$ played in the absence of these fields. Note, for example, how $(\mathbf{A} + \Phi)$ and $(\mathbf{G} + \Phi)$ of Eqs. (15-29) and (15-30) below play the same part "normally" given to $\mathbf{A}$ and $\mathbf{G}$ [cf. Eqs. (15-27) and (15-28)]. Some authors have invented ingenious symbols for such generalized functions; e.g., Guggenheim* uses inverted capital letters in this context. Confusion arises only if the *same* symbols, $\mathbf{A}$ and $\mathbf{G}$, are used to denote *different* thermodynamic functions, as their need arises. In this book $\mathbf{A}$ and $\mathbf{G}$ shall always refer to the properties defined by Eqs. (15-10) and (15-11), respectively.

(e) To end this section we present a critical discussion of the validity of the fundamental equation

$$d\mathbf{E} = T\,d\mathbf{S} - P\,d\mathbf{V}. \tag{15-12}$$

In passing we note that the conclusions we shall reach are also pertinent to the expressions (15-13a), (15-13b), and (15-13c) for $d\mathbf{A}$, $d\mathbf{G}$, and $d\mathbf{H}$, since Eq. (15-12) formed the basis for their derivation.

As we mentioned before, it is easy to see that Eq. (15-12) is obtained by combining the First and the Second Laws for any closed system (permitting only $P,\mathbf{V}$ work) *if* we restrict ourselves to *reversible* processes. It is far less obvious that Eq. (15-12) is still valid for a large class of *irreversible* processes, which such systems might undergo.† To understand this, we first write this equation in the alternative form:

$$d\mathbf{E} = \left(\frac{\partial \mathbf{E}}{\partial \mathbf{S}}\right)_{\mathbf{V}} d\mathbf{S} + \left(\frac{\partial \mathbf{E}}{\partial \mathbf{V}}\right)_{\mathbf{S}} d\mathbf{V}, \tag{15-12a}$$

with the obvious identifications $T = (\partial\mathbf{E}/\partial\mathbf{S})_{\mathbf{V}}$ and $-P = (\partial\mathbf{E}/\partial\mathbf{V})_{\mathbf{S}}$. While we are inclined to look upon Eq. (15-12) as describing a *physical process*, the form Eq. (15-12a) also represents a *geometrical relationship*, a displacement on the $\mathbf{E}(\mathbf{S},\mathbf{V})$ surface. Gibbs was the first one to emphasize geometrical in-

---

* E. A. Guggenheim, reference 6 of Chapter 4.

† For the time being, we exclude all those *irreversible* processes which involve changes in composition of our system. In this case $\mathbf{E}$ is no longer a function of $\mathbf{S}$ and $\mathbf{V}$ alone, but also of a set of composition variables (see section 19.1).

terpretations of this type.* As long as $\mathbf{E}(\mathbf{S},\mathbf{V})$ is a well-defined, properly behaved function, the partial derivatives of $\mathbf{E}$ with respect to $\mathbf{S}$ and $\mathbf{V}$ are also well defined. In turn, this is tantamount to saying that P and T are well defined. Incidentally, this was, of course, already implied in the *definition* of the free energy functions. (In fact, the requirement that T is well defined was mentioned already in the second sentence of this chapter.) Under these circumstances, the value of $d\mathbf{E}$ is completely determined by Eq. (15-12a), hence by Eq. (15-12), and we do not even need to differentiate between reversible and irreversible transitions. Thus we conclude that Eq. (15-12) is not only valid for reversible processes, but also for those irreversible processes that occur so "smoothly" that P and T are well defined throughout. In section 19.1 we shall generalize Eqs. (15-12) and (15-12a) to include spontaneously occurring chemical reactions.†

To clarify a point that is frequently misunderstood, let us relate the validity of Eq. (15-12) to that of the First Law (for closed systems):

$$d\mathbf{E} = dQ - dW. \tag{9-2}$$

As before, we limit ourselves to processes involving only $P,\mathbf{V}$ work. We distinguish two cases:

1. For *reversible* processes dQ equals $T\,d\mathbf{S}$ and dW equals $P\,d\mathbf{V}$. Once more, the equivalence of Eqs. (9-2) and (15-12) is obvious.
2. For *irreversible* processes, $dQ < T\,d\mathbf{S}$ and $dW < P\,d\mathbf{V}$. The First Law, in its "primary form," Eq. (9-2), is still valid; it does not differentiate between reversible and irreversible processes at all. If the conditions for the validity of Eq. (15-12), as discussed above, are also satisfied, it "simply" means that dQ is smaller than $T\,d\mathbf{S}$ *by exactly the same amount* as dW is smaller than $P\,d\mathbf{V}$. Thus the crucial point is that $d\mathbf{E}$ can be equal to $(dQ - dW)$ *and to* $(T\,d\mathbf{S} - P\,d\mathbf{V})$, even when $dQ \neq T\,d\mathbf{S}$ and $dW \neq P\,d\mathbf{V}$!

15.3 Alternative Conditions for Equilibrium and Stability

Following up the analysis of the preceding subsection, we shall continue to assume that during any irreversible process *P and T are well defined.*

(a) As our starting point, we take the relation

$$\int_{A}^{B} \frac{dQ}{T} \leqq \Delta\mathbf{S} \equiv \mathbf{S}_{B} - \mathbf{S}_{A}, \tag{15-2}$$

* For an elaboration of this point of view, we must refer the readers to a series of papers by Frank Weinhold. He gives a good summary, with additional references, in an article entitled "Thermodynamics and Geometry," *Physics Today* **29**, 23 (1976).

† If the reaction mixture is at (macroscopic) *equilibrium*, there is no change in composition variables, $\mathbf{E} = \mathbf{E}(\mathbf{S},\mathbf{V})$, and Eq. (15-12) requires no modification.

and we remind the readers that the inequality sign refers to irreversible processes and the equality sign to the limiting case of reversible processes. At *constant temperature* this equation can be integrated to yield

$$\frac{1}{T} Q_{A \to B} \leqq \mathbf{S}_B - \mathbf{S}_A,$$

or, by the First Law,

$$\frac{1}{T}(\mathbf{E}_B - \mathbf{E}_A + W_{A \to B}) + \mathbf{S}_A - \mathbf{S}_B \leqq 0.$$

Remembering the definition of the Helmholtz free energy, **A** [Eq. (15-10)], this equation reads

$$[(\mathbf{E}_B - T\mathbf{S}_B) - (\mathbf{E}_A - T\mathbf{S}_A)]_{T\ \text{constant}} + W_{A \to B} \leqq 0,$$

or simply [cf. Eq. (15-17)]

$$\Delta_T \mathbf{A} + W \leqq 0. \tag{15-26}$$

Once more we shall focus our attention on the situation in which *only P,***V** *work is possible.* Two cases are of particular interest:

1. **V** *(as well as T) is constant.* With these constraints $W = 0$, and (15-26) becomes

$$\Delta_{\mathbf{V},T} \mathbf{A} \leqq 0, \tag{15-27}$$

the *minimum theorem of the Helmholtz free energy.*

2. *P (as well as T) is constant.* With these constraints $W = \Delta(P\mathbf{V})$. Since $\mathbf{G} \equiv \mathbf{A} + P\mathbf{V}$, Eq. (15-26) becomes

$$\Delta_{P,T} \mathbf{G} \leqq 0, \tag{15-28}$$

the minimum theorem of the Gibbs free energy.

Exercise 15-8

Draw diagrams, corresponding to Fig. 17, to illustrate (1) the minimum theorem of **A** and (2) the minimum theorem of **G**.

In practice, most processes are carried out at constant P and T. This is certainly true for most chemical reactions in the laboratory and in industry. Therefore, the minimum theorem of the Gibbs free energy is of the greatest utility.

In section (11.1) we briefly mentioned the (incorrect) "Berthelot Principle," according to which a chemical reaction occurs spontaneously if and only if it is exothermic. We can now compare this criterion with the Gibbs free energy theorem for chemical reactions at constant P and T (only P,**V** work allowed). To this purpose we cast the two criteria in the form*

$$\Delta \mathbf{H} \leqq 0 \qquad \text{(Berthelot)}$$

* For the derivation of $Q_p = \Delta \mathbf{H}$ [Eq. (10-41)], see section 10.4.

and

$$\Delta\mathbf{G} = \Delta\mathbf{H} - T\Delta\mathbf{S} \leqq 0 \qquad \text{(correct)}.$$

It is obvious that these two give the same predictions if $\Delta\mathbf{H}$ and $\Delta\mathbf{G}$ have the *same* sign. This, in turn, would *always* be the case if $\Delta\mathbf{H}$ and $\Delta\mathbf{S}$ have *opposite* signs. Unfortunately, for most chemical reactions, $\Delta\mathbf{H}$ and $\Delta\mathbf{S}$ have the *same* sign, as we shall illustrate in several examples below. Whether or not the Berthelot Principle is still valid now becomes a question of the relative magnitudes of the terms $|\Delta\mathbf{H}|$ and $T|\Delta\mathbf{S}|$. In this situation the value of T is of vital importance. Let us briefly consider some representative examples, a phase transition, a dissociation, and a well-known industrial process.

1. $H_2O(\ell) \rightarrow H_2O(g)$ at 1 atm, *above* 100°C. (Of course, at 1 atm, 100°C, water and steam are in equilibrium; $\Delta\mathbf{G} = 0$.) Here $\Delta\mathbf{H} > 0$, but the process is spontaneous nevertheless, since $\Delta\mathbf{S}$ is also positive *and* T is large enough to make $\Delta\mathbf{G}$ negative.
2. At very high temperatures typical dissociation reactions, e.g.,

$$Cl_2(g) \rightarrow 2Cl(g) \qquad \text{at 1 atm},$$

occur spontaneously even though, of course, $\Delta\mathbf{H}$ is positive. The explanation is entirely analogous to that of case 1.
3. For the famous "water-gas" reaction,

$$C(s) + H_2O(g) = CO(g) + H_2(g) \qquad \text{at 1 atm},$$

$\Delta\mathbf{H}$ is on the order of 130 kJ mole^{-1} and $\Delta\mathbf{S}$ on the order of 130 J K^{-1} mole^{-1} (strictly speaking, both are temperature functions). At temperatures above about 1000 K, this reaction, which has been of considerable practical importance, occurs with (reasonable rates and) good yields, even though it is endothermic.

In all three examples $\Delta\mathbf{H}$ and $\Delta\mathbf{S}$ are positive. An example where both these terms are negative is provided by the formation of ammonia from the elements, at 1 atm pressure. At about 470 K, and above, $\Delta\mathbf{G}$ for this reaction is positive, and NH_3 yields are low, despite the fact that $\Delta\mathbf{H}$ is negative.*†

(b) From the discussion in section 15.2(d), and the derivations given in subsection (a) above, it should be evident that, under special conditions, the criteria (15-27) and (15-28) may appear in a slightly revised form, specifically tailored for applications in these circumstances. We shall not attempt to summarize all possible modifications, but digress briefly in order to mention one important case, which arises if a system is subjected to a *conservative* external force field.‡ In this case the equations remain quite simple, at least in a formal sense, because $\mathbf{E}$ just has to be replaced everywhere by $\mathbf{E} + \Phi$,

---

* See E. L. King, *Chemistry*, Painter Hopkins, Sausalito, Calif., 1979, pp. 110–111.

† Ammonia yields become significant at high pressures; see section 24.2.

‡ In a *conservative* force field, the forces are independent of time and velocity (hence frictional forces have to be excluded) and can be expressed as the negative gradient of a potential energy, which is a function of the coordinates only.

in which Φ is the potential energy function. With our resolve [cf. the discussion in subsection 15.2(d)] to leave the definitions of $\mathbf{A}$ and $\mathbf{G}$ unchanged, the conditions (15-27) and (15-28) are replaced by

$$\Delta_{\mathbf{V},T}(\mathbf{A} + \Phi) \leqq 0 \qquad (15\text{-}29)$$

and

$$\Delta_{P,T}(\mathbf{G} + \Phi) \leqq 0, \qquad (15\text{-}30)$$

respectively. It is important to realize that the maximum theorem of the entropy, in its original form,

$$\Delta_{\text{ad}}\mathbf{S} \geqq 0, \qquad (15\text{-}3)$$

does not require modification. However, in the conditions (15-4a′) and (15-4b′) $\mathbf{E}$ has to be replaced by $\mathbf{E} + \Phi$. For example, $\delta_{\mathbf{E},\mathbf{V}}\mathbf{S}$ now becomes $\delta_{\mathbf{E}+\Phi,\,\mathbf{V}}\mathbf{S}$, and so on.

(c) After this digression we return to the situation where no other than $P,\mathbf{V}$ work is possible and external fields are absent. It is convenient to combine the "old" results for the first- and second-order variation of the entropy, Eqs. (15-4), with their newly developed analogues for the free energy functions (Table 1). The combination of any of these equilibrium conditions with the *corresponding* stability condition is often referred to as a *condition for stable equilibrium.*

Table 1

Summary of Equilibrium and Stability Conditions[a]
(Only $P,\mathbf{V}$ work possible; no external fields)

Equilibrium	Stability
$\delta_{\mathbf{E},\mathbf{V}}\mathbf{S} = 0$	$\delta^2_{\mathbf{E},\mathbf{V}}\mathbf{S} < 0$
$\delta_{\mathbf{V},T}\mathbf{A} = 0$	$\delta^2_{\mathbf{V},T}\mathbf{A} > 0$
$\delta_{P,T}\mathbf{G} = 0$	$\delta^2_{P,T}\mathbf{G} > 0$

[a] Additional conditions may be written down in terms of the variation in $\mathbf{E}$ at constant $\mathbf{S}$ and $\mathbf{V}$, or the variation in $\mathbf{H}$ at constant $\mathbf{S}$ and $\mathbf{P}$, but these will not be used in this book. The constraints associated with each variation should be carefully noted.

Actually, these conditions do not differentiate between *stable* and *metastable* (sometimes called *local**) equilibrium. In addition, if the first- and second-order variations were all zero, we would have *neutral* equilibrium, and if the signs of the second-order

* We avoid this usage because "*local* equilibrium" may also refer to the equilibrium in a small volume element of a nonuniform fluid (cf. section 17.2) or to an equilibrium subjected to special constraints.

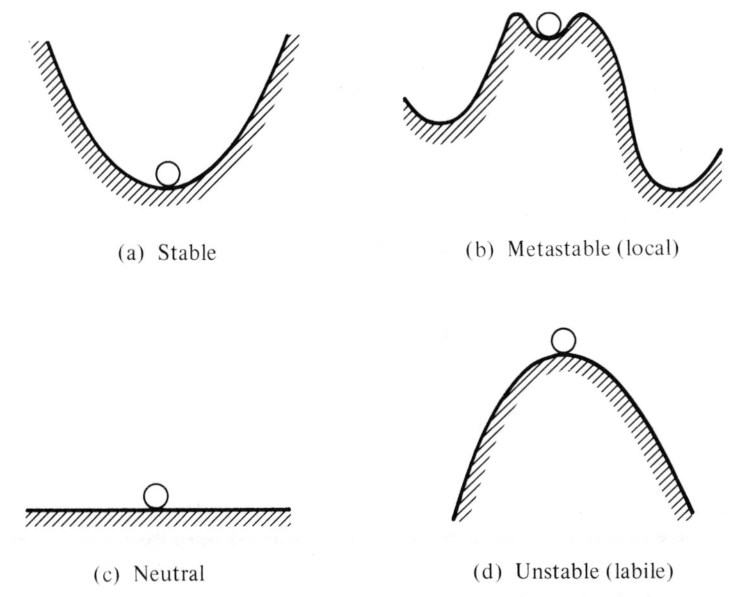

Fig. 18 Types of equilibria illustrated for a purely mechanical system. If the four diagrams are considered to represent $V(r)$ curves, then (a) corresponds to $(d^2V/dr^2) > 0$; (b) likewise; (c) to $(d^2V/dr^2) = 0$; and (d) to $(d^2V/dr^2) < 0$. In all cases $(dV/dr) = 0$.

variations were reversed (with respect to those in Table 1), we would be dealing with *unstable* (sometimes called *labile*) equilibrium. Most of these terms are borrowed from classical mechanics, where they were first given precise mathematical definitions. For our purpose, the illustrations given in Fig. 18 will suffice. It should be stressed that *unstable* equilibrium can never be realized in nature, because it would be upset by any infinitisimal perturbation, which is always present. The other three types of equilibria all play important roles in the physical sciences. Some very simple examples are:

Stable:	Water at 10°C, 1 atm
Neutral:	Water and ice at 0°C, 1 atm
Metastable:	(Supercooled) water at −10°C, 1 atm

We shall return repeatedly to these concepts in future chapters. (In particular, see sections 18.3 and 18.4.)

(d) A final comment. One occasionally finds statements such as: "If a system is at equilibrium at constant P and T, $\Delta\mathbf{G} = 0$." This is an unfortunate way of stating one of the general equilibrium conditions. A more meaningful presentation reads: "If a system at equilibrium is subjected to any conceivable small ("virtual") perturbation at constant P and T, $\delta\mathbf{G} = 0$." If the *same* system at equilibrium were subjected to such a perturbation at constant **V** and T, $\delta \mathbf{A}$ would vanish, and so on. In other words, since at equilibrium (as defined in phenomenological thermodynamics) *all* macroscopic properties and variables have to remain constant, *the three conditions, summarized in Table 1, will be satisfied simultaneously.* Any one is sufficient and the question of *which* conditions one ought to *use* in a particular case, and why, is discussed in detail in the third part of this book; we refer readers to Chapters 17 and 18. Those who skip the first, rather advanced, of those two chapters, will find an important illustration of the principles involved in section 18.1.

15.4 Electrochemical Corollary; Galvanic Cells and the Gibbs Free Energy

(a) The application of thermodynamics to chemical processes occurring in galvanic cells has become a standard topic in general chemistry courses. We shall assume that readers have acquired a working knowledge of simple notations, such as "cell schematics" of the type*

$$\text{Pt, } H_2(p)|\text{HCl}[\text{aq}(m)]\|\ \text{AgCl, Ag}|\text{Pt}$$

and are familiar with the most important kinds of electrodes that are used in the construction of these cells. It is our purpose to analyze some of the essential concepts critically and to put the thermodynamics involved on a sound basis. We shall not attempt to give a comprehensive account of all possible ramifications, since this would lead us well beyond the "equilibrium thermodynamics" to which this book is confined.

One of the most important properties of a galvanic cell is its *electromotive force* (EMF), $\mathscr{E}$, already encountered in Chapter 4. Its definition is based on the fundamental equation of electric circuit analysis:

$$\Delta u = \mathscr{E} - IR, \tag{15-31}$$

in which Δu is the electrostatic potential difference between the electrodes, I the electric current flowing through the system, and R its resistance. Apparently, $\mathscr{E}$ is the potential difference as $I \to 0$, which can be measured by means of a potentiometer bridge (as originally devised by Poggendorff). It represents an "equilibrium value," sometimes referred to as the "open-circuit value," of Δu.

Associated with each galvanic cell is a *cell reaction.* For example, for the cell schematically represented above, it is

$$H_2(g) + 2AgCl(s) = 2Ag(s) + 2H^+(aq) + 2Cl^-(aq).$$

* Several variations are possible. The p and m give the pressure of the H_2 gas and the molality of the HCl solution, respectively. These are needed, together with the temperature, to characterize the cell.

The basic equation for the thermodynamic study of galvanic cells relates $\mathscr{E}$ to $\Delta\mathbf{G}$ of this reaction:

$$\Delta\mathbf{G} = -n\mathscr{F}\mathscr{E}. \tag{15-32}$$

In this equation [derived in subsection (c) below]:

> Δ has the specific meaning introduced in Eq. (8-50a), i.e., it is equivalent to the operator $(\partial/\partial\xi)_{P,T}$,*
>
> $\mathscr{F}$, the faraday (=96,486 coulomb), is the absolute value of the charge of 1 mole of electrons, and
>
> n is the number of moles of electrons that flow through the cell if molar quantities are converted according to the stoichiometry of the cell reaction. (In our example, $n = 2$, as readers should readily verify, e.g., by writing out the two appropriate "half-reactions.")

In later parts of this book we shall write $\Delta\mathbf{G}$ for the reaction in terms of thermodynamic properties of the individual reactants and products. If all these species are in their, yet to be defined, *standard states*, Eq. (15-32) takes on the special form

$$\Delta\mathbf{G}^{\ominus} = -n\mathscr{F}\mathscr{E}^{\ominus}. \tag{15-32'}$$

Readers will recall from general chemistry how the *standard potentials*, $\mathscr{E}^{\ominus}$, may be computed from tabulated values of standard potentials for the half-reactions involved. In the rest of this chapter these matters will be left aside; instead, we shall concentrate primarily on the origin, meaning, and some of the consequences of the fundamental equation (15-32).

Evidently, $\mathscr{E}$, as it is associated with a cell reaction, can be both positive and negative. When $\mathscr{E}$ is positive, $\Delta\mathbf{G}$ is negative, and the cell will *discharge* as the cell reaction proceeds spontaneously from left to right. The reverse reaction, for which $\Delta\mathbf{G}$ is positive and $\mathscr{E}$ negative, can at best be "forced" to occur by *charging* the cell through the imposition of an external potential greater than and opposed to $\mathscr{E}$.

(b) Before concerning ourselves with the derivation of Eq. (15-32), it is absolutely essential to discuss the *reversibility* of electrochemical processes. The difficulties already associated with the reversibility concept *in general* as presented in Chapter 5 are compounded by the frequent use of this term *in a specific electrochemical sense.* To explain this, we consider two cases:

1. We return once more to the example used in subsection (a) above. In this system, when the cell is charged, we *merely reverse the chemical reaction* that occurs when the cell discharges. Such a cell is often called *electrochemically reversible.*

---

* In equations such as (15-32) and (15-32′), the constancy of P and T is always *implied.*

2. Next, consider the cell

$$Zn|H_2SO_4[aq(m)]|Pt.$$

Upon discharge, the reaction

$$Zn(s) + 2H^+(aq) \rightarrow H_2(g) + Zn^{2+}(aq)$$

occurs spontaneously.* But if we impose an external voltage larger than the cell EMF, in the opposite direction, the net chemical reaction will be quite different, namely, the electrolysis of water (H_2 evolves at the Zn electrode, and O_2 at the Pt electrode):

$$2H_2O(\ell) \rightarrow 2H_2(g) + O_2(g).$$

This cell is *electrochemically irreversible.* We next have to compare the *electrochemical* reversibility, thus introduced, to the *thermodynamic* reversibility, discussed in Chapter 5. It should be quite evident that a cell which is electrochemically irreversible can *never* be made to approach thermodynamic reversibility. When such a cell discharges infinitely slowly, by subjecting it to the appropriate counter EMF, it gives us a splendid example of a "pseudo-static" process, in Buchdahl's terminology (see section 5.2). On the other hand, it seems equally obvious that a cell which is electrochemically reversible *need not* operate reversibly in a thermodynamic sense, but *can* be made to approach thermodynamic reversibility, if it is subjected to a counter EMF that will allow the cell reaction to proceed infinitely slowly.

Unfortunately, if we study the construction and operation of galvanic cells more closely, we may detect several other sources of irreversibility. One such source is the presence of a *liquid junction*, although its effect can usually be minimized by the use of a "salt bridge."† Perfectionists will argue that, strictly speaking, *no* galvanic cells can *ever* operate reversibly, since there are always concentration gradients, which cause inherently irreversible diffusion processes to take place. Nevertheless, the author joins those who are convinced that in many electrochemically reversible cells the irreversible aspects can be sufficiently minimized so that the cell process approaches a thermodynamically reversible operation as closely as *any* real process can. If one wishes to carry out a chemical reaction under reversible conditions, galvanic cells offer usually our *best* hope, and many times our *only* hope. Under *all* circumstances, a (thermodynamically) reversible process remains an idealization.

(c) We now turn to the derivation of

$$\Delta \mathbf{G} = -n\mathcal{F}\mathcal{E}. \qquad (15\text{-}32)$$

---

* If the sulfuric acid is sufficiently strong, it will attack the Zn electrode even if the circuit is not closed. This circumstance does not interfere with the characterization of this cell.

† For the details we refer to the appropriate electrochemical literature. See, e.g., A. J. Bard and L. R. Faulkner, *Electrochemical Methods*, Wiley, New York, 1980, section 2.3, pp. 62ff.

This basic equation, or its analogue under standard conditions, (15-32′), can be found in most textbooks on general chemistry. Its validity is usually established by combining the relations:

$$\Delta\mathbf{G} = -W_{\text{electric, max}} \tag{15-19}$$

and

$$W_{\text{electric, max}} = n\mathscr{F}\mathscr{E}. \tag{15-33}$$

(In all these equations constant P and T is *implied.*) Readers were asked to justify Eq. (15-19) in Exercise 15-5, and Eq. (15-33) follows immediately from the definition of $\mathscr{E}$ as an "equilibrium value," given in subsection (a) above.

In Exercise 15-5 readers were told specifically to carry out the derivation of Eq. (15-19) along the lines by which $\Delta_T\mathbf{A} = -W_{\text{max}}$ was derived in the text, that is, *under the assumption of thermodynamic reversibility*. But all that we really need is the validity of [cf. Eq. (4-17d)]

$$d\mathbf{E} = T\,d\mathbf{S} - P\,d\mathbf{V} + \mathscr{E}\,dq. \tag{15-34}$$

This can be established for a large class of irreversible processes by exactly the same type of reasoning that led us to accept

$$d\mathbf{E} = T\,d\mathbf{S} - P\,d\mathbf{V} \tag{15-12}$$

outside the reversible domain [see section 15.2(e)]. The extra term in Eq. (15-34) corresponds, of course, to the electrical work. With P, T, and $\mathscr{E}$ fixed, during an infinitely slow discharge, the $\mathbf{E}(\mathbf{S}, \mathbf{V}, q)$ surface is well defined, and so on. Once we can accept Eq. (15-34), we combine it with the total differential of $\mathbf{G}$ ($\equiv \mathbf{E} - T\mathbf{S} + P\mathbf{V}$):

$$d\mathbf{G} = d\mathbf{E} - T\,d\mathbf{S} - \mathbf{S}\,dT + P\,d\mathbf{V} + \mathbf{V}\,dP$$

to obtain

$$d\mathbf{G} = -\mathbf{S}\,dT + \mathbf{V}\,dP + \mathscr{E}\,dq. \tag{10-35}$$

Integration, corresponding to the infinitely slow *discharge* of n Faradays ($\int dq = -n\mathscr{F}$), at constant P, T, and $\mathscr{E}$, yields the desired $\Delta\mathbf{G} = -n\mathscr{F}\mathscr{E}$, Eq. (15-32), under less stringent conditions than imposed previously.

(d) From the general discussion in sections 8.3 and 8.4, we conclude that Eq. (15-21) immediately leads to

$$\Delta\mathbf{S} = -\left(\frac{\partial\,\Delta\mathbf{G}}{\partial T}\right)_P, \tag{15-35}$$

and Eq. (15-24) to

$$\Delta\mathbf{H} = \Delta\mathbf{G} - T\left(\frac{\partial\Delta\mathbf{G}}{\partial T}\right)_P = -T^2\left(\frac{\partial\Delta\mathbf{G}/T}{\partial T}\right)_P = \left(\frac{\partial\Delta\mathbf{G}/T}{\partial 1/T}\right)_P. \tag{15-36}$$

$\Delta\mathbf{J}$ ($\mathbf{H}$, $\mathbf{G}$, or $\mathbf{S}$), as in the preceding section, refers to a chemical reaction at constant P and T; i.e., Δ is the operator $(\partial/\partial\xi)_{P,T}$.

By using the basic equation (15-32), these equations can be written "in electrochemical form." For example, we get

$$\Delta \mathbf{S} = n\mathscr{F}\left(\frac{\partial \mathscr{E}}{\partial T}\right)_P \tag{15-37}$$

and

$$\Delta \mathbf{H} = n\mathscr{F}\left[T\left(\frac{\partial \mathscr{E}}{\partial T}\right)_P - \mathscr{E}\right]. \tag{15-38}$$

Hence if we measure the EMF of a galvanic cell, as well as its temperature coefficient, $(\partial\mathscr{E}\ \partial T)_P$, we can compute $\Delta\mathbf{H}$, $\Delta\mathbf{S}$, and $\Delta\mathbf{G}$ of the associated cell reaction by means of Eqs. (15-38), (15-37), and (15-32), respectively. The expressions (15-36) and (15-38) for $\Delta\mathbf{H}$, and several related expressions, are all referred to in the literature as Gibbs–Helmholtz equations.*

All equations in this subsection have their "standard state" equivalents, in terms of the $\Delta\mathbf{J}^{\ominus}$ of a chemical reaction. One simply replaces each $\mathbf{J}$ ($\mathbf{H}$, $\mathbf{G}$, or $\mathbf{S}$) by the corresponding $\mathbf{J}^{\ominus}$, and each $\mathscr{E}$ by $\mathscr{E}^{\ominus}$ [compare Eq. (15-32′) in subsection (a) above.]

If the second derivative of $\mathscr{E}$ with respect to T can be measured with sufficient accuracy, we can obtain $\Delta\mathbf{C}_p$ of a cell reaction. For, by differentiation of (15-37),

$$\Delta \mathbf{C}_p = n\mathscr{F} T\left(\frac{\partial^2 \mathscr{E}}{\partial T^2}\right)_P. \tag{15-39}$$

15.5 Appendix: Availability and Related Concepts

(a) In this section we introduce, very briefly, some generalizations of Gibbs's ideas on "useful" work, as presented in section 15.2(b). These generalized concepts are not widely discussed in the current literature on thermodynamics,† although they appear to have considerable appeal to some authors, in relation to certain technological applications.‡

Consider a system at constant P and T, to be denoted by (I). It undergoes a reversible change (possibly, but not necessarily, a chemical reaction),

---

* Guggenheim attributes the forms (15-36) to Gibbs, and their electrochemical corollaries, such as Eq. (15-38), to Helmholtz. See E. A. Guggenheim, *Thermodynamics*, 3rd ed., North-Holland, Amsterdam, 1957, p. 387.

† The author is indebted to Joseph O. Hirschfelder for bringing this subject to his attention.

‡ For an extensive discussion, see, e.g., J. H. Keenan, *Thermodynamics*, Wiley, New York, 1957, Chapter XVII; or George N. Hatsopoulus and J. H. Keenan, *Principles of General Thermodynamics*, Wiley, New York, 1965, pp. 64ff.

from an initial state, 1, to a final state, 2. It is convenient to identify temporarily all thermodynamic quantities associated with this change by the superscript (I). To maintain its pressure and temperature, at P and T, respectively, the system must, in general, exchange a certain amount of heat, $|Q^{(I)}|$, with its environment. Since the process takes place reversibly,

$$Q^{(I)} = T\,\Delta \mathbf{S}^{(I)}. \tag{15-40}$$

In deriving the usual maximum work theorems in section 15.2(b) it was implied that the relevant heat exchange took place between the system at hand and a thermostat at the same temperature, T. Instead, we shall now assume that $Q^{(I)}$ is provided by a Carnot cycle, operating between a reservoir at T and the (surface of the earth and the) atmosphere, the latter considered to be a permanent reservoir at a constant pressure, P_o, and a constant temperature, T_o.* Let the heat flow *into* the Carnot cycle at T_o be equal to Q_o; hence Q_o and $Q^{(I)}$ (as defined above) must have the same sign.

To see this, let $Q^{(I)}$ be negative, i.e., system (I) *liberates* heat during the process of interest. This heat must be *absorbed* by the Carnot cycle at T; hence this cycle must *liberate* heat at T_o, which means that Q_o is also negative.

Therefore, we can adapt Eq. (12-18) to read

$$\frac{Q_o}{T_o} = \frac{Q^{(I)}}{T}. \tag{15-41}$$

By Eqs. (15-40) and (15-41),

$$Q_o = T_o\,\Delta \mathbf{S}^{(I)}. \tag{15-42}$$

Next we ask for the *total* work done by a compound system consisting of (I) *plus* the specified Carnot cycle. By the First Law,

$$W_{\text{total}} = Q_o - \Delta \mathbf{E}. \tag{15-43}$$

Since all parts of this compound system operate reversibly, this is also the *maximum* total work obtainable if a particular change takes place in system (I). Hence, in lieu of Eq. (15-43), we may also write

$$W_{\text{total, max}} = Q_o - \Delta \mathbf{E}, \tag{15-44}$$

which, by virtue of Eq. (15-42), becomes

$$W_{\text{total, max}} = -(\Delta \mathbf{E} - T\,\Delta \mathbf{S}^{(I)}). \tag{15-45}$$

---

* Denbigh refers to this permanent reservoir as the "medium." See K. Denbigh, *The Principles of Chemical Equilibrium*, 4th Ed., Cambridge University Press, Cambridge, 1981, p. 70.

But since the internal energy change associated with the Carnot cycle vanishes, $\Delta\mathbf{E}$ equals $\Delta\mathbf{E}^{(I)}$, so that Eq. (15-45) reads

$$W_{\text{total, max}} = -(\Delta\mathbf{E}^{(I)} - T_o\Delta\mathbf{S}^{(I)}). \tag{15-46}$$

Finally, we *define*

$$W_{\text{useful, max}} \equiv W_{\text{total, max}} - P_o\Delta\mathbf{V}^{(I)}, \tag{15-47}$$

in which $P_o\Delta\mathbf{V}^{(I)}$ represents the minimum work "lost" to the atmosphere as the result of a possible change in $\mathbf{V}^{(I)}$. (Again, after a Carnot cycle has been completed, there is no resulting change in volume.) The combination of Eqs. (15-46) and (15-47) gives the result we set out to derive:

$$W_{\text{useful, max}} = -(\Delta\mathbf{E}^{(I)} - T_o\Delta\mathbf{S}^{(I)} + P_o\Delta\mathbf{V}^{(I)}). \tag{15-48a}$$

At this stage we may drop the superscripts (I) and, remembering that P_o and T_o are constant, write

$$W_{\text{useful, max}} = -\Delta(\mathbf{E} - T_o\mathbf{S} + P_o\mathbf{V}). \tag{15-48b}$$

The readers should realize that $W_{\text{useful, max}}$ as it is defined in this section is quite different, *in general*, from $W_{\text{useful, max}}$ as it was introduced in section 15.2(b). The latter referred exclusively to *other than* $P,\mathbf{V}$ work done by a system equivalent to the present system (I). The former, as it appears in Eqs. (15-47) and (15-48), includes, *in addition*, the work done by the Carnot cycle, which could be *entirely* of the $P,\mathbf{V}$ type. In the first case in subsection (b), we shall discuss the special situation in which the two quantities $W_{\text{useful, max}}$ become identical.

Call

$$\mathbf{E} - T_o\mathbf{S} + P_o\mathbf{V} \equiv \Gamma \tag{15-49}$$

a new state function describing the original system [previously identified as (I)] with special reference to its location in the atmosphere. This allows us to write our final result Eq. (15-48) in the simple form

$$-W_{\text{useful, max}} = \Delta\Gamma \equiv \Gamma_2 - \Gamma_1. \tag{15-50}$$

If 2 is an arbitrary state and 1 corresponds to the most stable state of the system in the atmosphere, we shall write $\Gamma_2 \equiv \Gamma$ and $\Gamma_1 \equiv \Gamma_{\text{min}}$ and define

$$\Lambda \equiv \Gamma - \Gamma_{\text{min}}. \tag{15-51}$$

The name *availability* (or *availability function*) is assigned by some authors to Γ, by others to Λ. It is evident that for any particular system–atmosphere combination, Γ and Λ differ only by a constant, so that $\Delta\Gamma = \Delta\Lambda$. The usefulness of the functions Γ and Λ is contingent on our ability to compute $\Delta\mathbf{E}$, $\Delta\mathbf{S}$, and $\Delta\mathbf{V}$ for any process of interest in any system of practical significance. Interested readers should consult the references given in the third footnote on p. 170, and in the footnote on p. 173.

(b) Our understanding of the availability concept will be enhanced by considering two special cases.

1. The system of interest has the same pressure and temperature as the atmosphere; $P = P_o$ and $T = T_o$. In this case the Carnot cycle is superfluous, Γ becomes identical with the Gibbs free energy, $\mathbf{G}$, and Eqs. (15-48) reduce to Eq. (15-19). [Incidentally, in this case Eqs. (15-46) and (15-17) also become identical.]
2. The system of interest performs no work *of any kind*; it simply acts as a heat reservoir (to the Carnot cycle). Of course, in this case, $\Delta\mathbf{V}^{(\mathrm{I})} = 0$ and $(\Delta\mathbf{E} =)\Delta\mathbf{E}^{(\mathrm{I})} = Q^{(\mathrm{I})}$. Hence, by Eqs. (15-44) and (15-47),

$$W_{\text{useful, max}} = W_{\text{total, max}} \equiv W_{\max} = Q_o - Q^{(\mathrm{I})}. \tag{15-52}$$

But $-Q^{(\mathrm{I})} \equiv Q$ is the heat absorbed by the Carnot cycle at the temtemperature T; hence

$$W_{\max} = Q_o + Q. \tag{15-53}$$

With $Q_o/T_o + Q/T = 0$ [put $Q^{(\mathrm{I})} = -Q$ into Eq. (15-41)], this gives us

$$W_{\max} = \frac{Q(T - T_o)}{T}. \tag{15-54}$$

As readers should readily verify, Eqs. (15-53) and (15-54) simply gives us the work done by the Carnot cycle operating between T and T_o. (For a clockwise operation, $T > T_o$, and both Q and W are positive.)

Rant* has proposed to call the quantity $Q(T - T_o)/T$ the *exergy*, but this designation does not appear to have been widely accepted. It gives us the maximum portion of Q that can be turned into work. Note that $(T - T_o)/T$ is just the "efficiency" (or "conversion factor"), ξ, of any Carnot cycle operating between T and T_o. The exergy differs from ξ by the factor Q, which is determined by the choice of the particular system (I) and the specific change it undergoes.

In recent years, the availability concept has been generalized further by R. S. Berry *et al*, in the so-called "*thermodynamics in finite time*." This latest development has some features in common with the "thermodynamics of irreversible processes," already mentioned on several occasions in this book, the most obvious one being its time-dependent approach. However, finite time thermodynamics takes an essentially different point of departure: it questions the usefulness of reversible processes (which, as we know, take an

---

* Z. Rant, *Forsch. Ingenieurwesens* **22**, 36 (1956). See also P. Grassmann, *Allg. Wärmetechn.* **9**, 79 (1959). These papers give a good set of references to the European literature on the subject.

infinite time to complete) as a realistic standard against which the performance characteristics of real machines should be evaluated. Berry *et al* succeed in defining a "*finite time availability*" which will tell us how much work a system can perform optimally, if besides some of the "ordinary" constraints (e.g., constant P and T), we impose an additional condition on the rate, or duration, of the process this system undergoes. For details, which fall beyond the scope of this book, we refer to the literature.*

PROBLEMS

Problem 15-1

Two blocks of metal, A and B, at initial temperatures T_1 and T_2 ($T_2 > T_1$), are placed in thermal contact and allowed to come to equilibrium. Assume that the respective heat capacities, $\mathbf{C}_A$ and $\mathbf{C}_B$, are constant in the range $T_1 \to T_2$.

(a) Obtain an expression for the entropy change, $\Delta\mathbf{S}$, of the system comprising the two blocks, in terms of $\mathbf{C}_A$, $\mathbf{C}_B$, T_1, and T_2.

(b) Compute $\Delta\mathbf{S}$ if each block is 1 kg of silver, $T_1 = 150$ K, and $T_2 = 300$ K. Assume that in this temperature range $C_{Ag} \approx 24$ J K^{-1} mole^{-1}, a constant.

Problem 15-2

Use the data given, or computed, in Problem 14-3 to obtain $\Delta\mathbf{G}$ for the reaction

$$1 \text{ mole water} \xrightarrow[\text{pressure}]{1 \text{ atm}} 1 \text{ mole ice}$$

(a) at 0°C (*predict* the answer before you verify it numerically), and

(b) at −10°C (*predict* the *sign* of $\Delta\mathbf{G}$ before obtaining the *numerical* value).

Problem 15-3

A mole of water is evaporated at 100°C, 1 atm. The heat of vaporization is 40.58 kJ mole^{-1}. Calculate Q, W, $\Delta\mathbf{E}$, $\Delta\mathbf{H}$, $\Delta\mathbf{S}$, $\Delta\mathbf{A}$, and $\Delta\mathbf{G}$. Assume that the vapor is ideal, and neglect V of the liquid with respect to that of the vapor.

Problem 15-4

(a) Derive the relation $(\partial\,\Delta\mathbf{G}/\partial P)_T = \Delta\mathbf{V}$ (1), in which Δ is the operator $(\partial/\partial\xi)_{P,T}$.

(b) For the reaction

$$1 \text{ mole C(diamond)} \xrightarrow[1 \text{ atm}]{25^\circ\text{C}} 1 \text{ mole C(graphite)},$$

$\Delta\mathbf{G} = -2.895$ kJ mole^{-1}, showing that under these conditions graphite is the stable modification. *Assuming* that the densities of diamond and graphite, 3.51 g/cm^3 and 2.25 g/cm^3, respectively, are independent of pressure, use Eq. (1) to *estimate* the pressure above which diamond becomes the stable modification at 25°C.

---

* For an introductory survey, see B. Andresen, P. Salamon, and R. S. Berry, *Physics Today* **37** (1984), 62. An extensive list of references is given at the end of this paper.

Problem 15-5

The cell

$$\mathrm{Cd}|\mathrm{CdSO_4(aq)}|\mathrm{Cd\text{-}Hg}\ (X_{\mathrm{Cd}})$$

is associated with the reaction

$$\mathrm{Cd(s)} = \mathrm{Cd\ in\ Cd\text{-}Hg}\ (X_{\mathrm{Cd}}). \qquad (1)$$

On the left we have a (solid) cadmium electrode, on the right a (liquid) cadmium amalgam electrode. The EMF depends only on the mole fraction of Cd in the amalgam, X_{Cd}, and, of course, the pressure and temperature. For a cell with $X_{\mathrm{Cd}} =$ 0.0357, we find experimentally, at 25°C (and 1 atm),

$$\mathscr{E} = 0.06448\ \mathrm{V}$$

$$\left(\frac{\partial \mathscr{E}}{\partial T}\right)_P = 1.54 \times 10^{-4}\ \mathrm{V\ K^{-1}}.$$

Compute $\Delta\mathbf{G}$, $\Delta\mathbf{H}$, and $\Delta\mathbf{S}$ for cell reaction (1) under these conditions.

Problem 15-6

As discussed in section 15.4(a), the cell

$$\mathrm{Pt}|\mathrm{H_2}(p)|\mathrm{HCl}[\mathrm{aq}(m)]|\mathrm{AgCl, Ag}|\mathrm{Pt}$$

is associated with the reaction

$$\mathrm{H_2(g) + 2AgCl(s) = 2Ag(s) + 2H^+(aq) + 2Cl^-(aq)}.$$

Experimentally one finds, between 0 and 90°C (and 1 atm),

$$\mathscr{E}^{\ominus} = 0.2366 - 4.8564 \times 10^{-4}t - 3.4205 \times 10^{-6}t^2 + 5.869 \times 10^{-9}t^3\ \mathrm{V},$$

in which t is the centigrade temperature. Compute $\Delta\mathbf{G}^{\ominus}$, $\Delta\mathbf{H}^{\ominus}$, $\Delta\mathbf{S}^{\ominus}$, and $\Delta\mathbf{C}_p^{\ominus}$ for the cell reaction at 25°C, 1 atm.

$\mathscr{E}^{\ominus}$ is the EMF for the cell if the "activities" of gaseous H_2 and HCl(aq) are both unity. Activities and standard states have not been discussed as yet. For all practical purposes we can consider p of hydrogen gas in its standard state to be 1 atm, while the molality of HCl in its standard state is roughly 2 mole kg^{-1}. (See also Problem 26-7.)

Chapter 16

The Third Law of Thermodynamics

16.1 Introduction

The *Third Law* presents a serious challenge, not only to those who believe, as the author does, that *phenomenological* thermodynamics is "self-contained" and can stand on its own merits, but equally to those who feel that this discipline should be provided, wherever possible, with a sound corpuscular basis. Although most physical scientists admit, explicitly or implicitly, that the *Third Law* is of a very different nature from the other two, and that, in some way, the *Nernst Heat Theorem* plays a central role in its formulation and some of its applications, this is where all agreement ends. Several scientists have made valiant attempts to find a satisfactory phenomenological basis for this law (in the sense discussed in section 1.2). Nevertheless, in 1979, one prominent author still felt compelled to write:* "In spite of seventy years' effort, no 'third law of thermodynamics,' *independent of statistical arguments* [the italics are the present author's], universally obeyed, and somehow incorporating Nernst's heat theorem, has yet been formulated." If we consider this statement jointly with the words of another prominent author in this field:† "We are still far from understanding the Nernst heat theorem *on a statistical basis* . . . " (once more, the italics are the author's), then it becomes clear that in dealing with the Third Law we still find ourselves on somewhat shaky ground.

In the light of the quotations above, it should come as no surprise that there is even no agreement as to what the Third Law "really" *is*. Some have reluctantly concluded that it should be looked upon as no more than an important *convention*, which forms the basis for an elaborate tabulation of certain thermodynamic properties. These data, in turn, permit certain types of "thermodynamic calculations," which are of practical utility to chemists as well as to chemical engineers. Historically, the search for a Third Law is tied up with the desire to carry out precisely such manipulations. Fortunately, this goal can be reached successfully in most cases without delving

* M. L. McGlashan, *Chemical Thermodynamics*, Academic Press, London, 1979, p. 237.

† A. Münster, *Chemical Thermodynamics*, E. S. Halberstadt, trans., Wiley-Interscience, New York, 1970, p. 158.

too deeply into all the subtle controversies alluded to above. Therefore, we shall only give a brief account of the relevant material, and we do not claim that this account is always rigorous.

16.2 Phenomenological Discussion; the Nernst Heat Theorem and the Unattainability of Absolute Zero

(a) In section 15.3 we confronted the (incorrect) "Berthelot Principle," according to which a chemical reaction (at constant P and T) should occur spontaneously if $\Delta \mathbf{H}$ is negative, with the correct thermodynamic criterion, which tells us that $\Delta \mathbf{G}$ should be negative instead. It is obvious that as $T \to 0$, $\Delta \mathbf{H}$ and $\Delta \mathbf{G}$ approach each other (hence the two criteria become identical), *provided that* $\Delta \mathbf{S}$ *remains finite.* In 1902 this was investigated experimentally by Richards.* He found, for several reactions occurring in galvanic cells, that as $T \to 0$, the $\Delta \mathbf{G}$ versus T curve approached a horizontal tangent. Since $\Delta \mathbf{S} = -(\partial \Delta \mathbf{G}/\partial T)_p$, this implied

$$\lim_{T \to 0} \Delta \mathbf{S} = 0, \tag{16-1a}$$

which is frequently written as

$$\Delta \mathbf{S}_0 = 0. \tag{16-lb}$$

Although Richards' data pointed to these results in a limited number of examples only, the *Nernst Heat Theorem* (1906) asserted that Eqs. (16-1) must be accepted as a universally valid postulate.† It may be looked upon as the phenomenological statement of the Third Law, but in our way of looking at thermodynamics this would require that it can be derived from a fundamental fact of experience (see section 1.2). Although the issue has not been settled conclusively to everyone's satisfaction, the *unattainability of the absolute zero* is generally accepted as the desired fundamental "principle." It is also due to Nernst (1912), and tells us that no system can be cooled down to $T = 0$ (by a finite number of processes).

(Nernst originally thought that he could derive the unattainability principle from the *First* and *Second* Laws. The flaws in this reasoning were first pointed out by Einstein.‡)

(b) Before we can illustrate the link between the Heat Theorem and this unattainability, we must discuss briefly some features concerning heat capacities. In the first place, these quantities are always positive. This is not only intuitively clear (when a system *absorbs* heat, its temperature will *increase*), but can also be proven explicitly (see Chapter 17). Second, for crystalline

---

* T. W. Richards, *Z. Physik. Chem.* **42**, 129 (1902).

† W. Nernst, *Nachr. Kgl. Ges. Wiss. Göttingen, Math. Phys. Kl.*, 1906, p. 1.

‡ For a very sophisticated discussion, see P. T. Landsberg, *Rev. Mod. Phys.* **28**, 363 (1956). A brief summary is given by A. B. Pippard, *Elements of Classical Thermodynamics*, Cambridge University Press, Cambridge, 1957, pp. 48–51.

solids, both theory and experiment indicate that as $T \to 0$, heat capacities become negligible. The pioneering experiments in this area were carried out by W. F. Giauque *et al.* in the 1930s. Since the measurements at very low temperatures are difficult to carry out and are not feasible below *some* temperature, theoretical extrapolation formulas to absolute zero are of considerable importance. The most famous one, $C = aT^3$, is due to Debye.

Its derivation, based on a specific model concerning lattice vibrations, is given in most textbooks on statistical mechanics. For metals, due to the presence of "free" electrons, a term bT must be added. Since, at these very low temperatures, C_p and C_v for solids are to all intents and purposes equal, the subscripts p and v are frequently omitted in this context.

We shall now show how the Nernst Heat Theorem can be derived from the unattainability principle. Let $(T_1, \alpha) \to (T_2, \beta)$ formally denote an arbitrary process that is accompanied by a temperature change from T_1 to T_2. Thus $\alpha \to \beta$ may refer to an expansion, a demagnetization, a phase transition, a chemical reaction, and so on. We start with the premise that *the adiabatic process*

$$(T, \alpha) \to (0, \beta)$$

is impossible along any pathway. It is implied that T *may be chosen arbitrarily close to zero.*

Since the unattainability principle tells us that $(T, \alpha) \to (0, \beta)$ is impossible along *any* path, nothing prevents us from considering an *adiabatic* one. Parenthetically, we note that *if* it were possible to cool down a system to absolute zero, an adiabatic process would be our *only* hope. All other processes would require a reservoir at a temperature already lower than the initial T.

Exercise 16-1

If we cool down a substance adiabatically by the process $\alpha \to \beta$, show on the basis of the (First and the) Second Law, that a *reversible* pathway would be the most efficient. *Hint*: The proof can best be given through the use of a T,**S** diagram such as that shown here, by reasoning analogous to that given in section 15.1(b). In this diagram the α and β curves could represent, e.g., the magnetized and demagnetized states of a paramagnetic crystal. Consider reversible and irreversible adiabatic cooling, originating from point 1.

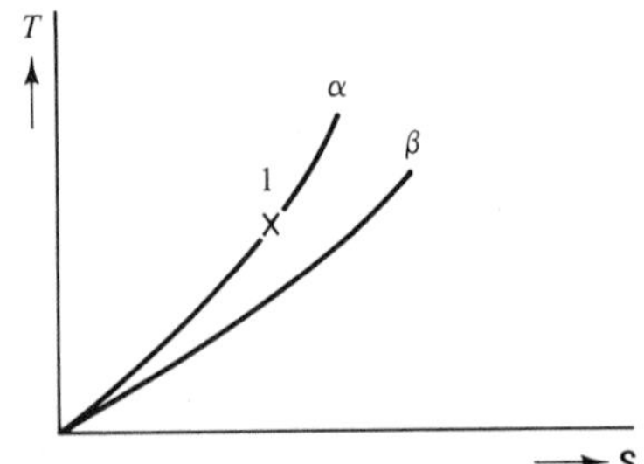

Carry out the thought process $(T, \alpha) \rightarrow (0, \beta)$ in two steps, $(T, \alpha) \rightarrow (0, \alpha)$ followed by $(0, \alpha) \rightarrow (0, \beta)$. Since the adiabatic overall process is postulated to be impossible, $\mathbf{S}$ must *decrease*; hence

$$\Delta \mathbf{S} = \int_T^0 \frac{\mathbf{C}^\alpha}{T}\, dT + (\mathbf{S}_0^\beta - \mathbf{S}_0^\alpha) < 0.$$

Since $(\mathbf{S}_0^\beta - \mathbf{S}_0^\alpha) \equiv \Delta \mathbf{S}_0$, this equation reads, upon interchanging the integration limits,

$$\Delta \mathbf{S}_0 < \int_0^T \frac{\mathbf{C}^\alpha}{T}\, dT. \tag{16-2}$$

From the discussion at the beginning of this subsection* it follows that the integral on the right-hand side of Eq. (16-2) must be finite and positive. We can make it arbitrarily small by letting T approach zero. But this necessarily implies that

$$\Delta \mathbf{S}_0 \leqq 0 \tag{16-3}$$

since if $\Delta \mathbf{S}_0$ were positive, we could always make T small enough to violate the inequality (16-2).

The unattainability principle also implies that *the adiabatic process*

$$(0, \alpha) \rightarrow (T, \beta)$$

is *irreversible*, hence associated with a *positive* entropy change. Proceeding as above, we readily obtain

$$\Delta \mathbf{S}_0 > -\int_0^T \frac{\mathbf{C}^\beta}{T}\, dT, \tag{16-4}$$

from which we conclude, by an argument entirely equivalent to the one that led us from Eq. (16-2) to (16-3), that

$$\Delta \mathbf{S}_0 \geqq 0. \tag{16-5}$$

From Eqs. (16-3) and (16-5) it follows that

$$\Delta \mathbf{S}_0 = 0. \tag{16-1b}$$

When the process is a chemical reaction, we recover the original Nernst Theorem, but our analysis is apparently more general.

---

* We refer to the contention that $\lim_{T \to 0} \mathbf{C} = 0$ and $\lim_{T \to 0} (\mathbf{C}/T)$ remains finite. Insofar as this is based on extrapolation formulas, obtained from statistical mechanics, it is evidently *not* a phenomenological statement. Consequently, some authors introduce it, within the framework of a phenomenological discussion, as an *additional* postulate.

(c) A derivation of this type can be found, with minor modifications, in many textbooks. It can also be given "in reverse"; if we *accept* the Nernst Theorem, we can *derive* the unattainability principle. For this reason, several authors (e.g., Pippard, Kirkwood and Oppenheim) consider these as representing two alternative (phenomenological) statements of the Third Law. However, this strict equivalence has been challenged by, among others, Münster,* who gives a very critical discussion of these matters. This, in turn, is the basis for McGlashan's poignant statement quoted on the first page of this chapter. The present author joins those who consider the issues involved far from settled.

16.3 Planck's Postulate and Statistical Mechanics

(a) The Nernst Theorem deals exclusively with entropy *changes* accompanying certain *processes* at absolute zero. In 1911, Planck proposed a more sweeping statement according to which the entropies of all *substances*, at absolute zero, should be set equal to zero *individually*. By analogy with the formulation of the Nernst Theorem in Eqs. (16-1a) and (16-1b), this implies, per mole of substance,

$$\lim_{T \to 0} S = 0, \tag{16-6a}$$

or simply

$$S_0 = 0. \tag{16-6b}$$

In Planck's own words:† "... as the temperature diminishes indefinitely, the entropy of a chemical homogeneous body of finite density approaches indefinitely near to the value zero." Questions immediately arose as to whether this statement was too general. To many scientists, a more satisfactory formulation was offered by G. N. Lewis, in 1923. In slightly simplified form, it reads:‡ "*The entropy for all pure, crystalline, perfectly ordered substances, at absolute zero, is zero, and the entropy of all other substances is positive.*" To add to the verbal confusion, some authors refer to one or more of *these* statements as "the" Third Law of Thermodynamics. However, Eqs. (16-6) do not fit into a phenomenological framework at all. Instead, they are linked to a famous equation from statistical mechanics:

$$S_T = k \ln \Omega_T \quad \text{or simply} \quad S = k \ln \Omega, \tag{16-7}$$

---

* A. Münster, p. 154 of the reference given in the second footnote on p. 176 of this book.

† Max Planck, *Treatise on Thermodynamics*, 5th ed., A. Ogg, trans., Dover, New York, 1917, p. 244.

‡ G. N. Lewis and M. Randall, *Thermodynamics and the Free Energy of Chemical Substances*, 1st ed., McGraw-Hill, New York, 1923, p. 448.

due to Boltzmann. In these equations k is the Boltzmann constant ($=R/N_{Av}$). The quantity Ω can be interpreted most "simply" within a quantum mechanical framework, namely, as the *degeneracy* of the energy state in question. Of course, Boltzmann introduced Eq. (16-7) in a classical context, in which Ω (originally denoted by W) is a *state density* in "phase space."

For more than a century scientists have attempted to give a qualitative interpretation of Ω. Classically, it could be looked upon as the number of different accessible microscopic ways in which a given macroscopic state of an *isolated* body can be realized. Quantum mechanically, it gives the number of accessible independent eigenstates of such a system. Either way, one has argued that the higher this number, the more *probable* the state of the system; in fact, Boltzmann undoubtedly chose the symbol W because it is the first letter of "Wahrscheinlichkeit," the German word for probability. Even if one accepts such a picture, it still does not justify the familiar tendency to associate Ω, and hence S, with the degree of "randomness" or "disorder" in the system. Only in a few special instances can one establish the necessary connection unambiguously, e.g., when we talk about the "configurational entropy" associated with the allocation of a number of different kinds of atoms to a set of lattice sites. According to the foregoing interpretation of the entropy, only a crystal at absolute zero could exhibit "perfect order" (whence the name "perfect crystal"), with $\Omega_0 = 1$, and S at its minimum value, $S_0 = 0$. However, it is not at all clear what constitutes a "perfect crystal" unless one *defines* it as a crystal in a state with $\Omega = 1$, and then the argument becomes circular. Moreover, in many cases, the correlation of Ω, or S, with a degree of "randomness" or "disorder" is not so obvious at all. For example, the entropy of mixing of equimolar mixtures of H_2O and $(C_2H_5)_2NH$ at 322.25 K is -8.78 J K^{-1}!* Only a very complex analysis, involving several types of hydrogen bonding, could *perhaps* relate this fact to a "disorder" argument.

Formal appearances notwithstanding, Eqs. (16-7) do not yield *absolute entropies* of individual substances, hence cannot rigorously justify the validity of Eqs. (16-6) at absolute zero. For even if we had a perfect crystal, in the sense that the system is in a nondegenerate ground state considering the configurational, rotational, vibrational, and electronic degrees of freedom, and if we could also take into account possible complications due to isotope mixing, even then there could still be further degeneracies due to the states of the nuclei, about which we often know little or nothing. Therefore, to obtain any numerical values of S, one has to start by adopting some *convention*, and several authors refer to the data thus obtained as *practical entropies* or *conventional entropies.* With an eye toward the computation of $\Delta \mathbf{S}$ for chemical reactions, it is customary to put the unknown factor in Ω (due to isotope mixing and all possible nuclear effects) equal to unity for all substances. It

* J. L. Copp and D. H. Everett, *Discuss. Faraday Soc.* **15**, 164 (1953) as quoted by M. L. McGlashan; see footnote on p. 176, §7.13. McGlashan gives a particularly lucid criticism of fallacious interpretations of the entropy concept.

is a reasonable expectation that the features thus "hidden" would make the same contribution for each atom, as it occurs on both sides of the reaction equation. Hence we get a cancellation of errors* and our scheme is entirely *consistent.*

(b) With these qualifications in mind, two different sets of numerical entropy data can be obtained, at least in principle. They are known in the literature as *caloric entropies* (or *thermal entropies*) and *spectroscopic entropies*, respectively. Strictly speaking, only the evaluation of the former requires the use of the Third Law. For comparison purposes it is of interest to discuss spectroscopic entropies briefly in this same section (see also section 16.4).

To obtain caloric entropies per mole of a substance, S_T^{cal}, we employ empirical heat capacities (and latent heats of phase transitions) in the range extending from the temperature of interest, T, down to the lowest temperature at which the relevant experiments have been carried out, T^* say. To be somewhat more specific, suppose that at T^* we have a nonmetallic crystal, so that we can use the theoretical Debye equation, $C = aT^3$, in order to extrapolate our data to absolute zero. This assumes that no phase changes occur between T^* and 0, and that in fact all that happens during the cooling in *this* temperature interval is the "damping out" of the lattice vibrations (to which the Debye equation applies). Provided that we can accept the "Third Law" (?) in the form Eq. (16-6b), we thus obtain

$$S_T^{\text{cal}} = \frac{1}{3}aT^{*3} + \int_{T^*}^{T} \frac{C_p}{T}\,dT + \sum_i \frac{\Delta H_{\text{trans},i}}{T_i} \tag{16-8}$$

in which the last term includes all phase transitions occurring between T^* and T. The readers should carefully review all the restrictions and assumptions that went into the derivation of this frequently used equation.

For a system of noninteracting molecules, as in an ideal gas, entropies can be computed by means of the methods of statistical mechanics, if an important sum, the *molecular partition function*,

$$\sum_i \omega_i e^{-\varepsilon_i/kT}, \tag{16-9}$$

can be obtained. In Eq. (16-9), the ε_i are the molecular energy levels ($i = 0$ is the lowest), the ω_i the corresponding degeneracies. *In principle*, the latter quantities can be computed using quantum mechanics. *In practice*, one obtains the energy differences, $\varepsilon_i - \varepsilon_o$, directly from the appropriate spectrum of the gaseous substance, and the ω_i by a judicious combination of theoretical arguments and analysis of spectral details. The entire procedure requires a

---

* Of course, in isotope-selective reactions (used, e.g., in chemical isotope separations), such a cancellation would not occur, and the corresponding difference could be important.

familiarity with statistical mechanics, quantum mechanics, and molecular spectroscopy. A more elaborate discussion would fall entirely outside the scope of this book. Using this method, one has been able to obtain entropies for a fair number of relatively small molecules. Since spectroscopic measurements play a vital part in the procedure, the resulting data are known as "spectroscopic entropies."

(c) It should be apparent that it is somewhat misleading to refer to either caloric or spectroscopic entropies as "experimental"; the methods by which each of these is obtained contain certain assumptions and approximations. Therefore, it is of considerable interest to compare the two sets of values for those molecules for which both have been obtained.* Generally, the agreement is excellent; the relevant molar entropies at 298 K are typically on the order of a few hundred J K^{-1}, and the difference between S^{spectr} and S^{cal} rarely exceeds the experimental error. There are notable exceptions, e.g., H_2, CO, N_2O, and H_2O, for which S^{spectr} is too large by about 4 to 6 J K^{-1} $mole^{-1}$, or about 1 to 1.5 cal K^{-1} $mole^{-1}$. Such discrepancies are frequently referred to as *exceptions to the Third Law*, a most ambiguous designation. In the author's opinion, the source of this anomalous behavior has been properly identified in a general way, but the more quantitative, detailed, explanations still fall short of their goal.

It is easy to account for these discrepancies in a purely formal way. We merely postulate that, for the crystals involved, Ω_0 is larger than the "perfect order" value of unity. This leads to a "residual entropy" per mole of $R \ln x$ $(x > 1)$, which should be added to S^{cal}. The big question remains how x should be rationalized. Let us analyze this question for the case of carbon monoxide. The "perfect crystal" corresponds to the configuration in which all dipoles are arranged in a unique regular array. To explain why Ω_0 is larger than unity, we note that the dipole moment of a CO molecule is very small† (only 0.12 D) so that other arrangements of these molecules in the crystal cannot be excluded. Several authors postulate that each carbon monoxide can have two distinct orientations in the lattice, namely CO or OC, so that $x = 2$. This is associated with a residual molar entropy of $R \ln 2$, which is 5.77 J K^{-1}. This slightly over accounts for the discrepancy; actually, $(S^{spectr} - S^{cal})$ equals 4.6 J K^{-1}. Unfortunately, there is some doubt as to whether such a twofold configurational degeneracy actually exists at absolute zero, *at least in a crystal in stable equilibrium.* Some will argue instead that, at these very low temperatures (including the lowest ones at which heat capacity measurements are still carried out), the CO molecules

---

* Most data are due to W. F. Giauque and collaborators. For an extensive tabulation, with many literature references, see p. 588 of J. G. Aston's chapter on the Third Law in Vol. I of *A Treatise on Physical Chemistry* by H. S. Taylor and S. Glasstone; 4th printing of the 3rd edition, published by Van Nostrand, New York, in 1946.

† The C≡O bond moment is opposed by a contribution from the lone pair in the nonbonding carbon *sp* hybrid.

rotate infinitely slowly, so that we are in effect dealing with *metastable crystals*; the time needed for the orientation of the CO molecules far exceeds the time scale of the relevant caloric experiments, the residual entropy is "frozen in."

The entire scheme is consistent in the sense that if one or more of the exceptional substances partake in a chemical reaction, the Nernst Heat Theorem for this process will also break down. Frequently, therefore, tabulated values of S^{cal} are corrected empirically for the relatively small anomalies involved, so that they are no longer of any real concern in practical applications.

16.4 Applications to Chemical Reactions

(a) Entropies obtained from caloric measurements and the Third Law, or from spectroscopic analysis and statistical mechanics, can be used to obtain values of $\Delta\mathbf{G}$ and $\Delta\mathbf{G}^{\ominus}$ for chemical reactions. The importance of knowing $\Delta\mathbf{G}$ and $\Delta\mathbf{G}^{\ominus}$ was mentioned in Chapter 15, and will be reemphasized throughout much of the second half of this book. The relation between $\Delta\mathbf{G}^{\ominus}$ and the equilibrium constant will not be derived until Chapter 19, although it should already be familiar to any student who has taken a course in general chemistry (intended for science majors). In such a course, applications are largely confined to model systems, e.g., the "ideal mixture of ideal gases" (see Chapter 23) for which a "good" equilibrium constant (in the sense that it is a function of T only), K_p, may be written in terms of partial pressures. We then have (section 19.2)

$$\Delta\mathbf{G}^{\ominus} = -RT \ln K_p,$$

and the standard state of each component corresponds effectively to its having a partial pressure of 1 atm.* To simplify the ensuing discussion, we shall assume for the remainder of this section that we are indeed interested in $\Delta\mathbf{G}^{\ominus}$ for this model system.

If we wanted to know $\Delta\mathbf{G}^{\ominus}$ for, say, a reaction in an ionic solution, several complications would have to be dealt with, and these could only confuse the main issues we wish to raise *at this stage*.

Of course, we can write

$$\Delta\mathbf{G}_T = \Delta\mathbf{H}_T - T\Delta\mathbf{S}_T$$

and

$$\Delta\mathbf{G}_T^{\ominus} = \Delta\mathbf{H}_T^{\ominus} - T\Delta\mathbf{S}_T^{\ominus}.$$

---

* The detailed discussion of these standard states will be deferred until Chapter 19; see, in particular, section 19.3(a).

It would be expedient if we could obtain $\Delta \mathbf{H}_T^\ominus$ and $\Delta \mathbf{S}_T^\ominus$, and hence $\Delta \mathbf{G}^\ominus$, at any desired temperature, exclusively from caloric measurements of reaction heats and heat capacities. These $\Delta \mathbf{G}^\ominus$ would then allow the calculation of equilibrium constants in many situations in which it is very difficult to carry out equilibrium measurements directly. It is nearly always possible to evaluate $\Delta \mathbf{H}_{298.15}^\ominus$ from known thermal data, such as the heats of formation of reactants and products at 25°C. This, once more, is a familiar type of problem assigned in general chemistry courses. If we also know the relevant heat capacities, we can obtain $\Delta \mathbf{H}_T^\ominus$ at any temperature employing an equation such as (10-48), which we rewrite in the form

$$\Delta \mathbf{H}_T^\ominus = \Delta \mathbf{H}_{298.15}^\ominus + \int_{298.15}^{T} \Delta \mathbf{C}_p^\ominus \, dT. \qquad (16\text{-}10)$$

If we knew $\Delta \mathbf{S}^\ominus$ at *any one* particular temperature, T', we could similarly compute $\Delta \mathbf{S}^\ominus$ at *any other* temperature T from

$$\Delta \mathbf{S}_T^\ominus = \Delta \mathbf{S}_{T'}^\ominus + \int_{T'}^{T} \frac{\Delta \mathbf{C}_p^\ominus}{T} \, dT. \qquad (16\text{-}11)$$

Around the turn of the century it became evident that unlike $\Delta \mathbf{H}^\ominus$, $\Delta \mathbf{S}^\ominus$ could not be obtained at *any one* temperature from caloric measurements combined with the first two laws of thermodynamics. This is what led scientists to the search for the "Third Law." It is immediately evident that the Nernst Theorem, discussed in section 16.2, allows us to write Eq. (16-11) in the form

$$\Delta \mathbf{S}_T^\ominus = \int_{0}^{T} \frac{\Delta \mathbf{C}_p^\ominus}{T} \, dT. \qquad (16\text{-}12)$$

The problems associated with obtaining the relevant $\mathbf{C}_p^\ominus$ (for reactants and products) down to absolute zero were addressed earlier in this chapter. In practice, one rarely uses Eq. (16-12) as such, but instead utilizes the Third Law in the form postulated by Planck (section 16.3), to arrive initially at values of $\mathbf{S}_{298.15}^\ominus$ for the individual substances. These "Third Law entropy values" have been tabulated extensively. In turn, they allow us to compute $\Delta \mathbf{S}_{298.15}^\ominus$ as $\sum_i \nu_i \mathbf{S}_{i,298.15}^\ominus$, after which we can use Eq. (16-11) with $T' = 298.15$.

(b) A second method of obtaining the relevant thermodynamic quantities utilizes spectroscopic measurements on dilute gases, and statistical mechanics. This procedure bypasses the Third Law with all its controversies, but it is customarily mentioned in this context, just as it was appropriate to discuss "spectroscopic entropies" in conjunction with "caloric entropies" in section 16.3(b). Two advantages of this alternative method are:

1. It is unnecessary to use theoretical extrapolations of empirical heat capacity data, between T^* (the lowest temperature at which caloric measurements have been carried out) and absolute zero, and
2. spectroscopic measurements at *one* convenient temperature yield all the $(\varepsilon_i - \varepsilon_o)$ needed to compute the partition functions (16-9) at *this* temperature *and at any other temperature as well.*

Of course, the entire procedure is confined to those relatively simple gaseous species for which the molecular spectrum can be resolved and analyzed with sufficient accuracy.

In this scheme one obtains $\Delta\mathbf{H}_T^\ominus$ and $\Delta\mathbf{G}_T^\ominus$ ($\Delta\mathbf{S}_T^\ominus$ as such is usually disregarded) from

$$\Delta\mathbf{H}_T^\ominus = \Delta\mathbf{H}_0^\ominus + \Delta(\mathbf{H}_T^\ominus - \mathbf{H}_0^\ominus) \tag{16-13a}$$

and

$$\Delta\mathbf{G}_T^\ominus = \Delta\mathbf{H}_0^\ominus + \Delta(\mathbf{G}_T^\ominus - \mathbf{H}_0^\ominus). \tag{16-13b}$$

This requires some explanation. Spectroscopy and statistical mechanics give us the quantities $(\mathbf{H}_T^\ominus - \mathbf{H}_0^\ominus)_k$ and $(\mathbf{G}_T^\ominus - \mathbf{H}_0^\ominus)_k$ for the individual species k at any desired T. The presence of the terms $\mathbf{H}_{0k}^\ominus$ is due to the fact that the energy levels i for all species k are obtained from the spectral analysis as $(\varepsilon_i - \varepsilon_o)_k$. When constructing tables of the $(\mathbf{H}_T^\ominus - \mathbf{H}_0^\ominus)_k$ and $(\mathbf{G}_T^\ominus - \mathbf{G}_0^\ominus)_k$ thus computed, at regular temperature intervals, it appeared that these data did not lend themselves to accurate interpolation. For this reason it is much better to tabulate these quantities divided by T, or the dimensionless analogues obtained if we divide them by RT. Thus, in lieu of Eqs. (16-13), the readers may come across such mysterious-looking alternatives as

$$\Delta\mathbf{H}_T^\ominus = \Delta\mathbf{H}_0^\ominus + \Delta\left(\frac{\mathbf{H}_T^\ominus - \mathbf{H}_0^\ominus}{RT}\right)RT \tag{16-14a}$$

and

$$\Delta\mathbf{G}_T^\ominus = \Delta\mathbf{H}_0^\ominus + \Delta\left(\frac{\mathbf{G}_T^\ominus - \mathbf{H}_0^\ominus}{RT}\right)RT. \tag{16-14b}$$

Lewis and Randall, and some of their followers, call the individual $(\mathbf{H}_T^\ominus - \mathbf{H}_0^\ominus)_k/T$ and $(\mathbf{G}_T - \mathbf{H}_0^\ominus)_k/T$ the *enthalpy function* and the *free energy function* (of substance k), respectively. This nomenclature is not generally accepted and somewhat ambiguous, since several authors refer to **A** and **G** as "the" *free energy functions* (compare the heading and initial paragraphs of section 15.2).

To complete the procedure, we still have to know $\Delta\mathbf{H}_0^\ominus$. This term has a simple meaning:

$$\Delta\mathbf{H}_0^\ominus = \Delta\mathbf{E}_0^\ominus = \sum_k \nu_k \varepsilon_{ok}, \tag{16-15}$$

the energy change of the reaction if it could be carried out in an ideal mixture of ideal gases (remember that all spectroscopic data refer presumably to isolated molecules!) at absolute zero.* To obtain $\Delta\mathbf{H}_0^\ominus$ experimentally, we must, once more, know $\Delta\mathbf{H}_T^\ominus$ at one particular temperature [as in the alternative

* Of course, at absolute zero, energies, enthalpies, and free energies of ideal gases become identical. For a brief discussion of the difficulties associated with the concept of ideal gases and ideal gas mixtures at absolute zero, see also the appendix to Chapter 23.

method discussed in subsection (a) above]. We then compute $\Delta H_0^\ominus$ by using Eq. (16-14a) at *this* temperature.

As an example of what can be achieved by this method, consider the water-gas reaction,

$$CO + H_2O = CO_2 + H_2 \qquad \text{at 1000 K.}$$

Guggenheim* computed $K_p = 1.27$, while direct measurements† yield $K_p = 1.5$. Denbigh‡ expresses the opinion that the latter figure, based on gas analysis, may actually be less accurate than the former. On the other hand, it should be noted that owing to the logarithmic relationship between $\Delta G^\ominus$ and K_p, a relatively small inaccuracy in the former leads to a relatively large percentage error in K_p (see Problem 16-1).

(c) Approximate methods have been developed that are useful in estimating free energy changes of reactions involving complicated organic compounds, which cannot be subjected to either of the techniques discussed in the preceding subsections. Essentially, the procedures involved are based on the premise that certain thermodynamic properties (including the entropy) of individual molecules can be computed by adding together contributions from various constituent groups, types of bonds present, and so on. Empirical values for such contributions have been tabulated.§ These methods may, in some instances, successfully predict whether a reaction is feasible or not, but the upper limit to the error in $\Delta G^\ominus$ values (about 20 kJ mole^{-1}) is so large that the associated equilibrium constants may become quite unreliable.

PROBLEMS

Problem 16-1

For a chemical reaction occurring in a dilute gas, at 25°C, we determine $\Delta H^\ominus$ from heats of formation and $\Delta S^\ominus$ from "Third Law entropies." Typical experimental errors in these "caloric" quantities are 0.25 kJ and 0.85 J K^{-1}, respectively. Compute the percentage error in K_p, obtained from $\Delta G^\ominus = -RT \ln K_p$, due to *either* of these uncertainties.

Note:

1. The answers are independent of the magnitude of $\Delta G^\ominus$.
2. Depending on their *sign*, the two errors, *taken together*, could reinforce each other, or roughly cancel out.

---

* E. A. Guggenheim, *Thermodynamics*, 3rd ed., North-Holland, Amsterdam, 1957, section 7.10, p. 309.

† Data by Haber and Hahn, quoted by Bryant, *Ind. Eng. Chem.* **23**, 1019 (1931).

‡ K. Denbigh, *The Principles of Chemical Equilibrium*, 3rd ed., Cambridge University Press, Cambridge, 1971, p. 381.

§ For an excellent summary of these approximation methods, including tables, examples, and references, see I. M. Klotz and R. M. Rosenberg, *Chemical Thermodynamics*, 3rd ed., W. A. Benjamin, Menlo Park, Calif., 1972, section 12.2.

Problem 16-2

Compute $\Delta\mathbf{G}^{\ominus}$ and K_p, at 1000 K, for the reaction

$$N_2 + 3H_2 = 2NH_3$$

from the standard heat of formation of ammonia from the elements (at 25°C), which is -46.11 kJ mol^{-1}, and the dimensionless quantities tabulated below.

	$\dfrac{\mathbf{H}^{\ominus}_{298.15} - \mathbf{H}^{\ominus}_0}{298.15R}$	$\dfrac{\mathbf{G}^{\ominus}_{1000} - \mathbf{H}^{\ominus}_0}{1000R}$
H_2	3.43	-16.5
N_2	3.51	-23.8
NH_3	4.00	-24.5

Problem 16-3

Mixtures of halogens generally give interhalogen compounds of various types. For example, iodine and chlorine will react to give ICl (liquid or gas) and ICl_3 (solid) depending on the conditions. Look up the thermodynamic data in the appropriate literature [e.g., D. D. Wagman *et al.*, "Selected Values of Chemical Thermodynamic Properties," *NBS Technical Note 270-3* (1968)], and calculate $\Delta\mathbf{G}^{\ominus}$, as well as the equilibrium constants at 25°C, for the reactions

$$I_2(s) + Cl_2(g) \leftrightharpoons 2ICl(\ell) \quad (A)$$

$$I_2(s) + Cl_2(g) \leftrightharpoons 2ICl(g) \quad (B)$$

$$I_2(s) + 3Cl_2(g) \leftrightharpoons 2ICl_3(s) \quad (C)$$

$$3ICl(\ell) \leftrightharpoons ICl_3(s) + I_2(s) \quad (D)$$

$$ICl_3(s) \leftrightharpoons ICl(\ell) + Cl_2(g) \quad (E)$$

$$ICl(\ell) \rightarrow ICl(g) \quad (F)$$

The answers to this problem will be used in Problem 19-4.

Problem 16-4

Compute the standard entropy of carbon suboxide, O=C=C=C=O, at 230 K and 1 atm pressure, assuming that under these conditions corrections for nonideality are negligible, from the following data [Lee A. McDougall and John E. Kilpatrick, *J. Chem. Phys.* **42**, 2311 (1965)]:

1. The heat of fusion at the melting point (160.962 K) is 5401.1 J mole^{-1}.
2. At 230 K, the heat of vaporization is 26,865.0 J mole^{-1}; the equilibrium pressure is 69.96 torr.
3. Between 15 and 230 K use the heat capacity data from the table on p. 189. Between 15 and 0 K use the Debye T^3 Law.

Molar Heat Capacity at Rounded Temperatures of Carbon Suboxide

T (K)	C (cal/deg · mole)	T (K)	C (cal/deg · mole)
15	1.392	135	16.43
20	2.610	140	16.78
25	4.022	145	17.14
30	5.374	150	17.50
35	6.629	155	17.87
40	7.706	160	18.25
45	8.667	160.962	18.32
50	9.550		
55	10.30	160.962	23.76
60	10.94	165	23.73
65	11.51	170	23.69
70	12.03	175	23.66
75	12.47	180	23.64
80	12.91	185	23.62
85	13.34	190	23.61
90	13.69	195	23.61
95	13.98	200	23.61
100	14.31	205	23.65
105	14.66	210	23.71
110	14.98	215	23.76
115	15.29	220	23.84
120	15.56	225	23.92
125	15.82	230	24.00
130	16.12	235	24.10
		240	24.19

REFERENCES

Owing to the controversial nature of the material involved, the definitive chapter on the Third Law has not been written as yet. Our recommendations for additional reading parallel the various references in our footnotes. "In order of appearance," these include chapters in the books by McGlashan (1979); Münster (1970); Landsberg (1956); Pippard (1957); Lewis and Randall, as revised by Pitzer and Brewer (1961); Aston's chapter in Taylor and Glasstone's *Physical Chemistry* (1946); Denbigh (1971); Guggenheim (1957); and Klotz-Rosenberg (1972).

For a proper understanding of some of these discussions, a working knowledge of statistical mechanics is essential, and a minimal familiarity with some quantum mechanical concepts is desirable.

Part III

Equilibrium and Stability

Chapter 17

Equilibrium and Stability in a Single-Component Gas

This entire chapter is of an advanced nature and may be disregarded in an introductory study, in particular by those who are primarily interested in "*chemical* thermodynamics." Section 17.1 deals with the relatively simple uniform* gas. In this case, equilibrium is a trivial matter; hence the discussion is confined to an application of the stability conditions introduced in Chapter 15. In section 17.2 we analyze a much more difficult problem, the consequences of the equilibrium and stability conditions for a *nonuniform* gas in a gravitational field. This is the only section in this book that deals with a system in which certain variables change continuously with position, by virtue of the presence of an external field. The analysis concerned requires a different formalism, which should be of interest to the advanced student. For concreteness sake, readers of this chapter may assume that our gas is always in *stable* equilibrium, although the mathematical treatment would be no different in the *metastable* case [cf. section 15.3(c)].

17.1 The Stability Conditions for a Uniform Gas

(a) Let the gas under consideration consist of n moles. For its *total* entropy we can write

$$\mathbf{S} = nS(E, V), \tag{17-1}$$

in which, as usual, S, E, and V represent the *molar* entropy, internal energy, and volume, respectively. We shall investigate the consequences of the *general* stability condition

$$\delta^2_{\mathbf{E},\mathbf{V}}\mathbf{S} < 0, \tag{17-2}$$

---

* Readers are urged to reacquaint themselves with our nomenclature, as introduced in section 4.1. Of particular relevance is the paragraph in small print toward the end of that section.

which was first obtained in section 15.1(c). This inequality tells us that if we subject our system to a small (virtual) perturbation at constant *total* energy and volume, the second-order variation of the entropy is negative. We effect such a perturbation by letting δn moles of gas adopt a slightly different *molar* energy and volume:

$$E' = E + \delta' E \tag{17-3a}$$

and

$$V' = V + \delta' V. \tag{17-3b}$$

The molar entropy of these δn moles is changed accordingly:

$$S'(E', V') = S + \delta' S. \tag{17-3c}$$

For the *total* internal energy and volume to be unchanged, the remaining $(n - \delta n)$ moles of gas must also undergo a slight change. Let us characterize the modified molar energy and volume of *this* portion of the gas by

$$E'' = E + \delta'' E \tag{17-4a}$$

and

$$V'' = V + \delta'' V, \tag{17-4b}$$

so that the corresponding molar entropy can be written

$$S''(E'', V'') = S + \delta'' S. \tag{17-4c}$$

Thus we have a perturbation resulting in an entropy change:

$$\delta \mathbf{S} = (n - \delta n)\, \delta'' S + \delta n\, \delta' S, \tag{17-5}$$

subject to the conditions

$$\delta \mathbf{E} = (n - \delta n)\, \delta'' E + \delta n\, \delta' E = 0 \tag{17-6a}$$

and

$$\delta \mathbf{V} = (n - \delta n)\, \delta'' V + \delta n\, \delta' V = 0. \tag{17-6b}$$

(b) We develop $S'(E', V')$ in a Taylor series, as follows:

$$\begin{aligned} S' = S &+ \left[\left(\frac{\partial S}{\partial E} \right)_V \delta' E + \left(\frac{\partial S}{\partial V} \right)_E \delta' V \right] \\ &+ \frac{1}{2} \left[\left(\frac{\partial^2 S}{\partial E^2} \right)_V (\delta' E)^2 + 2 \frac{\partial^2 S}{\partial E\, \partial V} \delta' E\, \delta' V + \left(\frac{\partial^2 S}{\partial V^2} \right)_E (\delta' V)^2 \right] \\ &+ \cdots . \end{aligned} \tag{17-7}$$

A similar expression can be written down for S'', by replacing all primes by double primes. In combination with Eqs. (17-4c) and (17-5), these expressions yield the desired expression for $\delta \mathbf{S}$.

Exercise 17-1

Verify that the first-order terms in $\delta\mathbf{S}$ add up to zero, as they should (why?).

When the second-order terms are added up, and Eqs. (17-6) are used to eliminate $\delta''E$ and $\delta''V$, we obtain

$$\delta^2_{\mathbf{E,V}}\mathbf{S} = \frac{1}{2}\frac{n\,\delta n}{n-\delta n}\left[\left(\frac{\partial^2 S}{\partial E^2}\right)_V (\delta' E)^2 + 2\left(\frac{\partial^2 S}{\partial E\,\partial V}\right)\delta' E\,\delta' V + \left(\frac{\partial^2 S}{\partial V^2}\right)_E (\delta' V)^2\right]. \tag{17-8}$$

The condition (17-2) requires that since $[n\,\delta n/(n-\delta n)]$ is positive, the expression in brackets has to be negative *for arbitrary values of* $\delta' E$ and $\delta' V$. This can only be true if

$$\left(\frac{\partial^2 S}{\partial E^2}\right)_V < 0 \tag{17-9a}$$

and

$$\left(\frac{\partial^2 S}{\partial E^2}\right)_V \left(\frac{\partial^2 S}{\partial V^2}\right)_E > \left(\frac{\partial^2 S}{\partial E\,\partial V}\right)^2. \tag{17-9b}$$

For a justification of these inequalities, the readers are referred to the appropriate textbooks on linear algebra, under the heading: "Positive (or Negative) Definiteness of Quadratic Forms."* Note that

$$\left(\frac{\partial^2 S}{\partial V^2}\right)_E < 0 \tag{17-9c}$$

follows from the conditions (17-9a) and (17-9b), hence yields no additional information.

Starting with

$$dS = \frac{1}{T}\,dE + \frac{P}{T}\,dV,$$

we obtain, after considerable but straightforward manipulations,

$$\left(\frac{\partial^2 S}{\partial E^2}\right)_V = -\frac{1}{T^2 C_v}, \tag{17-10a}$$

$$\left(\frac{\partial^2 S}{\partial V^2}\right)_E = \frac{1}{T}\left(\frac{\partial P}{\partial V}\right)_T - \frac{1}{T^2 C_v}\left[T\left(\frac{\partial P}{\partial T}\right)_V - P\right]^2, \tag{17-10b}$$

---

* See, e.g., Gilbert Strang, *Linear Algebra and Its Applications*, Academic Press, New York, 1976, Chapter 6.

and

$$\frac{\partial^2 S}{\partial V\, \partial E} = \frac{1}{C_v T^2}\left[T\left(\frac{\partial P}{\partial T}\right)_V - P\right]. \tag{17-10c}$$

Exercise 17-2

Derive Eqs. (17-10).

With Eq. (17-10a), we immediately conclude from Eq. (17-9a) that

$$C_v > 0. \tag{17-11a}$$

Using all three Eqs. (17-10), the condition (17-9b) yields

$$\left(\frac{\partial P}{\partial V}\right)_T < 0. \tag{17-11b}$$

Exercise 17-3

Derive Eqs. (17-11).

The resulting inequalities (17-11a) and (17-11b) are known as the conditions of *thermal* and *mechanical stability*, respectively. Of course, these conditions make eminent sense intuitively, and were anticipated in earlier sections of this book.

The condition of thermal stability was used in section 16.2(b). The condition of mechanical stability allowed us to derive (see Exercise 10-5) that C_p is larger than C_v (with rare exceptions, in which case $C_p = C_v$). Hence we can also conclude that C_p is always positive.

(c) The inequality (17-11b), *by itself*, could have been obtained much faster if we had based our analysis on the Helmholtz free energy criterion,

$$\delta^2_{\mathbf{V},\mathbf{T}}\mathbf{A} > 0, \tag{17-12}$$

rather than on the entropy criterion, Eq. (17-2) [see Table 15.1 and the accompanying discussion in section 15.3(c)]. The readers should have no problem in obtaining the inequality

$$\left(\frac{\partial^2 A}{\partial V^2}\right)_T > 0, \tag{17-13}$$

from which the condition of *mechanical* stability is readily recovered.

Exercise 17-4

Convince yourselves of the validity of the inequality (17-13), and use it to derive Eq. (17-11b).

The condition of thermal stability, Eq. (17-11a), cannot be obtained from the inequality (17-12). The reason is that this Helmholtz free energy criterion is less powerful than the entropy criterion, Eq. (17-2), because it permits a more restricted class of perturbations only. In using Eq. (17-2), variations only had to leave the *total* energy and volume of the system constant, allowing for *two* types of *local* perturbations. If, on the other hand, we use Eq. (17-12), we can only "play this game" with the volume; the temperature has to remain the same *throughout the system.* A similar situation will arise in section 17.2, as well as in subsequent chapters: All criteria for stable equilibrium are valid *simultaneously*, but they do not necessarily yield the same information. It should make eminent sense that the *amount of* such information increases as the selected criterion permits a more *flexible* type of perturbation inside the system. Obviously, this will be associated with a greater complexity of the mathematical treatment involved.

17.2 Equilibrium and Stability Conditions for a Nonuniform Gas in a Gravitational Field

(a) Let $\rho(x, y, z)\,dx\,dy\,xz \equiv \rho(\tau)\,d\tau$ represent the *molar density*, the number of moles of gas in the volume element $d\tau$. The molar volume, $V(\tau)$, is just the inverse of ρ:

$$V(\tau) = \frac{1}{\rho(\tau)}. \tag{17-14}$$

Of course,

$$\int \rho(\tau)\,d\tau = n, \tag{17-15}$$

the total number of moles of gas. The integral in Eq. (17-15), and all subsequent integrals in this section, extend over the entire volume, **V**.

For the potential energy of the entire gas, due to its presence in a gravitational field, we write

$$\Phi = \int \phi(\tau)\rho(\tau)\,d\tau, \tag{17-16}$$

in which $\phi(\tau)$ is the potential energy per mole, in the volume element $d\tau$. If the Z axis is chosen to coincide with the direction of the field, the function ϕ has the familiar form

$$\phi(z) = \phi(0) + Mgz, \tag{17-17}$$

in which M is the weight of 1 mole of the gas and g represents the gravitational constant. This special form of ϕ will not be introduced until later in this section; for the time being we shall keep our analysis more general and work in terms of a more arbitrary $\phi(\tau)$. It is important to realize that, disregarding the displacement of the system *as a whole*, the only way we can vary Φ is by a (virtual) redistribution of the gas molecules, hence by varying $\rho(\tau)$ at constant n; $\phi(\tau)$ is to be taken as fixed.

(b) We shall first derive the equilibrium conditions by means of the Helmholtz free energy criterion,

$$\delta_{\mathbf{V},T}(\mathbf{A} + \Phi) = 0, \tag{17-18}$$

as introduced in section 15.3(b). Since **A** is given by

$$\mathbf{A} = \int \rho(\tau) A(\rho, T)\, d\tau,$$

and Φ by the integral (17-16), the condition (17-18) becomes

$$\delta \int \rho(\tau)[A(\rho, T) + \phi(\tau)]\, d\tau = 0. \tag{17-19}$$

In carrying out the variation, we note that the constancy of T is "built in" from the outset [compare the alternative derivation in subsection (c) below], and the constancy of **V** is assured by taking all integrals over a fixed volume. Equation (17-19) thus becomes

$$\int \left[A + \phi + \rho \left(\frac{\partial A}{\partial \rho} \right)_T \right] \delta\rho\, d\tau = 0. \tag{17-20}$$

By Eq. (17-14), the last term in brackets can be written in terms of V:

$$\rho \left(\frac{\partial A}{\partial \rho} \right)_T = -V \left(\frac{\partial A}{\partial V} \right)_T. \tag{17-21}$$

But from $dA = -S\,dT - P\,dV$ [Eq. (15-13a)], it follows that $P = -(\partial A/\partial V)_T$ [Eq. (15-15b)], so that Eq. (17-21) becomes

$$\rho \left(\frac{\partial A}{\partial \rho} \right)_T = PV. \tag{17-22}$$

Substitution of Eq. (17-22) into Eq. (17-20) yields

$$\int (A + \phi + PV)\, \delta\rho\, d\tau = 0, \tag{17-23}$$

or, since $G \equiv A + PV$,

$$\int (G + \phi)\, \delta\rho\, d\tau = 0. \tag{17-24}$$

This has to be true subject to the auxiliary condition

$$\int \delta\rho\, d\tau = 0, \tag{17-25}$$

which follows immediately from Eq. (17-15) and expresses the fact that the *mean* molar density (hence also the *mean* molar volume, $\mathbf{V}/n$) has to remain unaffected by the variation. For Eqs. (17-24) and (17-25) to be true simultaneously, for an *arbitrary* distribution of $\delta\rho$, $(G + \phi)$ must be constant,* i.e., $(G + \phi)$ *must have the same value in all volume elements.*

When there is no external field, this result simply tells us that at equilibrium, G is the same throughout the system. If this were not the case, there would be a net flow of gas molecules from regions of high G to regions of low G [cf. the discussion of the flow of matter, as determined by "chemical potentials," in section 19.1(c)]. When, on the other hand, an external field *is* present, gradients in G and ϕ have to compensate each other. The implications are seen more clearly by returning to the special case of a gravitational field, with ϕ given by Eq. (17-17). In this situation the constancy of $(G + \phi)$ requires that

$$\left(\frac{\partial}{\partial z}(G + \phi)\right)_T = \left(\frac{\partial G}{\partial z}\right)_T + \frac{d\phi}{dz} = 0. \tag{17-26}$$

But

$$\left(\frac{\partial G}{\partial z}\right)_T = \left(\frac{\partial G}{\partial P}\right)_T \left(\frac{\partial P}{\partial z}\right)_T = V\left(\frac{\partial P}{\partial z}\right)_T = \frac{1}{\rho}\left(\frac{\partial P}{\partial z}\right)_T$$

and

$$\frac{d\phi}{dz} = Mg,$$

so that Eq. (17-26) yields the important result

$$\left(\frac{\partial P}{\partial z}\right)_T = -\rho(z)Mg. \tag{17-27}$$

This tells us that the pressure gradient in our gas is balanced by the effect of the external force; the system is in *hydrostatic equilibrium.* For an application, see Problem 17-1.

(c) We shall now treat the same problem by means of the entropy criterion:

$$\delta_{\mathbf{E}+\mathbf{\Phi},\mathbf{V}}\mathbf{S} = 0. \tag{17-28}$$

(See section 15.3.) Of course, Eq. (17-15) is still true (the total number of moles of gas, n, is constant) and, as before, all integrations extend over the fixed volume, $\mathbf{V}$. Since

$$\mathbf{S} = \int \rho S(E, \rho)\, d\rho, \tag{17-29}$$

---

* This may be clear "by inspection" but can be proven formally by the use of Lagrange's *method of undetermined multipliers.* This method will be illustrated in subsection (c) below, for a somewhat more complicated case.

the variation (17-28) takes the form

$$\int \left\{ \left[S + \rho \left(\frac{\partial S}{\partial \rho} \right)_E \right] \delta\rho + \rho \left(\frac{\partial S}{\partial E} \right)_\rho \delta E \right\} d\tau = 0. \tag{17-30}$$

But

$$\rho \left(\frac{\partial S}{\partial \rho} \right)_E = -V \left(\frac{\partial S}{\partial V} \right)_E = -\frac{PV}{T} \tag{17-31}$$

and

$$\left(\frac{\partial S}{\partial E} \right)_\rho = \left(\frac{\partial S}{\partial E} \right)_V = \frac{1}{T}, \tag{17-32}$$

so that Eq. (17-30) becomes

$$\int \left[\left(S - \frac{PV}{T} \right) \delta\rho + \frac{\rho}{T} \delta E \right] d\tau = 0. \tag{17-33}$$

The auxiliary conditions can be written

$$\int \delta\rho \, d\tau = 0 \tag{17-25}$$

and

$$[\delta(\mathbf{E} + \Phi) =] \int [(E + \phi)\delta\rho + \rho\,\delta E]\, d\tau = 0. \tag{17-34}$$

Remember that $\phi(\tau)$ is fixed (Φ can only be varied by a redistribution of gas molecules) and that the constancy of $\mathbf{V}$ is already built in. The standard procedure for incorporating these two auxiliary conditions, (17-25) and (17-34), into the main variation (17-33) is to use the aforementioned *method of undetermined multipliers*,* due to Lagrange. In brief, we first multiply Eq. (17-25) by c_1 and Eq. (17-34) by c_2 (these constants are the "undetermined multipliers") and add the resulting equations to Eq. (17-33):

$$\int \left\{ \left[S - \frac{PV}{T} + c_1 + c_2(E + \phi) \right] \delta\rho + \left(\frac{\rho}{T} + c_2 \rho \right) \delta E \right\} d\tau = 0. \tag{17-35}$$

Now that we *have* incorporated the two auxiliary conditions, the variations of ρ and E can be truly arbitrary, and Eq. (17-35) can be valid only if the coefficients of $\delta\rho$ and δE vanish individually:

$$S - \frac{PV}{T} + c_1 + c_2(E + \phi) = 0 \tag{17-36}$$

---

* See, e.g., S. M. Blinder. *Advanced Physical Chemistry*, reference 3 of Chapter 6, p. 14, or any appropriate textbook on mathematics.

and

$$\frac{\rho}{T} + c_2\rho = 0. \tag{17-37}$$

The last equation tells us that T *is constant* (the same throughout the gas). If we substitute $c_2 = -1/T$ in Eq. (17-36), we get

$$S - \frac{PV}{T} - \frac{1}{T}(E + \phi) = -c_1 \qquad \text{a constant.}$$

But since $G \equiv E - TS + PV$, this tells us that $(G + \phi)$ *is also constant*, our "old" result [see subsection (b)]. As in section 17.1 (see, in particular, its final paragraph), the more flexible entropy criterion forced us to do more work, but yielded more information than the less powerful Helmholtz free energy criterion, in which we had to assume the constancy of T (throughout the system) from the outset.

(d) From the analysis given in section 17.1, we immediately anticipate that:

1. The condition of *mechanical* stability can be obtained *easiest* from $\delta^2_{\mathbf{V},\mathbf{T}}(\mathbf{A} + \Phi) > 0$.
2. The condition of *thermal* stability can be obtained *only* from $\delta^2_{\mathbf{E}+\Phi,\,\mathbf{V}}\mathbf{S} < 0$.

We shall give the first derivation only, and leave the second one as an exercise for interested readers.

The second variation in $(\mathbf{A} + \Phi)$ can be written as

$$\delta^2_T \int (A + \phi)\rho\, d\tau > 0, \tag{17-38}$$

in which, as before, the constancy of $\mathbf{V}$ is assured by taking the integral over a fixed volume. Since

$$\left(\frac{\partial^2 \rho\phi}{\partial\rho^2}\right)_T = \phi\,\frac{d^2\rho}{d\rho^2} = 0,$$

Eq. (17-38) immediately reduces to

$$\int \delta^2_T(\rho A)\, d\tau > 0$$

or

$$\int \left(\frac{\partial^2 \rho A}{\partial \rho^2}\right)_T (\delta\rho)^2\, d\tau > 0. \tag{17-39}$$

This can only be true for arbitrary variation in ρ if

$$\left(\frac{\partial^2 \rho A}{\partial \rho^2}\right)_T > 0 \tag{17-40}$$

in all volume elements. Since

$$\left(\frac{\partial^2 \rho A}{\partial \rho^2}\right)_T = -V^3\left(\frac{\partial P}{\partial V}\right)_T, \tag{17-41}$$

we recover the condition of mechanical stability:

$$\left(\frac{\partial P}{\partial V}\right)_T < 0, \tag{17-11b}$$

valid throughout the gas.

Exercise 17-5

Derive Eq. (17-41).

Exercise 17-6

Combining the techniques developed throughout this chapter, derive the condition of thermal stability,

$$C_v > 0, \tag{17-11a}$$

from the entropy criterion, $\delta^2_{\mathbf{E}+\Phi,\mathbf{V}}\mathbf{S} < 0$.

PROBLEMS

Problem 17-1

The *Barometric Formula,*

$$\frac{P_z}{P_o} = e^{-Mgz/RT}, \tag{17-42}$$

expresses the variation in pressure of a gas with altitude z. It can be derived in many ways, e.g., by means of statistical mechanics from the so-called *Boltzmann distribution law.*

(a) Show that Eq. (17-42) follows immediately from our condition for hydrostatic equilibrium, Eq. (17-27), if we assume that we are dealing with an *ideal* gas.

(b) Compute $P_{\text{top}}/P_{\text{bottom}}$ for O_2 gas (assumed to be ideal) in a cylinder with a height of 50 cm at 20°C.

(c) Compute the barometric pressure 1 km above sea level ($P_o = 1$ atm), considering the air as a single-component ideal gas with a molecular weight of 28.5, at a uniform temperature of 20°C. (For a somewhat more realistic version of this part of the assignment, see Problem 18-3.)

(d) A typical barometer reading in the Boulder–Denver (Colorado) area is 630 torr. Compare this number with what you compute on the basis of the barometric formula, Eq. (17-42), with the same approximations as in part (c).

Chapter 18

Two-Phase Equilibria in One-Component Systems

Throughout this chapter the following assumptions will be made:

1. The presence of external fields can be disregarded, and the individual phases are uniform.*
2. We can neglect all surface and boundary effects, and the only possible work is of the $P,\mathbf{V}$ type.
3. All interphase boundaries must be heat conducting, deformable, and permeable to the exchange of matter between the phases.

18.1 Derivation of the Equilibrium Conditions; the Clapeyron Equation

(a) As we emphasized in section 15.3, we can base our derivation on any one of a number of general equilibrium conditions. Customarily, one selects

$$\delta_{P,T}\mathbf{G} = 0, \tag{18-1a}$$

i.e., one *accepts*, from the outset, that the two phases at equilibrium have the same (constant) P and T. The entire derivation is then reduced to the contents of subsection (c) below. Instead, we shall take as our starting point

$$\delta_{\mathbf{E},\mathbf{V}}\mathbf{S} = 0. \tag{18-1b}$$

As we shall see, this forces us to carry out a somewhat more complex analysis, but we shall be able to *demonstrate* the aforementioned equality of pressure

* See footnote, p. 192.

and temperature.* The utilization of a third criterion,

$$\delta_{\mathbf{V},T}\mathbf{A} = 0, \tag{18-1c}$$

will be left as an exercise for the reader.

For the relevant *total* extensive property $\mathbf{J}$ (i.e., $\mathbf{V}$, $\mathbf{E}$, $\mathbf{S}$, $\mathbf{A}$, or $\mathbf{G}$) of a system of two phases, 1 and 2, we can write

$$\mathbf{J} = n_1 J_1 + n_2 J_2, \tag{18-2}$$

with the obvious constraint that the total number of moles in our one-component system,

$$n = n_1 + n_2, \tag{18-3}$$

has to be constant.

(b) Carrying out the variation (18-1b) yields initially

$$\delta\mathbf{S} = S_1\,\delta n_1 + S_2\,\delta n_2 + n_1\left[\left(\frac{\partial S_1}{\partial E_1}\right)_{V_1}\delta E_1 + \left(\frac{\partial S_1}{\partial V_1}\right)_{E_1}\delta V_1\right]$$
$$+ n_2\left[\left(\frac{\partial S_2}{\partial E_2}\right)_{V_2}\delta E_2 + \left(\frac{\partial S_2}{\partial V_2}\right)_{E_2}\delta V_2\right] = 0, \tag{18-4}$$

subject to the auxiliary conditions

$$\delta\mathbf{E} = E_1\,\delta n_1 + E_2\,\delta n_2 + n_1\,\delta E_1 + n_2\,\delta E_2 = 0, \tag{18-5a}$$

$$\delta\mathbf{V} = V_1\,\delta n_1 + V_2\,\delta n_2 + n_1\,\delta V_1 + n_2\,\delta V_2 = 0, \tag{18-5b}$$

$$\delta n = \delta n_1 + \delta n_2 = 0. \tag{18-5c}$$

Equations (18-5) give us three relations between δn_1, δn_2, δE_1, δE_2, δV_1, and δV_2, so that only three of these variations can be independent. The standard way to take this into account is to use Lagrange's method of undetermined multipliers, as demonstrated in Chapter 17. However, the simplest way to proceed with the current derivation is to solve Eqs. (18-5) for δn_2, δE_2, and δV_2 in terms of δn_1, δE_1, and δV_1, and substitute the results in Eq. (18-4):

$$\left[(S_1 - S_2) - (E_1 - E_2)\left(\frac{\partial S_2}{\partial E_2}\right)_{V_2} - (V_1 - V_2)\left(\frac{\partial S_2}{\partial V_2}\right)_{E_2}\right]\delta n_1$$
$$+ n_1\left[\left(\frac{\partial S_1}{\partial V_1}\right)_{E_1} - \left(\frac{\partial S_2}{\partial V_2}\right)_{E_2}\right]\delta V_1 + n_1\left[\left(\frac{\partial S_1}{\partial E_1}\right)_{V_1} - \left(\frac{\partial S_2}{\partial E_2}\right)_{V_2}\right]\delta E_1 = 0. \tag{18-6}$$

---

* This should come as no surprise to those readers who opted to study Chapter 17. To them, some of the comments in the present section will appear repetitive.

From the equation

$$dS = \frac{1}{T}\,dE + \frac{P}{T}\,dV \tag{15-14}$$

we have

$$\left(\frac{\partial S_i}{\partial E_i}\right)_{V_i} = \frac{1}{T_i} \tag{18-7a}$$

and

$$\left(\frac{\partial S_i}{\partial V_i}\right)_{E_i} = \frac{P_i}{T_i} \tag{18-7b}$$

for $i = 1$ or 2. By virtue of these relations, Eq. (18-6) simplifies to

$$\left[(S_1 - S_2) - \frac{E_1 - E_2}{T_2} - \frac{P_2(V_1 - V_2)}{T_2}\right]\delta n_1 + n_1\left(\frac{P_1}{T_1} - \frac{P_2}{T_2}\right)\delta V_1 + n_1\left(\frac{1}{T_1} - \frac{1}{T_2}\right)\delta E_1 = 0. \tag{18-8}$$

Since the auxiliary conditions are now "built into" Eq. (18-8), δn_1, δV_1, and δE_1 can be chosen *independently*. In order for this equation to hold for *all possible* variations (including $\delta n_1 = \delta V_1 = 0$, $\delta E_1 \neq 0$, and so on) the coefficients of δE_1, δV_1, and δn_1 have to vanish *individually*. Thus we obtain, in order,

$$T_1 = T_2, \tag{18-9a}$$

$$P_1 = P_2, \tag{18-9b}$$

$$G_1 = G_2. \tag{18-9c}$$

Exercise 18-1

Those readers who are familiar with Lagrange's method of undetermined multipliers should work out the *last part* of the derivation [starting with Eqs. (18-4) and (18-5)] by means of this systematic procedure.

Exercise 18-2

Carry out the *entire* derivation, using

$$\delta_{\mathsf{V},T}\mathbf{A} = 0. \tag{18-1c}$$

Do you expect to be able to derive *all three* equations (18-9)? Read (or reread) the closing paragraph of section 17-1 (which should be accessible even to those who did not study that section in its entirety)!

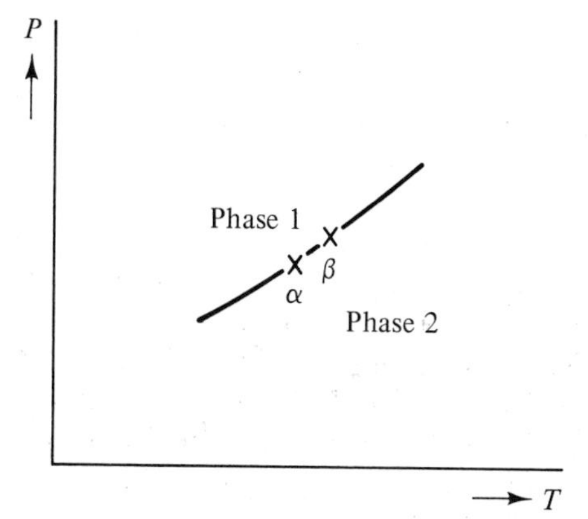

Fig. 19 Illustrating the derivation of the Clapeyron equation. (One should imagine α and β to be very close together.)

(c) The rest of the proof is "routine." In Fig. 19 we have drawn a small portion of a two-phase line, separating the two phases, 1 and 2. By Eq. (18-9c), the equilibria at points α and β require $G_{1\alpha} = G_{2\alpha}$ and $G_{1\beta} = G_{2\beta}$, respectively; hence $G_{2\beta} - G_{2\alpha} = G_{1\beta} - G_{1\alpha}$. If we consider α and β to be adjacent points, this can be written as

$$dG_1 = dG_2. \tag{18-10}$$

But, by Eq. (15-13b), $dG_i = -S_i\,dT + V_i\,dP$ ($i = 1, 2$); hence (18-10) becomes

$$-S_1\,dT + V_1\,dP = -S_2\,dT + V_2\,dP,$$

from which

$$\frac{dP}{dT} = \frac{S_2 - S_1}{V_2 - V_1} \equiv \frac{\Delta S}{\Delta V}. \tag{18-11a}$$

Since we are dealing with an equilibrium, at constant P and T, $\Delta S = \Delta H/T$, and the final result is

$$\frac{dP}{dT} = \frac{\Delta H}{T\Delta V}, \tag{18-11b}$$

the famous *Clapeyron equation* for the slope of *any* two-phase line in one-component systems. Thus Eq. (18-11b) covers boiling lines, melting lines, sublimation lines, and so on. *In general*, it is no simple matter to integrate this equation, since we usually do not have explicit expressions for both ΔH and ΔV as a function of P and T. The integration may become possible if we make simplifying assumptions, which frequently involves the introduction of models. This is illustrated in the next section.

18.2 Some Applications; the Clausius–Clapeyron Equation

(a) In Fig. 20 we have drawn the simplest phase diagram for a one-component system. $S + L$ represents the melting line, $L + G$ the boiling line, and $S + G$ the sublimation line. Tr is the triple point and Cr the critical point. We can use the Clapeyron equation to investigate the sign of the slope of these two-phase lines. Specifically, let us look at the melting line, for it is the most interesting one in this context. Since heat is always *required* to melt a solid, ΔH, if taken to represent $(H_L - H_S)$, is always positive. Consequently, Eq. (18-11b) tells us immediately that the sign of dP/dT is the same as that of the (small) quantity $(V_L - V_S)$. "Normally," substances expand upon melting, so that the slope of the melting line is positive. An important exception is provided by H_2O* ($V_{\text{water}} < V_{\text{ice}}$; ice floats on top of water!), whose melting line tilts slightly to the left. This gives us an excellent illustration of the power and limitations of phenomenological thermodynamics, as first discussed in section 1.5. Through Eq. (18-11b) we are able to *relate* two, at first sight unrelated, properties: the sign of the slope of the melting line and the sign of $(V_L - V_S)$.

This interrelation can also be interpreted if we adapt the mechanical stability condition, as introduced in Chapter 17, to the case at hand. The important feature is† that, at constant T, V has to decrease monotonically (continuously within each phase, discontinuously at the phase transition) with increasing P. Hence if we *increase* P across the $(S + L)$ line at constant T, ΔV must *always be negative.* Thus if the melting line

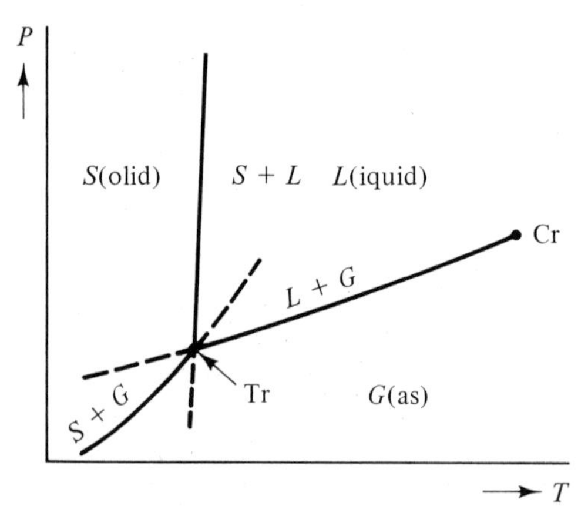

Fig. 20 Simple P,T diagram for a one-component system. ("Normal" situation; the melting line tilts slightly to the right.)

---

* Strictly speaking, this is true only for "ordinary" hexagonal ice (ice I) in equilibrium with water. See, e.g., D. Eisenberg and W. Kauzmann, *The Structure and Properties of Water*, Oxford University Press, London, 1969, Fig. 3.11, p. 93.

† John C. Wheeler, *J. Chem. Phys.* **61**, 4476 (1974).

slopes to the right (left), so that such an increase in P takes us from L to S (S to L), V_S must be smaller than V_L (V_L must be smaller than V_S). This reasoning appears to be unnecessarily sophisticated in this very simple instance, but it gives us a powerful, systematic way to get at important inequalities in less trivial situations.

Note that thermodynamics does *not* tell us *why* the particular substance H_2O shows this anomalous behavior. To answer *that* type of question, we have to investigate those corpuscular (molecular) features that determine the structures of the two phases involved, ice I and water.

(b) In order to apply Eq. (18-11b) to the boiling and sublimation lines, it is customary to convert it into an approximate integrated form, by making the following *three approximations* (X denotes L or S):

1. Assume that the volume of the condensed phase, V_X, is negligible with respect to V_G.
2. Assume that the gas is ideal.
3. Assume that $\Delta H_{X \to G} \equiv \Delta H_{\text{vap}}$, the molar enthalpy of vaporization, is constant over the temperature range of interest.

With these stipulations, Eq. (18-11b) can be integrated to yield

$$\ln P = -\frac{\Delta H_{\text{vap}}}{R}\frac{1}{T} + \text{const.} \tag{18-12}$$

Exercise 18-3

Derive Eq. (18-12).

For computational purposes it is convenient to use equivalent forms such as

$$\log \frac{P_2}{P_1} = -\frac{\Delta H_{\text{vap}}}{2.303\text{R}}\left(\frac{1}{T_2} - \frac{1}{T_1}\right) = \frac{\Delta H_{\text{vap}}}{2.303\text{R}}\frac{T_2 - T_1}{T_1 T_2}. \tag{18-13}$$

The result [Eq. (18-12)] and the alternative [Eq. (18-13)] are known as the *Clausius–Clapeyron equation.*

(c) In the final part of this section we confine our attention to the boiling curve. Plotting the logarithm of the *experimental* equilibrium vapor pressure versus $1/T$ nearly always yields a straight line, *all the way from the triple point to the critical point,* in accordance with Eq. (18-12). This is truly remarkable, since the three approximations used in *deriving* this equation are certainly invalid over this temperature range. To wit, at the higher pressures near the critical point, the vapor undoubtedly is no longer ideal, and as we approach this point, both $\Delta V \equiv V_G - V_L$ and ΔH_{vap} should go to zero! Very few authors mention this point or the elaborate rationalization of the experimental behavior, as given more than thirty years ago by Brown.* Following his

---

* Oliver L. I. Brown, *J. Chem. Educ.* **28**, 428 (1951). See also S. Waldenstrøm, K. Stegavik, and K. Razi Naqvi, *J. Chem. Educ.* **59**, 30 (1982).

analysis, we observe that Eq. (18-11b), for the *boiling line*, can be converted into the *exact form*

$$\frac{d \ln P}{d(1/T)} = \frac{-\Delta H_{\text{vap}}}{Z[1 - (V_L/V_G)]\mathrm{R}}, \tag{18-14}$$

in which $Z(P, T)$ is the *compressibility factor* of the gas:

$$Z(P, T) \equiv \frac{PV_G}{\mathrm{R}T}. \tag{14-17}$$

Exercise 18-4

Derive Eq. (18-14).

There is no way to turn the right-hand side of Eq. (18-14) into a true constant. Nevertheless, the experimental results suggest that ΔH_{vap} is approximately proportional to $Z[1 - (V_L/V_G)]$. This is not unreasonable, since in going from the triple point (normally at quite low pressures) to the critical point (usually at rather high pressures):

1. Z will vary from about unity (ideal gas value) to a value between 0.2 and 0.3.
2. $[1 - (V_L/V_G)]$ goes from near unity to zero, and
3. ΔH_{vap} decreases gradually from its low-pressure value to zero.

18.3 Intersection of Two-Phase Lines; the 180° Rule

(a) Thermodynamics cannot tell us whether or not two-phase lines intersect each other,* but *if* they do, a *triple point* is created, from which a third such line *must* originate. Thus in Fig. 20, the triple point is common to the boiling, melting, and sublimation lines. Note that *each* of the three stable phases (solid, liquid, and gas), occupies an angle less than 180° at the triple point; the situation depicted in Fig. 21 has never been observed. This is an illustration of the famous 180° *rule*: *No stable phase may occupy an angle of more than* 180° *at the triple point.*† In the simple example illustrated by Figs. 20 and 21, an entirely equivalent formulation of this rule states that the metastable extension of any two-phase line past the triple point (dashed in the drawings) has to be located in the area *between* the stable portions of the

---

* Phenomenological thermodynamics cannot explain either why the boiling line *always* ends in a critical point (unless, of course, the substance concerned decomposes before reaching critical conditions), but apparently the melting line *never* does.

† Whether a stable phase can occupy *exactly* 180° is a question we shall address briefly in subsection (d) below.

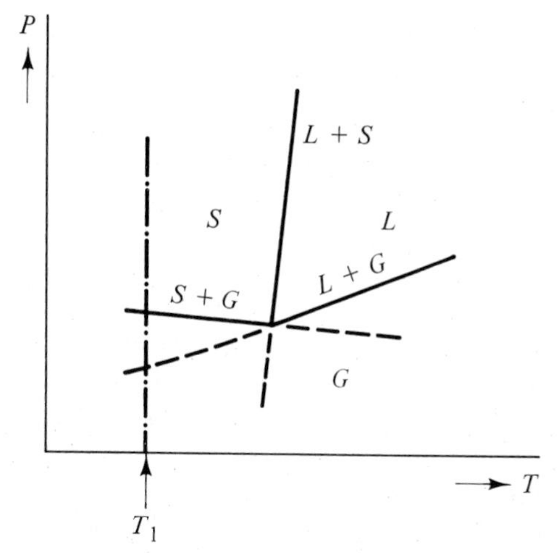

Fig. 21 A P,T diagram (for a one-component system) *violating* the 180° rule. (This situation has never been observed.)

other two two-phase lines. In this form the rule can be rationalized by means of a thought experiment, as illustrated in Fig. 22. If we open the valve between the two containers, gas will flow from the side with the highest to the side with the lowest vapor pressure. Consequently, the amount of condensed phase will decrease in the former (through vaporization) and increase in the latter (through condensation). If the situation depicted in Fig. 21 could be true, this would mean that at T_1 the stable solid could spontaneously convert to the metastable liquid, which of course does not happen [see also section 18.4(e), and compare Problem 14-3(d)].

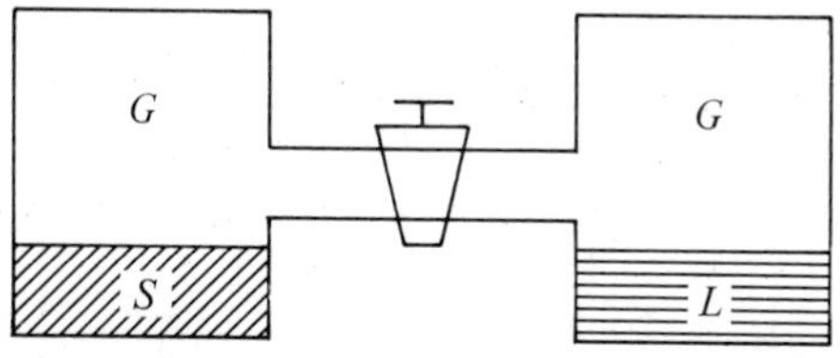

Fig. 22 Experimental set-up at T_1 (see Fig. 21), which gives a simple rationalization of the 180° rule as it pertains to the usual three-phase equilibrium in one-component systems.

(b) In 1974, Wheeler pointed out* that this familiar 180° rule, in one-component systems, is but one special manifestation of less well known, quite general, geometric constraints at triple points, which pertain to a large number of phase equilibria.

---

* John C. Wheeler, "Geometric Constraints at Triple Points," *J. Chem. Phys.* **61**, 4474 (1974).

He mentions examples, the best known of which is *Schreinemakers' rule* (in the X_1,X_2 plane, at constant P and T) for a ternary mixture, and gives an imposing list of violations as they appear in phase diagrams throughout the literature. Finally, he points out that these rules can be extended to more complex situations, such as the arrangement around *quadruple points* in binary systems and phase diagrams pertaining to *higher-order transitions*. For a proper understanding of these matters, considerable familiarity with the graphical representations of heterogeneous equilibria is required. Interested readers are urged to study this rigorous paper* in detail.

Wheeler's elegant analysis goes well beyond the scope of this book. His derivations treat the problems from a general, unified point of view, in which all reference to metastable states is avoided. The present author joins those who are perfectly willing to admit metastable states into the domain of phenomenological equilibrium thermodynamics, and has no serious objections to the proof associated with the thought experiment illustrated in Fig. 22. Nevertheless, we shall proceed to give an alternative proof, by adapting one of Wheeler's general methods to the case of the simple triple-point configuration of Fig. 20. This ought to provide interested readers with some new tools, as well as with some valuable insights. Introductory students may skip the rest of this section.

(c) For concreteness sake (this is not essential to the proof) we shall consider the "normal" case, in which the melting line has a positive slope. At the triple point,

$$\left(\frac{dP}{dT}\right)_{L+S} > \left(\frac{dP}{dT}\right)_{L+G}. \tag{18-15}$$

We have to prove that

$$\left(\frac{dP}{dT}\right)_{L+S} > \left(\frac{dP}{dT}\right)_{S+G} > \left(\frac{dP}{dT}\right)_{L+G}. \tag{18-16}$$

This is equivalent to proving that $(dP/dT)_{S+G}$ is a weighted sum of the other two slopes:

$$\left(\frac{dP}{dT}\right)_{S+G} = w_1\left(\frac{dP}{dT}\right)_{L+G} + w_2\left(\frac{dP}{dT}\right)_{L+S}, \tag{18-17}$$

in which w_1 and w_2 are positive and add up to unity.

To convince ourselves of this equivalence, let us take two arbitrary quantities, c and a, such that $c > a$. Assume that two *positive* numbers, w_1 and w_2, add up to unity, and that a third quantity b can be expressed as

$$b = w_1 a + w_2 c.$$

---

* Ibid.

Then, since $c > a$, $b > w_1a + w_2a = (w_1 + w_2)a = a$, and since $a < c$, $b < w_1c + w_2c = (w_1 + w_2)c = c$. Hence we have shown that $b > a$ and $b < c$, which yields, with the original assumption ($c > a$),

$$a < b < c. \quad \text{Q.E.D.}$$

Each of the three slopes is given by a Clapeyron equation in the form (18-11a)

$$\left(\frac{dP}{dT}\right)_{S+G} = \frac{S_G - S_S}{V_G - V_S}, \tag{18-18a}$$

$$\left(\frac{dP}{dT}\right)_{L+G} = \frac{S_G - S_L}{V_G - V_L}, \tag{18-18b}$$

and

$$\left(\frac{dP}{dT}\right)_{L+S} = \frac{S_L - S_S}{V_L - V_S}. \tag{18-18c}$$

Since

$$\begin{aligned}\left(\frac{dP}{dT}\right)_{S+G} &= \frac{S_G - S_S}{V_G - V_S} = \frac{S_G - S_L}{V_G - V_S} + \frac{S_L - S_S}{V_G - V_S} \\ &= \frac{S_G - S_L}{V_G - V_L}\frac{V_G - V_L}{V_G - V_S} + \frac{S_L - S_S}{V_L - V_S}\frac{V_L - V_S}{V_G - V_S},\end{aligned} \tag{18-19}$$

we have

$$\left(\frac{dP}{dT}\right)_{S+G} = \frac{V_G - V_L}{V_G - V_S}\left(\frac{dP}{dT}\right)_{L+G} + \frac{V_L - V_S}{V_G - V_S}\left(\frac{dP}{dT}\right)_{L+S}. \tag{18-20}$$

If we identify $(V_G - V_L)/(V_G - V_S)$ with w_1 and $(V_L - V_S)/(V_G - V_S)$ with w_2, then w_1 and w_2 are positive and add up to unity; hence Eq. (18-20) is of the desired form Eq. (18-17), and we have completed the proof.

(d) Let us look at three aspects of this proof somewhat more critically. First, the molar volumes and entropies appearing in the Clapeyron equations (18-18) have to be evaluated in the respective phases, near the triple point, *immediately adjacent to the two-phase lines concerned*. In order for Eq. (18-20) to follow from Eqs. (18-18) and (18-19), these molar quantities must have the same value no matter where they occur. For example, near the triple point, V_L and S_L must have the same values in Eqs. (18-18b) and (18-18c). Hence we are making the implicit reasonable assumption that these entropies and volumes approach single well-defined values in each of the three phases, as the triple point is approached *along any path* (in the P,T plane) within the phase of interest.

Next, in our simple example, none of the ΔS or ΔV vanish near the triple point. This guarantees that each phase actually occupies *less than* 180°, although the general rule, as formulated at the beginning of subsection (a), refers to *no more than* 180°. To see this, suppose that near the triple point V_L and S_L were equal to V_G and S_G, respectively. In that case, since $(dP/dT)_{L+G}$ is finite, it would follow from Eq. (18-20) that $(dP/dT)_{S+G} = (dP/dT)_{L+S}$; the solid phase occupies *exactly* 180° near the triple point. Actually, $\Delta V_{L\to G}$ and $\Delta S_{L\to G}$ are zero only at the *critical* point, which *never* coincides with the *triple* point *in one-component systems.*

Such coincidence would violate the *phase rule* (see Chapter 21). In binary systems it is not unusual for a "three-phase line" (solid component–L–G) to intersect the critical (L–G) line. Around such a "*critical end point*"* (of the three-phase equilibrium), the unusual situation described above *does* occur; one phase will occupy *exactly* 180°.

Finally, in identifying the coefficients of $(dP/dT)_{L+G}$ and $(dP/dT)_{L+S}$ in Eq. (18-20) with *positive* constants, w_1 and w_2, we made specific assumptions about the sign of the volume change accompanying each of the phase transitions. In our simple example, these signs, as well as those of the corresponding entropy changes, are more or less "obvious," but they can be justified systematically on the basis of the thermodynamic stability conditions. This has already been suggested in section 18.2(a), in which we pointed out the relation between the sign of $\Delta V_{L\to S}$ and the condition of "mechanical" stability.

Exercise 18-5

Similarly, the condition of "thermal" stability of section 17.1(b) can be adapted to read (see footnote, p. 209) that, at constant P, S increases monotonically with T. Convince yourselves that this leads to the correct sign of the entropy change accompanying each of the relevant phase transitions.

18.4 Graphical Considerations of Equilibrium and Stability; Maxwell's Rule; the Lever Rule

(a) Graphical representations in the thermodynamics of phase equilibria are due primarily to the work of Gibbs and van der Waals. Their methods are particularly useful when explicit analytical relationships are unavailable or become unmanageable. There is the added advantage that these graphs allow us to differentiate between *stable* and *metastable* (*locally stable*) equilibria, a distinction obscured in the usual mathematical formulations based on infinitesimal variations. (Compare the discussion in section 15.3 and the illustrations in Fig. 18.)

* See, e.g., J. Zernike, *Chemical Phase Theory*, A. E. Kluwer, Deventer, The Netherlands, 1955, p. 112.

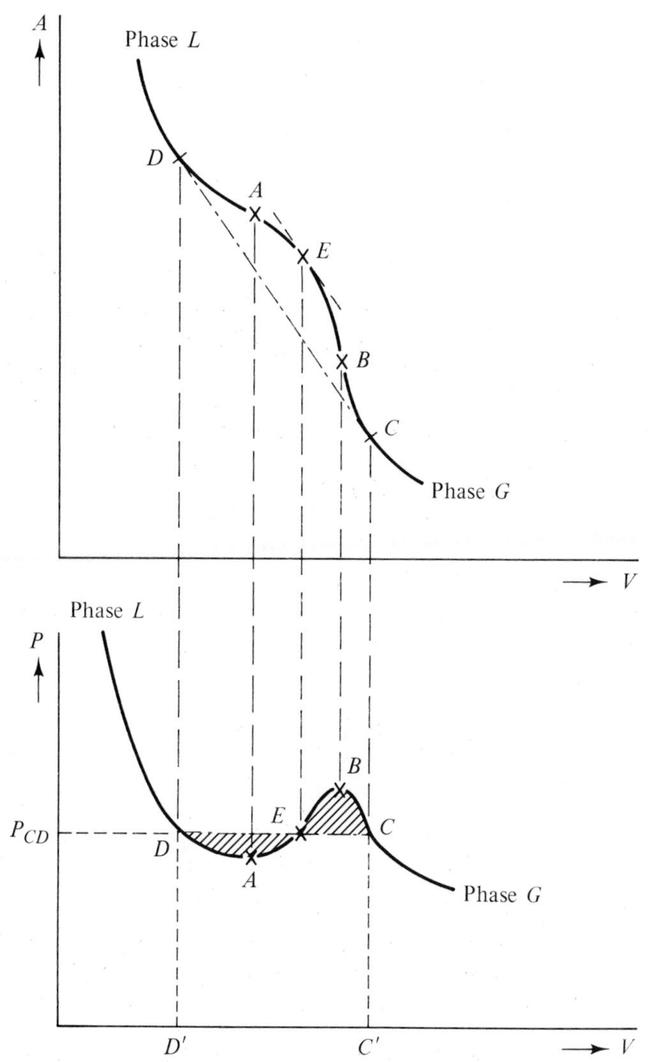

Fig. 23 *A*,*V* and *P*,*V* isotherms of a van der Waals *type* fluid.

Most of the discussion will be couched in terms of a van der Waals *type* fluid. By this we mean a system that reflects the general behavior of this model, as exemplified by the familiar shape of the continuous *P*,*V* isotherm (Fig. 23), without necessarily obeying the van der Waals equation of state *in a strict sense.* Since many fluids, in the accessible portion of the *P*,*V* diagram, are adequately described by the van der Waals equation, our analysis will not only offer a useful theoretical framework, but also reveal several features of practical importance. We remind readers that the stable liquid and gas portions of the *P*,*V* isotherm can be linked by a continuous curve

$DABC$, consisting of two metastable segments, DA and BC, and an unstable part, AB. By careful expansion or compression one has been able to realize parts of the metastable sections, but no point on the segment AB can ever be obtained experimentally. Note that along the latter segment $(\partial P/\partial V)_T > 0$; the condition of mechanical stability is violated.

(b) Of course the horizontal line CD represents the neutral equilibrium between the coexisting phases, (gaseous) C and (liquid) D, in the proportions dictated by the "lever rule" (see below). The phase rule (Chapter 21) tells us that at the temperature *chosen*, P_{CD} is entirely *fixed*. Van der Waals himself realized this, but he was unable to specify where exactly the line CD should be drawn. It was Maxwell who first pointed out that P_{CD} should be located in the P,V diagram (Fig. 23) in such a way that the shaded areas $DAED$ and $EBCE$ are equal (*Maxwell's Rule*).

Occasionally, one finds the following incorrect proof. Take 1 mole of fluid *reversibly* through the cycle $DCBEAD$. For *any* cycle $\Delta S = 0$; hence the assumption of reversibility also makes $Q = 0$. Moreover, since for *any* cyclic process ΔE also vanishes, the First Law tells us that for the cycle of interest $W = 0$ as well. When CD is drawn in the way suggested by Maxwell, we achieve just that; the areas $EBCD$ and $DEAD$ represent equal amounts of work, appearing with opposite signs.

Exercise 18-6

Criticize the proof given in small print above.

To prove this rule, we start with the *mathematical identity*

$$A_C - A_D = \int_{V_D \atop (DABC)}^{V_C} \left(\frac{\partial A}{\partial V}\right)_T dV = -\int_{V_D \atop (DABC)}^{V_C} P\,dV. \tag{18-21}$$

The validity of Eq. (18-21) is subject only to the condition that the function $A(V, T)$ be well defined in every point of the curve. The equilibrium condition (18-9c) requires $G_C = G_D$, or

$$A_C + P_{CD}V_C = A_D + P_{CD}V_D;$$

hence

$$A_C - A_D = P_{CD}(V_D - V_C). \tag{18-22}$$

Equating the right-hand sides of Eqs. (18-21) and (18-22) yields

$$\int_{V_D \atop (DABC)}^{V_C} P\,dV = P_{CD}(V_C - V_D). \tag{18-23}$$

This tells us that the area under the curve $DABC$ equals that of the rectangle $DCC'D'$, which completes the desired proof. Note that the explicit form of the van der Waals equation of state has not been used.

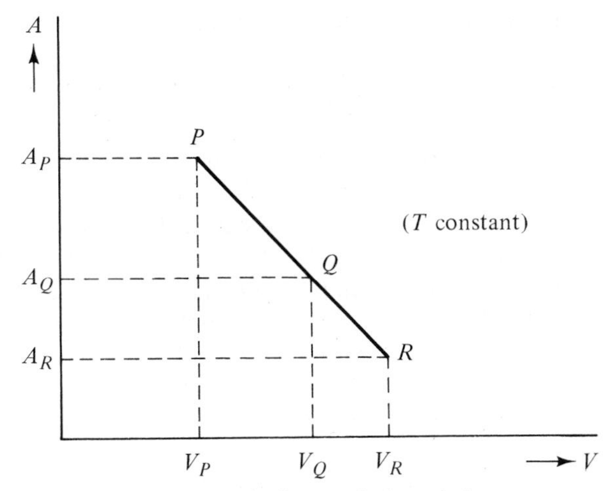

Fig. 24 The lever rule (see text).

(c) The *lever rule*, also referred to as the "center of gravity rule," relates to many types of phase diagrams and is of importance to chemists as well as chemical engineers. In accordance with the subject matter of this section, the rule will be introduced within the framework of an A,V diagram (Fig. 24).

Let us take a mixture of X_P moles of a phase P and X_R moles of a phase R, such that

$$X_P + X_R = 1. \tag{18-24}$$

With the positions of P and R given, we ask for the location of the point Q which will give us the (mean) molar Helmholtz free energy, A_Q, and the (mean) molar volume, V_Q, of this mixture. Since A and V are extensive properties,

$$A_Q = X_P A_P + X_R A_R \tag{18-25a}$$

and

$$V_Q = X_P V_P + X_R V_R. \tag{18-25b}$$

By (18-24) and the geometric relationships displayed in Fig. 24, these last two equations yield

$$X_P = \frac{A_Q - A_R}{A_P - A_R} = \frac{V_Q - V_R}{V_P - V_R} = \frac{QR}{PR} \tag{18-26a}$$

and

$$X_R = 1 - X_P = \frac{PQ}{PR}. \tag{18-26b}$$

This shows that Q is located on the line PR such that

$$\frac{PQ}{QR} = \frac{X_R}{X_P}, \tag{18-27}$$

which expresses the desired lever rule algebraically.

(d) In the top half of Fig. 23, the dotted-dashed line of coexisting phases appears as the double tangent to the A,V curve. Let us convince ourselves that this is indeed the proper way to locate points C and D. From the basic equilibrium conditions $G_2 = G_1$ and $P_2 = P_1$ [Eqs. (18-9b) and (18-9c)], remembering that $P = -(\partial A/\partial V)_T$, we immediately obtain

$$\left[\left(\frac{\partial A}{\partial V}\right)_T\right]_{\text{at } C} = \left[\left(\frac{\partial A}{\partial V}\right)_T\right]_{\text{at } D} \tag{18-28a}$$

and

$$\left[A - V\left(\frac{\partial A}{\partial V}\right)_T\right]_{\text{at } C} = \left[A - V\left(\frac{\partial A}{\partial V}\right)_T\right]_{\text{at } D}. \tag{18-28b}$$

Obviously, the first of these equations tells us that the A,V curve must have the *same slope* at C and D. Simple geometrical considerations (see Fig. 25) reveal that Eq. (18-28b) implies the tangent lines in C and D to have the *same intercept* with the A axis. Together, these two conclusions confirm the correctness of our *double tangent construction*. Its importance goes far beyond the specific case at hand. It is a prescription to locate coexisting phases in many different types of geometrical representations of heterogeneous equilibria (see also Chapters 22 and 27).

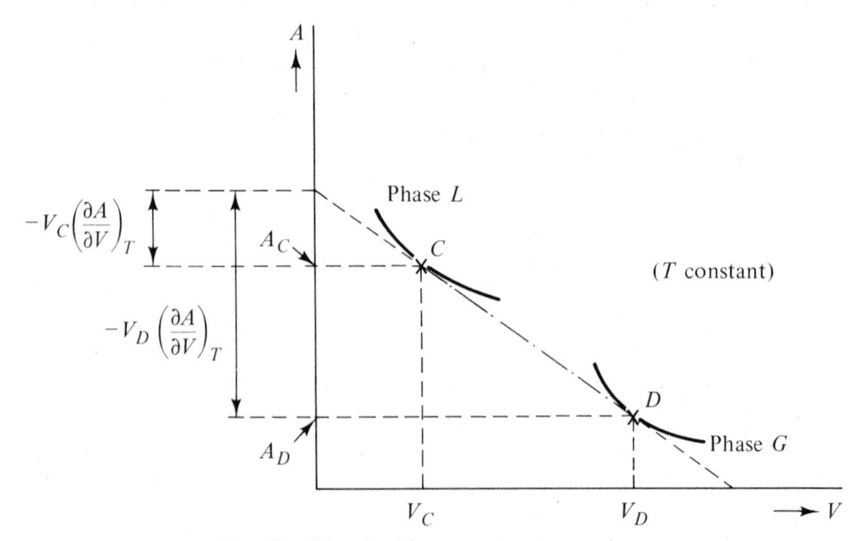

Fig. 25 The double tangent construction.

(e) We are now in a position to give Fig. 23 a closer look. Since $P = -(\partial A/\partial V)_T$, the A,V curve must have a negative slope everywhere. The dotted-dashed double tangent line locates the coexisting phases, C and D. All features of the A,V and P,V curves must be properly correlated. For example, since $(\partial P/\partial V)_T = -(\partial^2 A/\partial V^2)_T$, A and B appear as points of inflection on the upper curve. Moreover, the tangent line to this curve in E $(= -P_E)$ must be parallel to CD $(P_{CD} = P_E)$.

In Fig. 26 we have enlarged a portion of the A,V curve of Fig. 23.

1. Point p represents a *stable* gas; $[(\partial^2 A/\partial V^2)_T]_{\text{in } p}$ is positive. A perturbation at constant V (and T) may instantaneously give a mixture of r and s, but since the Helmholtz free energy of this mixture, represented by the point q, is higher than that of p, the perturbed system will immediately fall back from q to p.
2. An *unstable* fluid such as E, upon the slightest perturbation (which is always present in practice), will split initially into a mixture of, say, i and j, with the resultant Helmholtz free energy being that of point h, below E. The system will *not* return to E but, on the contrary, i and j each will split further into two phases with still lower free energy, and so on. Thus we can never realize fluid E; it will "cascade down" until we arrive at the situation represented by point k, which denotes

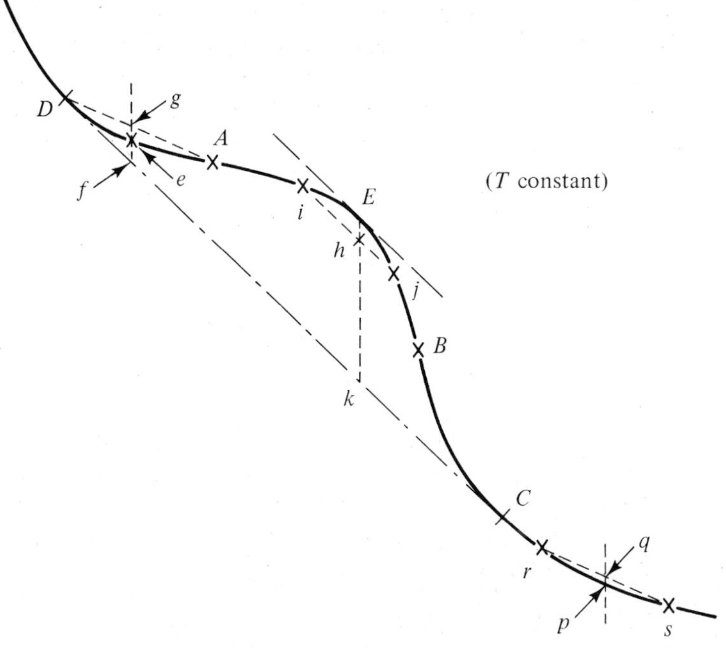

Fig. 26 Enlargement of the A,V curve of Fig. 23.

a mixture of the coexisting phases C and D, in the ratio kD/kC (lever rule). Note that throughout the unstable region, $(\partial^2 A/\partial V^2)_T < 0$; hence $(\partial P/\partial V)_T > 0$.

3. For the *metastable* segments, DA and BC, $(\partial^2 A/\partial V^2)_T > 0$, just as for the stable parts. It is possible to *realize* a metastable liquid such as e, even though a *sufficiently large* perturbation would bring it *down* to point f, representing the appropriate mixture of C and D. But a *very small* perturbation may bring the system *up* to point g (representing a mixture of two phases adjacent to e), from which it may return to e without further consequences. Thus the metastable liquid e is stable with respect to small perturbations, unstable with respect to large ones.

(f) The A,V curve for a solid always has the "narrow parabola" type of shape illustrated in Fig. 27. It is "narrow" because a very small change in V_S is associated with a very steep rise in A. Such a change can be brought about, either by subjecting the solid to a very high pressure (the volume will decrease) or to a very high tension (the volume will increase), corresponding to states at the left and the right of the minimum in the A,V curve (where $P = 0$), respectively. States with "negative pressures" (under tension) are not represented in the usual P,T diagrams, such as the one in Fig. 20. Since the double tangent construction [see subsection (d) above] has to apply to the coexistence of any two phases, there is a triple tangent line at the triple-point temperature (Fig. 27), and at this temperature only.

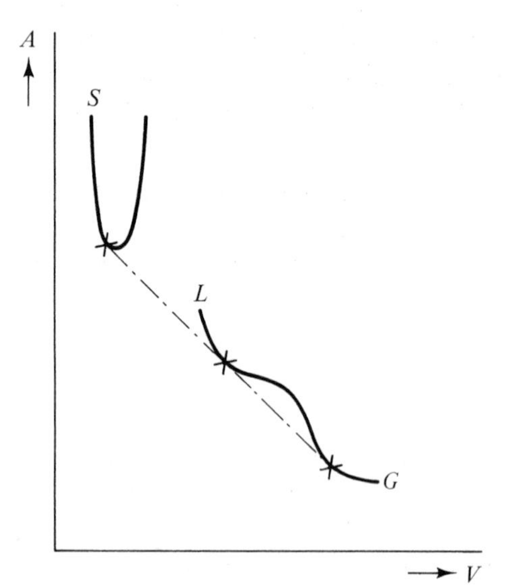

Fig. 27 A,V curve at the temperature of the triple point.

Exercise 18-7

(a) Draw A,V diagrams at temperatures (i) above and (ii) below the triple point, involving the three phases S, L, and G. Draw the appropriate double tangent lines for the stable as well as the metastable two-phase equilibria. Check that everything is consistent with the situation illustrated in Fig. 20.

(b) Draw a similar A,V diagram above the critical temperature. [Remember what happens to the usual (i.e., P,V) van der Waals type isotherm above the critical temperature.]

PROBLEMS

Problem 18-1

A typical barometer reading in Boulder, Colorado, is 63 cm mercury. Use the Clausius–Clapeyron equation to compute the boiling point of water under these circumstances. (The average heat of vaporization of water, over the temperature range involved, is 2260 J g^{-1}.)

Problem 18-2

According to *Trouton's rule*, the entropy change accompanying the boiling of a liquid is approximately 85 J K^{-1} mole^{-1}, at its normal boiling point, T_b. Use this relation to show that for a small change in pressure,

$$\frac{\Delta T_b}{T_b} \approx \frac{\Delta P \text{ (in atm)}}{10}. \tag{1}$$

Accepting the validity of Eq. (1), compute once more the boiling point of water* in Boulder, Colorado, and compare the result with that obtained in Problem (18-1). Equation (1) is known as *Craft's rule*.

Problem 18-3

At 0°C, 1 atm, the densities of water and ice are 1.00 g cm^{-3} and 0.915 g cm^{-3}, respectively. Under these conditions it takes 6010 J of heat to melt 1 mole of ice.

(a) Use the Clapeyron equation in the form Eq. (18-11b) to compute dP/dT at the normal freezing point.

(b) *Estimate*, on the basis that $\Delta P/\Delta T \approx dP/dT$ [hence on the assumption that the numbers used in part (a) are roughly constant over the range from 1 to 100 atm], the freezing point of ice at a pressure of 100 atm.

---

* Trouton's rule works best for *nonpolar* liquids, so for water we can only expect rough agreement with the Clausius–Clapeyron value. For a critical discussion of the validity of Trouton's rule, and the related rules of Hildebrand and Everett, see L. K. Nash, *J. Chem. Educ.* **61**, 981 (1984).

Chapter 19

Chemical Equilibrium

19.1 The Gibbs Equation; Chemical Potentials and Affinity

(a) In this discussion we shall make frequent use of a number of concepts and equations, which were introduced and derived in earlier parts of the book. Of particular importance are the entire contents of Chapter 8, as well as those parts of Chapter 15 which dealt with the validity of the equations

$$d\mathbf{E} = T\,d\mathbf{S} - P\,d\mathbf{V} \tag{15-12}$$

and

$$d\mathbf{E} = \left(\frac{\partial \mathbf{E}}{\partial \mathbf{S}}\right)_{\mathbf{V}} d\mathbf{S} + \left(\frac{\partial \mathbf{E}}{\partial \mathbf{V}}\right)_{\mathbf{S}} d\mathbf{V} \tag{15-12a}$$

for *closed* systems.

We start by writing down the generalization of Eq. (15-12a), when dn_1, $dn_2, \ldots, dn_r$ moles of substances $1, 2, \ldots, r$ are added to what obviously has to be an *open* system. Following Gibbs, we write

$$d\mathbf{E} = \left(\frac{\partial \mathbf{E}}{\partial \mathbf{S}}\right)_{\mathbf{V},n_k} d\mathbf{S} + \left(\frac{\partial \mathbf{E}}{\partial \mathbf{V}}\right)_{\mathbf{S},n_k} d\mathbf{V} + \sum_{k=1}^{r} \left(\frac{\partial \mathbf{E}}{\partial n_k}\right)_{\mathbf{S},\mathbf{V},n_j} dn_k \tag{19-1}$$

for the total differential of $\mathbf{E}$. For the time being we treat all dn_k as independent. In the first two partial differentials it is implied that *all* mole numbers are fixed, while for each term in the sum over k *all* those *except* n_k are considered constant. Obviously, $(\partial \mathbf{E}/\partial \mathbf{S})_{\mathbf{V},n_k}$ and $(\partial \mathbf{E}/\partial \mathbf{V})_{\mathbf{S},n_k}$ are the same as if the system were closed, i.e., T and $-P$, respectively. The terms in the sum *define* the *chemical potentials*, μ_k, of the various components in our mixture:

$$\mu_k \equiv \left(\frac{\partial \mathbf{E}}{\partial n_k}\right)_{\mathbf{S},\mathbf{V},n_j}, \tag{19-2}$$

so that Eq. (19-1) reads

$$d\mathbf{E} = T\,d\mathbf{S} - P\,d\mathbf{V} + \sum_{k=1}^{r} \mu_k\,dn_k. \tag{19-3}$$

This is the famous Gibbs relation. In the alternative form

$$d\mathbf{S} = \frac{1}{T}\,d\mathbf{E} + \frac{P}{T}\,d\mathbf{V} - \frac{1}{T}\sum_{k=1}^{r} \mu_k\,dn_k, \tag{19-3$'$}$$

it is one of the fundamental equations in the development of the (phenomenological) "thermodynamics of irreversible processes."* In fact, on the basis of kinetic theory and statistical mechanics, it can be made plausible that if the system is not too far from equilibrium, $\mathbf{E}$ is still an explicit well-defined function of $\mathbf{S}$, $\mathbf{V}$, and the n_k. Under these circumstances, the intensive thermodynamic variables P, T, and the μ_k remain also well defined, and thus Eqs. (19-3) and (19-3′) retain their validity outside the domain of thermostatic equilibrium.

With the definitions of $\mathbf{G}$, $\mathbf{A}$, and $\mathbf{H}$, Eq. (19-3) readily yields

$$d\mathbf{H} = T\,d\mathbf{S} + \mathbf{V}\,dP + \sum_{k=1}^{r} \mu_k\,dn_k, \tag{19-4}$$

$$d\mathbf{A} = -\mathbf{S}\,dT - P\,d\mathbf{V} + \sum_{k=1}^{r} \mu_k\,dn_k, \tag{19-5}$$

and

$$d\mathbf{G} = -\mathbf{S}\,dT + \mathbf{V}\,dP + \sum_{k=1}^{r} \mu_k\,dn_k. \tag{19-6}$$

Exercise 19-1

Derive

$$\mathbf{S}\,dT - \mathbf{V}\,dP + \sum_{k} n_k\,d\mu_k = 0. \tag{8-29a}$$

Obviously, the validity of these three equations is subject to similar conditions as those given for Eqs. (19-3) and (19-3′); for example, the very important relation (19-6) is valid if $\mathbf{G}$ is a well-defined function of the variables $P, T, n_1, \ldots, n_r$. We note that there are at least three alternative definitions for μ_k:

$$\mu_k \equiv \left(\frac{\partial \mathbf{H}}{\partial n_k}\right)_{\mathbf{S},P,n_j} = \left(\frac{\partial \mathbf{A}}{\partial n_k}\right)_{\mathbf{V},T,n_j} = \left(\frac{\partial \mathbf{G}}{\partial n_k}\right)_{P,T,n_j}. \tag{19-7}$$

The last expression confirms what we stated in section 8.2—that $\mu_k \equiv \bar{G}_k$; Gibbs's chemical potentials are in fact partial molar (Gibbs) free energies and have all the properties of partial molar quantities.

(b) To describe chemical equilibria, we have to identify the various substances k with the reactants and products of a given reaction (or, in the most

---

* See, e.g., reference 4 of Chapter 1.

general case, with those in a set of simultaneous reactions), to be written as

$$\sum_{k=1}^{r} \nu_k k = 0 \tag{19-8}$$

(see section 8.4). If all changes in the dn_k are due exclusively to this reaction, Eq. (8-45) must hold, whence

$$\sum_{k=1}^{r} \mu_k \, dn_k = \left(\sum_{k=1}^{r} \nu_k \mu_k \right) \frac{dn_i}{\nu_i}, \tag{19-9}$$

in which i is *any* reactant or product. By Eq. (8-46), dn_i/ν_i is just $d\xi$ (ξ is the *progress variable* or *extent of reaction*). If we define the *affinity*, $\mathbb{A}$, as the negative of $\Delta_{P,T}\mathbf{G}$ so that

$$\mathbb{A} \equiv -\sum_{k=1}^{r} \nu_k \mu_k, \tag{19-10}$$

then Eq. (19-9) simplifies to

$$\sum_{k=1}^{r} \mu_k \, dn_k = -\mathbb{A} \, d\xi. \tag{19-11}$$

With Eq. (19-11), Eqs. (19-3) through (19-6) take on the form

$$d\mathbf{E} = T \, d\mathbf{S} - P \, d\mathbf{V} - \mathbb{A} \, d\xi, \tag{19-12}$$

$$d\mathbf{S} = \frac{1}{T} \, d\mathbf{E} + \frac{P}{T} \, d\mathbf{V} + \frac{\mathbb{A}}{T} \, d\xi, \tag{19-13}$$

$$d\mathbf{H} = T \, d\mathbf{S} + \mathbf{V} \, dP - \mathbb{A} \, d\xi, \tag{19-14}$$

$$d\mathbf{A} = -\mathbf{S} \, dT - P \, d\mathbf{V} - \mathbb{A} \, d\xi, \tag{19-15}$$

and

$$d\mathbf{G} = -\mathbf{S} \, dT + \mathbf{V} \, dP - \mathbb{A} \, d\xi. \tag{19-16}$$

Hence

$$\mathbb{A} = -\left(\frac{\partial \mathbf{E}}{\partial \xi}\right)_{\mathbf{S},\mathbf{V}} = -\left(\frac{\partial \mathbf{H}}{\partial \xi}\right)_{\mathbf{S},P} = T\left(\frac{\partial \mathbf{S}}{\partial \xi}\right)_{\mathbf{E},\mathbf{V}} = -\left(\frac{\partial \mathbf{A}}{\partial \xi}\right)_{\mathbf{V},T} = -\left(\frac{\partial \mathbf{G}}{\partial \xi}\right)_{P,T}. \tag{19-17}$$

The last expression for $\mathbb{A}$ is in accordance with Eqs. (8-51) and (8-52); since for chemical reactions $-\mathbb{A} \equiv \Delta_{P,T}\mathbf{G}$.

If we consider a set of simultaneous reactions, each with its own extent of reaction, ξ_α, we define $\mathbb{A}_\alpha \equiv -(\partial \mathbf{E}/\partial \xi_\alpha)_{\mathbf{S},\mathbf{V},\xi_{\beta=\alpha}}$. In this case each $\mathbb{A} \, d\xi$ in Eqs. (19-12) through (19-16) has to be replaced by a sum $\sum_\alpha \mathbb{A}_\alpha \, d\xi_\alpha$.

Sometimes we shall need derivatives of $\mathbb{A}$ with respect to ξ; of particular interest are

$$\left(\frac{\partial \mathbb{A}}{\partial \xi}\right)_{\mathbf{S},\mathbf{V}} = -\left(\frac{\partial^2 \mathbf{E}}{\partial \xi^2}\right)_{\mathbf{S},\mathbf{V}} \tag{19-18a}$$

and

$$\left(\frac{\partial \mathbb{A}}{\partial \xi}\right)_{P,T} = -\left(\frac{\partial^2 \mathbf{G}}{\partial \xi^2}\right)_{P,T}. \tag{19-18b}$$

The convexity of **E**(**S**, **V**) and **G**(P, T) ensures that $\partial\mathbb{A}/\partial\xi$ is always negative. Figure 28 illustrates the situation if the independent variables P and T are

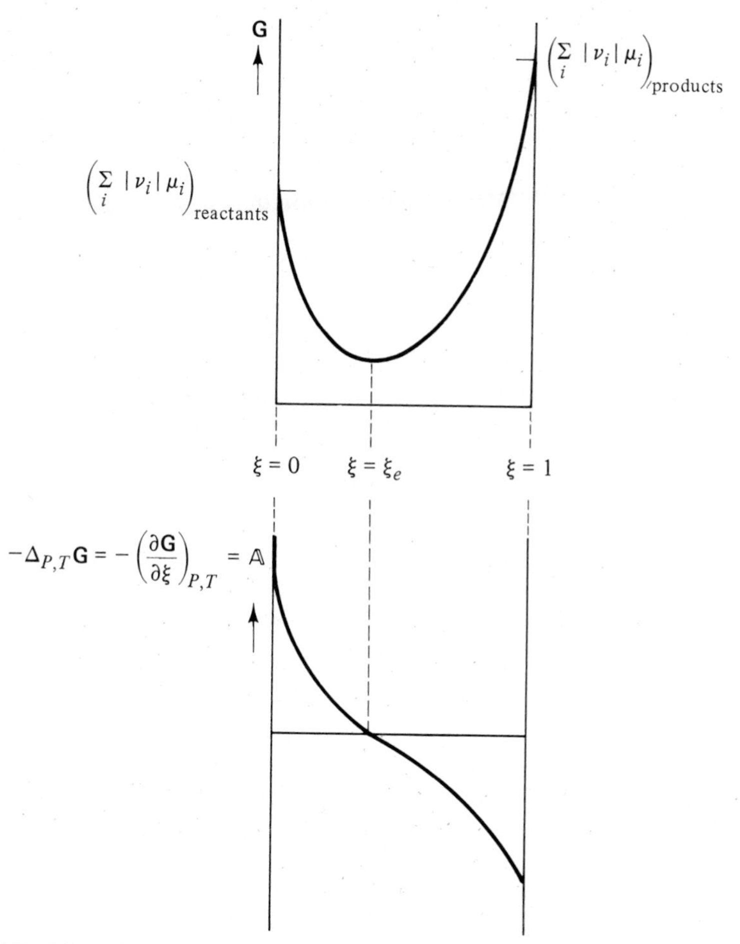

Fig. 28 **G**(ξ) and $\mathbb{A}(\xi)$ at constant P and T. The reactants are mixed in stoichiometric proportions, so that $\xi_{\text{max}} = 1$ (see section 8.4); e denotes equilibrium.

held constant. At constant **S** and **V**, the **E**(ξ) curve would look entirely similar to **G**(ξ) of Fig. 28, so that $\mathbb{A}(\xi)$ would also look qualitatively the same. At equilibrium

$$\delta_{\mathbf{S},\mathbf{V}}\mathbf{E} = \delta_{\mathbf{E},\mathbf{V}}\mathbf{S} = \delta_{\mathbf{S},P}\mathbf{H} = \delta_{\mathbf{V},T}\mathbf{A} = \delta_{P,T}\mathbf{G} = 0,$$

hence any one of the Eqs. (19-12) through (19-16) will tell us that $\mathbb{A}$ is zero:

$$-\mathbb{A} = \Delta_{P,T}\mathbf{G} = \sum_k \nu_k \mu_k = 0 \qquad \textit{at equilibrium.} \tag{19-19}$$

Note that we arrive at Eq. (19-19) by considering an *infinitesimal* ("virtual," if one wishes) displacement under equilibrium conditions, *not* by having to visualize how ν_1 moles of 1 actually react with ν_2 moles 2, etc., while *maintaining* such conditions. The latter procedure, still given in some books, necessitates the introduction of such ingenious, but unrealistic, devices as the van't Hoff reaction vessel, where the various reactants and products are reversibly pumped out of and into a very large container through a set of semipermeable membranes.

For a spontaneous reaction from left to right, we have, e.g., $\delta_{P,T}\mathbf{G} < 0$ or $\mathbb{A} > 0$ (since $\delta\xi > 0$). For the reaction to *reverse* spontaneously we still need $\delta_{P,T}\mathbf{G} < 0$, but now this corresponds to $\mathbb{A} < 0$ (since $\delta\xi < 0$). In this way the minimum **G** and vanishing $\mathbb{A}$ are approached from either direction. Comparison of the reaction equation (19-8) with the equilibrium condition (19-19) shows a very simple relation: Each molecule in the former is replaced by the corresponding chemical potential in the latter. For example, for the reaction

$$N_2 + 3H_2 = 2NH_3,$$

the thermodynamic equilibrium condition is

$$\mu_{N_2} + 3\mu_{H_2} = 2\mu_{NH_3}.$$

Note that *at equilibrium*, Eq. (19-12) reduces to the Gibbs equation for *closed* systems, Eq. (15-12), as it should. Summarizing our conclusions concerned from section 15.2(e) and the present analysis, we see that Eq. (15-12) is *valid*:

1. for any reversible (equilibrium) process in a closed system;
2. for irreversible processes in a closed system *provided* that **E** is a well-defined function of **S** and **V** only, i.e., no changes in composition are involved.

Equation (15-12) is *invalid* if a spontaneous chemical reaction occurs, but as long as **E** remains well defined (in this case as an explicit function of **S**, **V**, *and* ξ), we can replace it by Eq. (19-12).

(c) We end this section with some general comments on chemical potentials and the affinity. First, let us consider a very simple process (hardly a chemical reaction), that of the passage of a solute, *i*, between two immiscible solvents, I and II, at constant P and T. It is left to the reader to verify that for an irreversible transfer of i from I to II,

$$\mu_{i \text{ in I}} > \mu_{i \text{ in II}}, \tag{19-20a}$$

that for an infinitesimal passage of i between I and II under equilibrium conditions,

$$\mu_{i \text{ in I}} = \mu_{i \text{ in II}}, \tag{19-20b}$$

and that for an irreversible transfer of i from II to I,

$$\mu_{i \text{ in II}} > \mu_{i \text{ in I}}. \tag{19-20c}$$

Exercise 19-2

Derive Eqs. (19-20) from Eq. (19-6).

These criteria can be used to rationalize the name *chemical potential*: While in electricity (positive) *charge* flows from higher to lower *electrical* potential, *matter*, according to thermodynamics, apparently flows from a region of higher to one of lower *chemical* potential. The same simple example also illustrates why the μ_i are measures for what G. N. Lewis called the *escaping tendency* of these substances.

To extend this reasoning to the description of chemical reactions, we obviously need some modifications: The *individual* μ_i's are no longer decisive, but the reaction will proceed spontaneously from the side with the highest to the side with the lowest *sum* $\sum_i |\nu_i|\mu_i$, in which i is restricted to either reactants or products. Thus the affinity, which is just the difference between these two sums, is the more revealing property in this situation. Historically, chemists have talked about the affinity *of various substances for each other*. This goes back to the time when it was thought (Empedocles, Glauber) that only substances that "love each other" will react. Berthelot (1878) returned to the concept, but equated it incorrectly, in our nomenclature, with $-\sum_k \nu_k \bar{H}_k$ [compare section 11.1 and the analysis, in small print, at the end of section 15.3(a)]. While van't Hoff was probably the first one to have hit the right track, the modern definition is due to De Donder (1922).*

The most general definition of the affinity, $\mathbb{A}$, is given in terms of the "entropy production" or "uncompensated heat," mentioned briefly in section 15.1(e). We have the important relation

$$T\,d_i\mathbf{S} \equiv dQ' \equiv \mathbb{A}\,d\xi \geq 0, \tag{19-21}$$

* See, e.g., Th. De Donder and P. Van Rysselberghe, *The Thermodynamic Theory of Affinity*, Stanford University Press, Palo Alto, Calif., 1936, and I. Prigogine and R. Defay, *Chemical Thermodynamics*, D. H. Everett, trans., Longmans, Green, London, 1954.

in which, as usual, the equality sign refers to reversible and the inequality to irreversible (spontaneous) changes. It is readily seen that when dealing with chemical reactions, the various criteria given earlier in this section (such as Eq. (19-19) are recovered. When Eq. (19-21) is no longer confined to *chemical* processes, ξ takes on the more generalized meaning of "extent of change."*

19.2 Equilibrium Constants; Chemical Reactions in Ideal Mixtures of Ideal Gases

(a) To convert the general thermodynamic equilibrium condition (19-19) into an equation for the typical equilibrium constant of a chemical reaction, we have to obtain expressions for the μ_k of all reactants and products. For pedagogical purposes it is desirable to start our analysis with the consideration of a reaction in an *ideal mixture of ideal gases*,† even though a comprehensive discussion of this model system will not be given until Chapter 23. All we need at this stage is the expression for the chemical potential of a component k in such a mixture:

$$\mu_k = \mu_k^{\ominus}(T) + RT \ln \frac{p_k}{p_k^{\ominus}}, \qquad (19\text{-}22)$$

in which $\mu_k^{\ominus}(T)$ denotes the *standard chemical potential* of k, *a function of the temperature only*, and p_k is its *partial pressure*, *defined* by

$$p_k \equiv X_k P. \qquad (19\text{-}23)$$

By convention, $p_k^{\ominus}$ equals 1 atm, and the quotient $(p_k/p_k^{\ominus})$ is sometimes written as $|p_k|$, the *numerical* value of this partial pressure when it is expressed in units of atmospheres.

On a corpuscular level, an ideal mixture of ideal gases is *defined* as consisting entirely of noninteracting mass points. But *some* interactions must be allowed for; otherwise, no reaction could ever take place, and the system involved could never reach thermodynamic equilibrium. The crucial point is that such interactions should be weak enough as to have a negligible effect on the values of the thermodynamic properties of the components in the reaction mixture. In particular, the validity of Eq. (19-22) should not be impaired. The model can be approached empirically by considering the gas mixture at sufficiently low *total* pressure. In this way it appears as a generalization of the single-component ideal gas, discussed in Chapter 2. The essential Eq. (19-22) may be looked upon, for the time being, as a generalization of the equation for G of such a gas. [See also section 19.3(a), in particular Eq. (19-46).]

---

* Ibid.

† We refrain from calling it simply an *ideal gas mixture*, in order to distinguish this model from a different one, the *ideal mixture of real gases*, which we introduce in Chapter 24.

(b) For the reaction $\sum_k \nu_k k = 0$, Eq. (19-22) allows us to write

$$\Delta_{P,T}\mathbf{G}\left(=\sum_k \nu_k \mu_k\right) = \sum_k \nu_k \mu_k^{\ominus}(T) + RT \sum_k \nu_k \ln \frac{p_k}{p_k^{\ominus}}. \tag{19-24}$$

Let

$$\sum_k \nu_k \mu_k^{\ominus}(T) \equiv \Delta\mathbf{G}^{\ominus}(T), \tag{19-25}$$

the *standard free energy change* of the reaction, at the temperature of interest. This is what the free energy change of the reaction would be if at that temperature, all reactants and products had partial pressures of 1 atm. Define the *reaction quotient* (in terms of partial pressures) by

$$\mathfrak{Q}_p \equiv \prod_k p_k^{\nu_k}. \tag{19-26}$$

Note that $\mathfrak{Q}_p$ is defined without reference to equilibrium!

For the ammonia synthesis, written as

$$N_2 + 3H_2 = 2NH_3,$$

$\mathfrak{Q}_p$ becomes $p_{NH_3}^2 p_{N_2}^{-1} p_{H_2}^{-3}$; hence

$$\mathfrak{Q}_p = \frac{p_{NH_3}^2}{p_{N_2} p_{H_2}^3}.$$

By virtue of Eqs. (19-25) and (19-26), the expression (19-24) becomes

$$\Delta_{P,T}\mathbf{G} = \Delta\mathbf{G}^{\ominus}(T) + RT \ln \frac{\mathfrak{Q}_p}{\mathfrak{Q}_p^{\ominus}}. \tag{19-27}$$

Exercise 19-3

Derive Eq. (19-27). If the "abstract" compact formulation causes problems, the reader is advised to carry out this transformation first for some simple examples, such as the ammonia synthesis referred to in small print above.

Equation (19-27) *is valid whether the mixture is at equilibrium or not*, as long as $\mathbf{G}(P, T, \xi)$ is well defined (see section 19.1 above). At equilibrium, $\Delta_{P,T}\mathbf{G} = 0$; hence Eq. (19-27) yields

$$\Delta G^{\ominus}(T) = -RT \ln \frac{\mathfrak{Q}_p^e}{\mathfrak{Q}_p^{\ominus}}, \tag{19-28}$$

in which $\mathfrak{Q}_p^e$ is the reaction quotient when all reactants and products have their equilibrium value, p_k^e, in units of atmospheres. Upon defining the (*dimensionless*) equilibrium constant, K_p, by

$$K_p \equiv \frac{\mathfrak{Q}_p^e}{\mathfrak{Q}_p^{\ominus}} = \frac{\mathfrak{Q}_p^e}{(1\ \text{atm})^{\Delta\nu}}, \tag{19-29}$$

we recover the familiar relation

$$\Delta \mathbf{G}^{\ominus}(T) = -RT \ln K_p, \tag{19-30}$$

given in most textbooks on general chemistry.* Equation (19-30) was anticipated in section 16.4, as well as in Problems 16-1, 16-2, and 16-3. Our thermodynamic derivation shows clearly that K_p *is a function of the temperature only*. This is sometimes expressed by the statement that K_p is a "*good*" *thermodynamic equilibrium constant* for a reaction in an ideal mixture of ideal gases.

K_p and $\mathfrak{Q}_p^e$ are identical only for reactions with $\Delta\nu = 0$. As an example, consider

$$H_2 + Br_2 = 2HBr,$$

for which, at 298.2 K, $K_p = \mathfrak{Q}_p^e = 0.19 \times 10^{20}$. On the other hand, for the ammonia synthesis,

$$N_2 + 3H_2 = 2NH_3,$$

at 298.2 K, $\mathfrak{Q}_p^e = 5.3 \times 10^5$ atm^{-2} and $K_p = 5.3 \times 10^5$. Obviously, in general chemistry textbooks such "subtle" distinctions are rarely, if ever, made, although it is frequently pointed out that in these cases the numerical value of *the* equilibrium constant changes if pressures are expressed in different units. Thus the students may be told that K_p is either 5.3×10^5 atm^{-2}, or 0.92 torr^{-2} or 5.2×10^{-5} Pa^{-2}, and so on. These are really equivalent alternative $\mathfrak{Q}_p^e$ values. The important point is that if we compute K_p from $\Delta\mathbf{G}^{\ominus}(298.2) = -32$ kJ, as obtained from tabulated thermodynamic data (compare Problem 16-2), then we "automatically"† obtain the number 5.3×10^5. The readers should verify this.

(c) The equilibrium constant in terms of molarities, K_c, is defined by

$$K_c \equiv \frac{\mathfrak{Q}_c^e}{\mathfrak{Q}_c^{\ominus}} = \frac{\mathfrak{Q}_c^e}{(1 \text{ mole } \ell^{-1})^{\Delta\nu}}, \tag{19-31}$$

in which $\mathfrak{Q}_c$ is the reaction quotient in terms of molarities. For our model system, K_p and K_c are related by (Exercise 19-4)

$$K_p = K_c\left(\frac{RT}{1 \text{ atm } \ell \text{ mole}^{-1}}\right)^{\Delta\nu}. \tag{19-32}$$

In the terminology introduced in the text below Eq. (19-30), K_c is also a "good" thermodynamic equilibrium constant for a reaction in an ideal mixture of ideal gases.

---

* In *some* textbooks at that level, $\Delta\mathbf{G}^{\ominus}$ (or $\Delta\mathbf{G}^{\circ}$) refers to a temperature of 25°C, unless specified otherwise.

† Of course, this is a consequence of the *convention* adopted in tabulating these data.

Finally, it is left to the readers to show that

$$K_p = \mathfrak{Q}_X^e \left(\frac{P}{1 \text{ atm}} \right)^{\Delta \nu}, \tag{19-33}$$

in which $\mathfrak{Q}_X$ represents the reaction quotient in terms of mole fractions. This result shows that even for a reaction in an ideal mixture of ideal gases, $\mathfrak{Q}_X^e$ is in general a function of T *and* P. Hence, unless we are dealing with a reaction in which $\Delta \nu = 0$, it is misleading to write $\mathfrak{Q}_X^e$ as K_X and call it an equilibrium constant.

Exercise 19-4

Derive Eqs. (19-32) and (19-33).

(d) The explicit temperature dependence of K_p is of considerable interest. From Eq. (15-36), with all reactants and products in their standard states, we have

$$\left(\frac{\partial \, \Delta \mathbf{G}^{\ominus}/T}{\partial T} \right)_P = -\frac{\Delta \mathbf{H}^{\ominus}}{RT^2}, \tag{19-34}$$

Combined with Eq. (19-30), this immediately yields the well-known result

$$\frac{d \ln K_p}{dT} = \frac{\Delta \mathbf{H}^{\ominus}}{RT^2}. \tag{19-35}$$

In section 23.2 we shall show that *in an ideal mixture of ideal gases* ($\Delta \mathbf{E}^{\ominus}$ and) $\Delta \mathbf{H}^{\ominus}$ equal(s) ($\Delta \mathbf{E}$ and) $\Delta \mathbf{H}$; the heat of a reaction is independent of the partial pressure of the components. Hence, *for this model*,

$$\frac{d \ln K_p}{dT} = \frac{\Delta \mathbf{H}}{RT^2}, \tag{19-36}$$

an equation first obtained by van't Hoff and sometimes referred to as van't Hoff's *reaction isobar*.

Exercise 19-5

(a) In an ideal mixture of ideal gases, the partial pressures p_i are related to the molarities, c_i, as follows:

$$p_i = \frac{n_i RT}{\mathbf{V}} = c_i RT.$$

Convert the *reaction isobar*, Eq. (19-36), into the van't Hoff *reaction isochor*:

$$\frac{d \ln K_c}{dT} = \frac{\Delta \mathbf{E}}{RT^2}. \tag{19-37}$$

(b) Prove that for the model under consideration,

$$\left(\frac{\partial \ln \mathfrak{Q}_X^e}{\partial P}\right)_T = -\frac{\Delta v}{P}. \tag{19-38}$$

Suppose that $\Delta\mathbf{H}$ is positive, i.e., we are dealing with a chemical reaction which is endothermic when it proceeds from left to right. We see that in this case the right-hand side of Eq. (19-36) is positive; hence the same must be true for the left-hand side. Thus an increase in T results in a higher value of K_p; the equilibrium has shifted to the right. This proves van't Hoff's famous *Principle of Mobile Equilibrium* (1884):

> *An increase in temperature causes an equilibrium to be displaced in the direction of heat absorption (i.e., causes an endothermic reaction to take place).*

(e) The reaction isobar, Eq. (19-36), can be integrated if we know $\Delta\mathbf{H}$ as a function of T (compare the end of section 10.4 and Problem 10-4). For the special case that $\Delta\mathbf{H}$ is constant in a given temperature range ($T_1 \rightarrow T_2$), we obtain

$$\ln \frac{(K_p)_{T_2}}{(K_p)_{T_1}} = -\frac{\Delta\mathbf{H}}{R}\left(\frac{1}{T_2} - \frac{1}{T_1}\right) = \frac{\Delta\mathbf{H}}{R}\,\frac{T_2 - T_1}{T_1 T_2}. \tag{19-39}$$

In this situation a plot of $\log K_p$ versus $1/T$ should yield a straight line:

$$\log K_p = \frac{A}{T} + B. \tag{19-40}$$

Unfortunately, $\Delta\mathbf{H}$ is rarely constant, even for a reaction in an ideal mixture of ideal gases, but strangely enough, within experimental error, nearly *all* gaseous equilibria, for which K_p is a good equilibrium constant (i.e., when studied at sufficiently low pressures), show a straight line of the type (19-40) over a reasonable temperature range. Scheffer went so far as to state that if such a straight line *cannot* be drawn, we are confronting either serious experimental errors or an erroneous interpretation of the data (in the sense that, over the temperature range in question, we are dealing with more than one equilibrium).* Of course, we can always write

$$\log K_p = \frac{\Delta\mathbf{S}^\ominus}{2.303R} - \frac{\Delta\mathbf{H}^\ominus}{2.303RT}, \tag{19-41}$$

in which both $\Delta\mathbf{S}^\ominus$ and $\Delta\mathbf{H}^\ominus$ are temperature functions. According to Scheffer, the *approximate* validity of Eq. (19-40) must be rationalized by

---

* F. E. C. Scheffer, *Toepassingen van de thermodynamika op chemische processen*, Waltman, Delft, the Netherlands, 1945, pp. 36ff.

noting that the temperature-dependent parts of $\Delta \mathbf{S}^{\ominus}/R$ and $\Delta \mathbf{H}^{\ominus}/RT$, both determined essentially by the temperature dependence of $\Delta \mathbf{C}_p^{\ominus}$, will largely compensate each other.

Exercise 19-6

Consider the relation

$$\Delta \mathbf{S}^{\ominus} = \frac{\Delta \mathbf{H}^{\ominus}}{T}. \tag{19-42}$$

(a) Explain why Eq. (19-42) is not true *in general.*

(b) In what *exceptional situation* would Eq. (19-42) be valid? Note that in this case, Eq. (19-41) yields $\log K_p = 0$, hence $K_p = 1$. Convince yourselves that this is entirely consistent with the exceptional equilibrium conditions prevailing under these circumstances.

(f) In this section we have given a derivation of several equations, and a critical discussion of some related concepts, all of which should remind the readers of their general chemistry courses. Presently, we have to reach beyond the simple model system to which such elementary discussions are confined. As a first step we would like to turn our attention to chemical reactions in real gaseous systems. In order to appreciate the difficulties concerned, we first have to digress briefly and introduce a new concept, the *fugacity*, for the simple case of a single-component gas.

19.3 The Fugacity of a Single-Component Gas; Fugacities and Equilibrium in Real Gaseous Mixtures

(a) Let us start with the basic equation

$$V = \left(\frac{\partial G}{\partial P}\right)_T \tag{15-20}$$

for 1 mole of a (single-component) gas. Subtracting the identity

$$\left(\frac{\partial RT \ln (P/1 \text{ atm})}{\partial P}\right)_T = \frac{RT}{P},$$

in which P has to be expressed in atmospheres, yields

$$\left(\frac{\partial [G - RT \ln (P/1 \text{ atm})]}{\partial P}\right)_T = V - \frac{RT}{P}. \tag{19-43}$$

At constant T we integrate Eq. (19-43) between 0 and P:

$$\left(G - RT \ln \frac{P}{1 \text{ atm}}\right) - \lim_{P\to 0}\left(G - RT \ln \frac{P}{1 \text{ atm}}\right) = \int_{0\,(T)}^{P} \left(V - \frac{RT}{P}\right) dP.$$

If we define $G^{\ominus}(T)$, the meaning of which will become clear below, by

$$G^{\ominus}(T) = \lim_{P \to 0} \left(G - RT \ln \frac{P}{1 \text{ atm}} \right), \tag{19-44}$$

we have obtained the general result

$$G = G^{\ominus}(T) + RT \ln \frac{P}{1 \text{ atm}} + \int_{0\,(T)}^{P} \left(V - \frac{RT}{P} \right) dP. \tag{19-45}$$

If the gas were ideal, the integrand in the right-hand side would be zero, so that Eq. (19-45) reduces to the familiar

$$G = G^{\ominus}(T) + RT \ln \frac{P}{1 \text{ atm}}. \tag{19-46}$$

It is seen that $G^{\ominus}(T)$ is the Gibbs free energy of 1 mole of a *hypothetical* gas, namely, a gas that would still be ideal at a *pressure* of 1 atm.

For a real gas we implicitly *define* the *fugacity*, f, by the equation

$$G = G^{\ominus}(T) + RT \ln \frac{f}{1 \text{ atm}}. \tag{19-47}$$

We see that $G^{\ominus}(T)$ can also be identified with the Gibbs free energy of a *real* gas with a *fugacity* of 1 atm. Comparison of Eqs. (19-45) and (19-47) shows that

$$RT \ln \frac{f}{P} = \int_{0\,(T)}^{P} \left(V - \frac{RT}{P} \right) dP. \tag{19-48}$$

This gives us an explicit expression for the fugacity, f, if we know the equation of state of our gas.

Exercise 19-7

Show that for a gas obeying the van der Waals equation of state,

$$\ln |f| = \ln \left| \frac{RT}{V - b} \right| + \frac{b}{V - b} - \frac{2a}{RTV}. \tag{19-49}$$

There are a number of alternative ways to express the result Eq. (19-48). Define the *fugacity coefficient*, γ, by

$$\gamma \equiv \frac{f}{P}. \tag{19-50}$$

Then, obviously,

$$RT \ln \gamma = \int_{0\,(T)}^{P} \left(V - \frac{RT}{P} \right) dP. \tag{19-51}$$

In terms of the compressibility factor, Z, defined by

$$Z \equiv \frac{PV}{RT}, \tag{14-17}$$

Eq. (19-51) can be rearranged to read

$$\ln \gamma = -\int_{0\ (T)}^{P} \frac{1-Z}{P}\, dP. \tag{19-52}$$

Note that

$$\lim_{P\to 0} \frac{f}{P} = \lim_{P\to 0} \gamma = 1. \tag{19-53}$$

To obtain *exact* values of γ (and f), we have to integrate Eq. (19-51) or (19-52). Such an integration can be carried out analytically if we know the explicit equation of state (compare Exercise 19-7 and Problem 19-2), or graphically by the appropriate use of experimental data. If one is content to obtain reasonable *estimates*, a number of approximation methods can be used. The best known of these is based on the *Law of Corresponding States*, according to which Z (among other properties) is a "universal" function of the reduced state variables, $P_r(\equiv P/P_{Cr})$ and $T_r(\equiv T/T_{Cr})$. That is, if this law were strictly valid, $Z(P_r, T_r)$ would be the same for all gases. Consequently, Eq. (19-52) suggests that γ should also be such a universal function, which indeed turns out to be *approximately* true. Graphs giving $\gamma(P_r)$ at various values of T_r have been published in a number of technical journals.*

Some final comments on the fugacities of single-component gases can best be made after giving a few numerical values. For N_2, at 600 atm and 450°C, $f \approx 800$ atm ($\gamma \approx 1.3$), and for NH_3, under the same conditions of pressure and temperature, $f \approx 500$ atm ($\gamma \approx 0.8$). Note that f may be substantially larger *or* smaller than P (γ may be well above, *or* well below, unity). This behavior can only be rationalized by means of corpuscular theories. On a phenomenological level, fugacities are occasionally referred to as *effective pressures*. This does *not* mean that a pressure gauge, calibrated in atmospheres, would point at the numbers 800 (for N_2) and 500 (for NH_3), respectively. It *does* mean that *these* numbers, rather than 600, appear in the equation for G [compare Eqs. (19-46) and (19-47) and any consequences derived from this equation.

(b) For (real) gas *mixtures*, in analogy with Eq. (19-47), we implicitly *define* the fugacity of component k, f_k, by

$$\mu_k = \mu_k^{\ominus}(T) + RT \ln \frac{f_k}{1 \text{ atm}}, \tag{19-54}$$

* See, e.g., R. H. Newton, *Ind. Eng. Chem.* **27**, 302 (1935); or B. W. Gamson and K. M. Watson, *Natl. Petrol. News., Tech. Sec.* **36**, R623 (Sept. 6, 1944).

and the corresponding fugacity coefficient, in analogy with Eq. (19-50), by

$$\gamma_k \equiv \frac{f_k}{p_k} = \frac{f_k}{X_k P}. \tag{19-55}$$

An auxiliary requirement, analogous to Eq. (19-53),

$$\lim_{P\to 0} \frac{f_k}{p_k} = \lim_{P\to 0} \frac{f_k}{X_k P} = \lim_{P\to 0} \gamma_k = 1, \tag{19-56}$$

ensures that as $P \to 0$, we recover Eq. (19-22) for an ideal mixture of ideal gases (with $p_k^{\ominus} = 1$ atm). It is left as an exercise to the readers, starting with

$$\bar{V}_k = \left(\frac{\partial \mu_k}{\partial P}\right)_{T,\,\text{composition}}, \tag{19-57}$$

in lieu of Eq. (15-20), to retrace the analysis of the preceding subsection, with the appropriate minor modifications, and obtain the results

$$\ln \frac{f_k}{1\text{ atm}} = \ln \frac{p_k}{1\text{ atm}} - \frac{1}{RT} \underset{(T)}{\int_0^P} \left(\frac{RT}{P} - \bar{V}_k\right) dP \tag{19-58a}$$

and

$$\ln \gamma_k = -\frac{1}{RT} \underset{(T)}{\int_0^P} \left(\frac{RT}{P} - \bar{V}_k\right) dP. \tag{19-58b}$$

These equations are valid for all components k in the mixture.

Exercise 19-8

(a) Derive Eq. (19-57) from Eq. (15-20). (Compare the relevant paragraphs in section 8.3.)

(b) Derive Eqs. (19-58a) and (19-58b).

At this stage we can give the thermodynamic analysis of equilibria in *real* gas mixtures by proceeding along exactly the same lines as in section 19.2, with all partial pressures replaced by the corresponding fugacities. If $\mathfrak{Q}_f$ denotes the new reaction quotient (in terms of fugacities), $K_f \equiv \mathfrak{Q}_f^e/\mathfrak{Q}_f^{\ominus} = \mathfrak{Q}_f^e/(1\text{ atm})^{\Delta\nu}$ becomes the "good" thermodynamic equilibrium constant. That is, K_f is a function of the temperature only, by virtue of the relation

$$RT \ln K_f = -\Delta\mathbf{G}^{\ominus}(T). \tag{19-59}$$

In Eq. (19-59), $\Delta\mathbf{G}^{\ominus}$ represents the free energy change of the reaction if all reactants and products had a fugacity of 1 atm. Equation (19-34) is still true, but in lieu of Eq. (19-35) we obtain

$$\frac{d \ln K_f}{dT} = \frac{\Delta\mathbf{H}^{\ominus}}{RT^2}, \tag{19-60}$$

in which $\Delta\mathbf{H}^{\ominus}$ can no longer be replaced by $\Delta\mathbf{H}$.

Exercise 19-9

Derive Eqs. (19-59) and (19-60). Carefully note the analogies and differences with the case of an equilibrium in an ideal mixture of ideal gases. Convince yourselves that for a reaction in a real gaseous mixture, $\Delta\mathbf{H}^{\ominus}$ will, in general, no longer be equal to $\Delta\mathbf{H}$.

By virtue of the definition of γ_k, Eq. (19-55), we can write

$$K_f = \text{“}\mathfrak{Q}_\gamma^e\text{”}\,\frac{\mathfrak{Q}_p^e}{(1\ \text{atm})^{\Delta\nu}}. \tag{19-61}$$

This has the advantage of revealing K_f as the product of what *would be* the correct equilibrium constant *if* we were dealing with an ideal mixture of ideal gases, and a correction factor, "$\mathfrak{Q}_\gamma^e$," defined as $\Pi_k\,(\gamma_k^e)^{\nu_k}$ [compare Eq. (19-26) and the accompanying illustration], due to nonideality. Although "$\mathfrak{Q}_\gamma^e$" has the *form* of a reaction quotient, *it is really not a reaction quotient at all*! Sometimes Eq. (19-61) is written as

$$K_f = K_\gamma K_p, \tag{19-62}$$

which can easily lead to misinterpretations.

(c) In the preceding subsection, the problem of finding equilibrium constants for reactions in real gas mixtures has been solved *in a purely formal way*. Unfortunately, the entire presentation is useless unless we can obtain *numerical* values of all the f_k in the mixtures of interest. This would involve integration of Eq. (19-58a) or (19-58b). Since, in general, $\bar{V}_k(P,\,T,$ composition) data are not known, such integrations can hardly ever be carried out. For pressures that are too high to accept the ideal mixture of ideal gases, the only practical way out of this dilemma involves the introduction of better models for gas mixtures. Unlike the ideal mixture of ideal gases, these are not encountered by the majority of readers in more elementary science courses. The best known of such models is the ideal mixture of real gases, introduced by Lewis and Randall in 1923. Its intricacy requires a thorough analysis, which we shall give in section 24.1. A further discussion of chemical equilibria in real gas mixtures will be postponed accordingly (see section 24.2).

19.4 General Discussion of Equilibrium Constants; the Introduction of Activities

(a) It is easy to generalize the *formal* treatment given in section 19.3(b). We can retain the implicit definition of the fugacity of a component k, in an entirely arbitrary mixture, by writing

$$\mu_k = \mu_k^{\ominus} + RT\ln\frac{f_k}{f_k^{\ominus}}. \tag{19-63}$$

For the time being, we do not specify the standard state, nor the related functional dependence of $\mu_k^{\ominus}$. If we define the relative fugacity or *activity* of k, a_k, by

$$a_k \equiv \frac{f_k}{f_k^{\ominus}}, \tag{19-64}$$

Eq. (19-63) becomes

$$\mu_k = \mu_k^{\ominus} + RT \ln a_k. \tag{19-65}$$

Activities are dimensionless.

Of course in gaseous mixtures this general formalism has to reduce to the specific one of section 19.4. Thus, for gases $f_k^{\ominus} = 1$ atm, hence a_k becomes the numerical value of f_k (sometimes written $|f_k|$) when the latter is expressed in units of atmospheres. In this case $\mu_k^{\ominus}$ becomes a function of T only. In an ideal mixture of ideal gases f_k becomes identical with $p_k(\equiv x_k P)$, so that $a_k = |p_k|$, provided that the partial pressure is expressed in atmospheres.

With all μ_k given by Eq. (19-65), it is left as an exercise to the readers to retrace, once more, the various steps of the development first given in section 19.2, for the general reaction $\sum_k \nu_k k = 0$, at constant P and T. They should first obtain

$$\Delta\mathbf{G} = \Delta\mathbf{G}^{\ominus} + RT \ln \mathfrak{Q}_a, \tag{19-66}$$

in which $\mathfrak{Q}_a$ is the reaction quotient in terms of activities:

$$\mathfrak{Q}_a \equiv \prod_k a_k^{\nu_k}. \tag{19-67}$$

[See the clarification in conjunction with the definition of $\mathfrak{Q}_p$, Eq. (19-26), in section 19.2.] Unlike $\mathfrak{Q}_p$ and $\mathfrak{Q}_f$, $\mathfrak{Q}_a$ is *always* dimensionless.

In passing, for those interested in electrochemistry, we note that by combining Eq. (19-66) with the basic equations for galvanic cells,

$$\Delta\mathbf{G} = -n\mathcal{F}\mathcal{E} \tag{15-32}$$

and

$$\Delta\mathbf{G}^{\ominus} = -n\mathcal{F}\mathcal{E}^{\ominus}, \tag{15-32$'$}$$

we obtain a general form of the *Nernst equation*:

$$\mathcal{E} = \mathcal{E}^{\ominus} - \frac{RT}{n\mathcal{F}} \ln \mathfrak{Q}_a. \tag{19-68}$$

Most readers who took a course in general chemistry will have seen applications of this equation to model systems in which activities are replaced by molarities or molalities. (See the discussion of dilute solutions in Chapter 26.)

Equations (19-66) and (19-68) are true whether the reaction mixture is at equilibrium or not, with the usual stipulation that $\mathbf{G}(P, T, \xi)$ must be well defined at all stages of the reaction. Proceeding as in section 19.2, we obtain *at equilibrium*:

$$\Delta\mathbf{G}^{\ominus} = -RT \ln K_a, \tag{19-69}$$

in which the *equilibrium constant*, K_a, is defined as

$$K_a = \mathfrak{Q}_a^e. \tag{19-70}$$

From the general equation (19-34) we obtain the temperature dependence of K_a:

$$\left(\frac{\partial \ln K_a}{\partial T}\right)_P = \frac{\Delta\mathbf{H}^{\ominus}}{RT^2}. \tag{19-71}$$

The *partial* differential in the left-hand side [compare Eqs. (19-35) and (19-60)] indicates that K_a is no longer *necessarily* a function of the temperature *only*.

(b) When we are not dealing with gases exclusively, the specification of standard states becomes quite involved. We should remember at all times that the choices concerned are largely a matter of *convenience*. Thus, in solving different problems, we may find it desirable to choose *different* standard states for the *same* substance. Nevertheless, there are certain guidelines which are *usually*, if *not always*, followed. For example, for *pure liquids and solids*, it is customary to choose as a reference the state of 1 atm pressure, at the temperature of interest. Since small pressure changes have little effect on the chemical potentials of these substances, pressure effects are frequently ignored, and the activities of pure liquids and solids become unity, not only at all temperatures, but over a range of pressures. Thus for a heterogeneous equilibrium, such as

$$CaCO_3(s) = CaO(s) + CO_2(g),$$

K_a is simply equated with $a_{CO_2} = |f_{CO_2}|$. Since at given T, fugacities of single-component gases are uniquely related to the pressure [see section 19.3(a)], we immediately obtain the result that the dissociation pressure, P_{CO_2}, is completely determined by the temperature. This familiar conclusion can also be drawn on the basis of the *Phase Rule* (Chapter 21), and is confirmed experimentally.

In *solutions*, we first have to distinguish between solvent and solute(s). For the *solvent*, which we shall always denote by the subscript 1 (compare section 8.1), the standard state is usually taken to be the pure substance at the pressure and temperature of interest. Hence, in this case,

$$a_1 = \frac{f_1}{f_1^{\bullet}(P, T)} \tag{19-72}$$

and

$$\mu_1 = \mu_1^{\bullet}(P, T) + RT \ln a_1. \tag{19-73}$$

Exercise 19-10

The quantity a_1, as defined by Eq. (19-72), can never be *greater than* unity. Can you understand this? [*Hint*: See the discussion in section 19.1(c).] When is it *equal to* unity?

From the point of view of chemical equilibria, the activities of the *solutes* are of more direct interest to us. For the specification of the standard states involved, different conventions are used, depending on, among other things, the way in which we express the concentration of these solutes. Hence the associated $\mu_i^{\ominus}$ and $\Delta\mathbf{G}^{\ominus}$ of the reaction of interest are also dependent on these concentration units. In addition, they are functions of P, T, and the choice of solvent. Obviously, these matters become very complicated. As in the case of gaseous mixtures, the conventions involved are closely tied to some appropriate limiting cases—models such as the ideally dilute solution—which will be introduced in the last part of this book. Once more, the discussions concerned will have to be postponed accordingly.

PROBLEMS

Problem 19-1

In this problem we deal with the reaction

$$N_2 + 3H_2 = 2NH_3.$$

This ammonia synthesis is a favorite subject of discussions and exercises in most textbooks on general chemistry, physical chemistry, and chemical thermodynamics. We have used it in many sections of this book and shall refer to it again in Chapter 20 on the principle of Le Chatelier. It represents an example of great technical importance, in which the equilibrium position is markedly dependent on the conditions of temperature and pressure. In all relevant parts of this problem we shall assume that, initially, nitrogen and hydrogen are mixed in stoichiometric proportions.

Part I: In this first part we shall treat our system as an ideal mixture of ideal gases.

(a) With $\Delta\mathbf{G}_{300}^{\ominus} = -33.0$ kJ mole^{-1}, compute K_p at 300 K.
(b) With $\Delta\mathbf{H}_{300} = -92.2$ kJ mole^{-1}, compute K_p at 400 K,
(i) assuming that $\Delta\mathbf{H}$ is constant over the temperature range concerned, and
(ii) assuming the temperature dependence of $\Delta\mathbf{H}$ can be taken into account by using the following heat capacity data (in units of J K^{-1} mole^{-1}):

$$\bar{C}_{p,H_2} \approx C_{p,H_2}^{\bullet} = 27.28 + 3.26 \times 10^{-3}T + 5.0 \times 10^{4}T^{-2},$$

$$\bar{C}_{p,N_2} \approx C_{p,N_2}^{\bullet} = 28.58 + 3.76 \times 10^{-3}T - 5.0 \times 10^{4}T^{-2},$$

and

$$\bar{C}_{p,\mathrm{NH_3}} \approx C^{\bullet}_{p,\mathrm{NH_3}} = 29.75 + 25.10 \times 10^{-3}T - 15.5 \times 10^{4}T^{-2}.$$

(c) Verify that the results of the preceding calculations are in accordance with van't Hoff's *principle of mobile equilibrium.* Apparently, the lower the temperature, the higher the yield of ammonia. Unfortunately, at low temperatures the reaction is too slow, even with the introduction of the most effective catalyst(s). Therefore, one is forced to compromise and carry out the synthesis at about 800 K.

(d) With $\Delta\mathbf{G}^{\ominus}_{800} = 68.2$ kJ mole^{-1} (notice, and rationalize, the change in sign, as compared to that for the reaction at 300 K), compute:

(i) K_p at 800 K,

(ii) the yield of ammonia in mole percent if the reaction is carried out at $P =$ 10 atm, and

(iii) the same yield if P is raised to 685 atm. Apparently, as could have been predicted on the basis of Eq. (19-33), the reaction should be carried out at high pressure.

Part II: At high pressure we can be confident in stating that the ideal mixture of ideal gases (used throughout Part I) is no longer a viable model. Therefore, K_p computed in (d) above is really K_f. By means of procedures and approximations to be discussed in Chapter 24, a reasonable estimate for $\mathfrak{Q}^{e}_{\gamma}$ at $T = 800$ K, $P = 685$ atm, is 0.28. Compute a more realistic yield of ammonia [compared to that obtained in part (d) (iii) above] by taking into account this "correction factor."

Problem 19-2

Use the van der Waals equation of state, and the result of Exercise 19-7, to compute the fugacities and fugacity coefficients of He and H_2 at 300 K, and a pressure of 200 atm. Repeat the calculations with a pressure of 400 atm. For the relevant van der Waals constants, see Problem 10-2.

Problem 19-3

The chemical reaction associated with the Daniell cell is

$$\mathrm{Zn} + \mathrm{Cu}^{2+} = \mathrm{Zn}^{2+} + \mathrm{Cu}.$$

At 25°C, $\mathscr{E}^{\ominus}$ of this cell is 1.10 V. Obtain the equilibrium ratio $(a_{\mathrm{Zn}^{2+}}/a_{\mathrm{Cu}^{2+}})$ at this temperature.

Problem 19-4

In this assignment we return to the reactions listed in Problem 16-3. If 1 mole of I_2 and 2 moles of Cl_2 are placed in a 1-ℓ flask at 25°C, decide what species, ICl_3(s), ICl(ℓ), and/or I_2, will be present in the condensed phase and calculate the pressure (in torr), of ICl and Cl_2 in the gas phase at equilibrium. Use the data you obtained in Problem 16-3. Assume that the gas phase can be considered an ideal mixture of ideal gases, and neglect the solubility of any of the species present in liquid ICl.

Chapter 20

The Principle of Le Chatelier

20.1 Introduction; Historical

In the discussion of chemical equilibria, nearly all textbooks of general chemistry, and many textbooks of physical chemistry, give a *principle of moderation*, which can be traced back to H. L. Le Chatelier. In 1884, in a very brief note, he wrote:*

> *Any system in chemical equilibrium undergoes, as a result of a variation in one of the factors governing the equilibrium, a compensating change in a direction such that, had this change occurred alone, it would have produced a variation of the factor considered in the opposite direction.* (*I*)

Le Chatelier acknowledged that he was inspired by the writings of van't Hoff (cf. his *principle of mobile equilibrium*, section 19-2) and by G. Lippmann's ideas in the field of electricity. In turn, the latter referred to Lenz and his famous law:

> *When a force acting on a primary electric current induces a secondary current, the direction of the latter is such that its electrodynamical action opposes the acting force.*

The most striking feature of Le Chatelier's paper is that he saw no need to subject his rather loose formulation to thermodynamic scrutiny, nor to give a proof. The first one to take that route was F. Braun (1887, 1888), but his "proof" turned out to be incorrect. Nevertheless, it earned him a place in history, for the modern descendants of Le Chatelier's rule are frequently referred to under the heading "The Principle of Le Chatelier and Braun."

It would be of considerable interest for some historians of science to trace the modifications which the form (I) underwent in the course of half a century. The end result, as we still encounter it today, is an even more loosely

* H. L. Le Chatelier, *Compt. Rend.*, **99**, 786 (1884), as translated by D. H. Everett, reference 5 of Chapter 1, p. 262.

expressed assertion along the following lines:

> *If a system in chemical equilibrium is subjected to a perturbation (stress), the equilibrium will be shifted (a reaction will occur) such as to partially undo this perturbation (relieve/oppose the stress). (II)*

This current formulation has acquired an important characteristic which seems to be lacking in the original: It has an appealing simplicity within an almost *anthropomorphic* framework. Or, as the author has pointed out, it has obtained a *metaphysical* flavor, in the sense attributed to that term by Philipp Frank;* it attempts to interpret some basic principle of science in terms of "common sense" or "everyday experience." These circumstances have contributed much to its persisting popularity.

Amazingly enough, as early as 1909, P. Ehrenfest and, independently, M. C. Raveau expressed serious doubts as to the validity of a principle in the form (I) or (II). In 1911, the former gave an extensive thermodynamic discussion which, although not flawless, has formed the basis for all subsequent critique. We shall return to Ehrenfest's contributions throughout this chapter. His point of view has been reemphasized, in one form or another, by, among others, Schottky, Ulich, and Wagner (1929), Bijvoet (1933), Planck (1934), Verschaffelt (1935), Epstein (1939), Scheffer (1945), Prigogine and Defay (1955), and the author (1957).† Nevertheless, principle (II) can still be found in most general chemistry textbooks. Frequent attempts have been made to use it in analyzing equilibrium states in a variety of physical systems. Some authors have even gone so far as to claim its applicability in the psychological, economic, and sociological fields! This is another example of a type of extrapolation against which we already warned in section 1.3, and which we also encountered in section 15.1(d). Occasionally, the question has been raised whether we should even *attempt* to give a proof, or simply accept the principle as a generalization based on experience.

In tackling these issues, our point of departure must be that *all relevant information regarding the position of—and possible shifts in—stable equilibria must be contained in the second law of thermodynamics.* We shall accept no principles, or even Principles, unless they can be derived rigorously from the equilibrium and stability conditions of Table 1 [section 15.3(b)], or any equivalent alternative formulation thereof. Our challenge is to bridge the gap between these very general, very basic, thermodynamic criteria on the one hand, and the various "principles of moderation" on the other

---

* Philipp Frank, *Philosophy of Science* **15**, 275 (1947).

† J. de Heer, reference 1 at the end of the chapter. See this paper for a list of further references.

hand. As we already stated above, this challenge was first met head-on by Ehrenfest. It seems advisable to continue this chapter with a qualitative (or, at best, semiquantitative) discussion, which will put us in a better frame of mind to reproduce some of the complex mathematical analysis involved.

20.2 Qualitative Discussion; the Ehrenfest Dichotomy

Suppose that a gas reaction has come to equilibrium at a certain P and T. Introduce a "perturbation" ("stress," if one prefers) by placing this mixture in a thermostat at temperature $T + \delta T$, while maintaining a constant pressure. Then, in due time, the gas will have to adopt the new temperature and, *in a direct sense*, no equilibrium shift can, even partially, "undo" this. For a "principle" of type (II) to have any significance at all, we must be able to identify the proper *indirect* response. When pursuing this matter, we are immediately struck by another issue which, at first sight, seems to be purely semantic: In the context of vague formulations such as (II) above, *opposing* and *relieving* an action are frequently considered synonymous, but in the normal interpretation of these terms they refer to two quite *different* reactions. For example, if someone makes a frontal approach and tries to push us, we can react by (1) retreating, i.e., *relieving* the situation, or (2) pushing back, thus *opposing* the intrusion. If this were indeed a problem of semantics *only*, the confusion could readily be eliminated by agreeing to use *one* of these terms, the most "logical" one, exclusively or, preferably, switch back to a more neutral terminology as in the original formulation (I). But unfortunately, the issue concerned goes much deeper; it was one of the most fascinating discoveries of Ehrenfest that *both* types of response occur in the "real" world of science. In an indirect sense, yet to be specified, systems may *either oppose* a "perturbation" (they are what Ehrenfest called "widerstandfähig," that is, capable of resistance) *or relieve* it (in Ehrenfest's terminology they are "anpassungsfähig," i.e., capable of adaptation), depending on the nature of the system *and* the perturbation under consideration. This is what the author has referred to as the "Ehrenfest Dichotomy" and it strips the Le Chatelier principle of its simplistic mystique. As Planck put it so aptly:

> *The idea that nature has a certain interest to preserve an equilibrium state at all cost is wrong. Nature is essentially indifferent; in certain cases it reacts in one sense, but in other cases in the opposite sense.**

---

* M. Planck, *Ann. Physik* **19**, 759 (1934).

Let us illustrate all this by some examples. First we return to the rise in temperature of a gaseous reaction mixture at constant pressure. Assume for the moment that we could prevent any reaction from taking place (keep ξ constant), i.e., "freeze" the equilibrium in its original position. Then, in acquiring the temperature $T + \delta T$, an amount of heat, $|\delta Q_o|$, would have been absorbed by the system from the thermostat. But, of course, in reality the equilibrium *does* shift (ξ changes), and with it is associated a certain heat effect. Consequently, after thermal equilibrium is reestablished, the gas mixture will have taken up an amount of heat from the thermostat, $|\delta Q|$, which in general is different from $|\delta Q_o|$. The relevant question is whether $|\delta Q|$ is larger than or smaller than $|\delta Q_o|$. The answer is given by van't Hoff's Principle of Mobile Equilibrium [section 19-2(d)]: The equilibrium shift is in the endothermic direction, the mixture absorbs *more* heat from the reservoir than would have been the case if all gases had been inert. Hence, in our present notation,

$$|\delta Q| > |\delta Q_o|; \tag{20-1}$$

the system *adapted* itself, so to speak, to the "perturbation."

As a "counterexample," consider a gaseous reaction mixture under adiabatic conditions. We impose a "stress" by increasing the volume from $\mathbf{V}$ to $\mathbf{V} + \delta\mathbf{V}$. Again, no process can ever take place to "undo" this change in a direct sense. The meaningful question to ask here is whether, because of an induced chemical reaction, the actual pressure lowering, $|\delta P|$, is larger than or smaller than the pressure lowering, $|\delta P_o|$, that would result from this volume increase if all gases were inert. As we shall see below, the answer is

$$|\delta P| < |\delta P_o|; \tag{20-2}$$

the equilibrium shift is such as to *go against* the "perturbation."

These two examples clearly demonstrate the dichotomy first noted by Ehrenfest. As an illuminating, but somewhat different type of illustration, consider the favorite example of Chapter 19:

$$N_2 + 3H_2 \rightleftharpoons 2NH_3$$

at constant temperature. We ask what will happen if we add some nitrogen to an equilibrium mixture while the *total* pressure is also kept constant. Those accustomed to apply "blindly" the Le Chatelier principle in form (II), may immediately answer: The equilibrium will shift to the right. The correct answer is: It all depends on the initial composition of the reaction mixture, as determined by the conditions we impose. We shall show below [section 20.3(e)] that if the original mixture contains *more than* 50* mole % N_2, the

---

* The number 50 is obtained, as we shall see, if the reaction mixture is considered to be an ideal mixture of ideal gases. If it is *not*, the critical percentage would be somewhat different, but the general situation is unchanged.

addition will lead to the *decomposition* of some NH_3, under the formation of even more N_2. If, however, under similar conditions (constant T and *total* P), we add N_2 to an equilibrium mixture containing *less than* 50 mole % of this gas, the result will be the *formation* of some NH_3. Thus the Ehrenfest dichotomy shows up again, and the situation appears to be very complex indeed.

20.3 Rigorous Thermodynamic Treatment

(a) We first return to the second example of the preceding section. We can write down the predicted result (20-2) in a more explicit form as follows:

$$-\left(\frac{\partial P}{\partial \mathbf{V}}\right)_{\mathbf{S},\mathbb{A}} \delta\mathbf{V} < -\left(\frac{\partial P}{\partial \mathbf{V}}\right)_{\mathbf{S},\xi} \delta\mathbf{V}. \tag{20-3}$$

The term on the right does indeed correspond to what was called $|\delta P_o|$; it gives the pressure lowering that would result if (besides **S**) ξ is kept constant, i.e., the equilibrium is frozen. The term on the left, on the other hand, corresponds to $|\delta P|$; it gives the pressure lowering when (besides **S**) $\mathbb{A}$ is kept constant (at its equilibrium value of zero!), i.e., accompanying the equilibrium shift. Thus, if we prove the validity of Eq. (20-3), we have proven the original contention, Eq. (20-2).

Since we have an *adiabatic* process,* we can write

$$dP = \left(\frac{\partial P}{\partial \mathbf{V}}\right)_{\mathbf{S},\xi} d\mathbf{V} + \left(\frac{\partial P}{\partial \xi}\right)_{\mathbf{S},\mathbf{V}} d\xi,$$

from which

$$-\left(\frac{\partial P}{\partial \mathbf{V}}\right)_{\mathbf{S},\mathbb{A}} = -\left(\frac{\partial P}{\partial \mathbf{V}}\right)_{\mathbf{S},\xi} - \left(\frac{\partial P}{\partial \xi}\right)_{\mathbf{S},\mathbf{V}} \left(\frac{\partial \xi}{\partial \mathbf{V}}\right)_{\mathbf{S},\mathbb{A}}. \tag{20-4}$$

Under adiabatic conditions, $d\mathbb{A}$ can also be expressed in terms of $d\mathbf{V}$ and $d\xi$:

$$d\mathbb{A} = \left(\frac{\partial \mathbb{A}}{\partial \mathbf{V}}\right)_{\mathbf{S},\xi} d\mathbf{V} + \left(\frac{\partial \mathbb{A}}{\partial \xi}\right)_{\mathbf{S},\mathbf{V}} d\xi,$$

from which

$$\left(\frac{\partial \xi}{\partial \mathbf{V}}\right)_{\mathbf{S},\mathbb{A}} = -\frac{(\partial \mathbb{A}/\partial \mathbf{V})_{\mathbf{S},\xi}}{(\partial \mathbb{A}/\partial \xi)_{\mathbf{S},\mathbf{V}}}. \tag{20-5}$$

Finally, from Eq. (19-12), we extract one of the Euler reciprocal relations:

$$\left(\frac{\partial P}{\partial \xi}\right)_{\mathbf{S},\mathbf{V}} = \left(\frac{\partial \mathbb{A}}{\partial \mathbf{V}}\right)_{\mathbf{S},\xi}. \tag{20-6}$$

---

* In *general*, P can be written as $P(\mathbf{S}, \mathbf{V}, \xi)$.

Upon substitution of Eqs. (20-5) and (20-6) into Eq. (20-4), we obtain

$$-\left(\frac{\partial P}{\partial \mathbf{V}}\right)_{\mathbf{S},\mathbb{A}} = -\left(\frac{\partial P}{\partial \mathbf{V}}\right)_{\mathbf{S},\xi} + \frac{[(\partial \mathbb{A}/\partial \mathbf{V})_{\mathbf{S},\xi}]^2}{(\partial \mathbb{A}/\partial \xi)_{\mathbf{S},\mathbf{V}}}. \tag{20-7}$$

Of course, both partial differentials of P with respect to $\mathbf{V}$ are *negative.* In section 19.1 the crucial sign of $(\partial \mathbb{A}/\partial \xi)_{\mathbf{S},\mathbf{V}}$ was found to be *negative* as well. Remember that this is a consequence of the convexity property of $\mathbf{E}(\mathbf{S}, \mathbf{V})$, or, to put it in other words, of the fundamental thermodynamic stability conditions. In the most general terms it is *here* that *we introduce the decisive implications of the second law of thermodynamics.* Once we know these signs, Eq. (20-7) immediately tells us that

$$-\left(\frac{\partial P}{\partial \mathbf{V}}\right)_{\mathbf{S},\mathbb{A}} < -\left(\frac{\partial P}{\partial \mathbf{V}}\right)_{\mathbf{S},\xi}, \tag{20-8}$$

which completes the desired proof of the inequality (20-3). Evidently, Eq. (20-8) also implies

$$-\left(\frac{\partial \mathbf{V}}{\partial P}\right)_{\mathbf{S},\mathbb{A}} > -\left(\frac{\partial \mathbf{V}}{\partial P}\right)_{\mathbf{S},\xi};$$

hence

$$-\left(\frac{\partial \mathbf{V}}{\partial P}\right)_{\mathbf{S},\mathbb{A}} \delta P > -\left(\frac{\partial \mathbf{V}}{\partial P}\right)_{\mathbf{S},\xi} \delta P. \tag{20-9}$$

While, in the terminology introduced in section 20.2, Eq. (20-3) shows that the system "opposes" volume changes, Eq. (20-9) proves that it "relieves" pressure changes. Hence the Ehrenfest dichotomy is rigorously justified on the basis of thermodynamics.

(b) Next we consider the more general situation,* in which $d\mathbf{E}$ is given in the form

$$d\mathbf{E} = \sum_i Y_i \, dy_i. \tag{20-10}$$

We shall refer to the Y_i and y_i as pairs of conjugate intensive and extensive variables (see section 4.2). To avoid possible complications, that are not essential in the present context, we shall assume that our system is homogeneous (see section 4.1).

In the example of subsection (a), we had three pairs of variables:

$$Y_1 = T,\ y_1 = \mathbf{S};\quad Y_2 = -P,\ y_2 = \mathbf{V};\quad Y_3 = -\mathbb{A},\ y_3 = \xi.$$

Since T is neither a coordinate nor a force (section 4.5), we cannot use the conjugate force and coordinate terminology, and therefore we avoid the X_i, x_i notation.

* The beginning student may wish to skip this derivation and simply accept formulation (III) below.

Let $y_3, y_4, \ldots$ all be constant. Then the total differential of Y_1 can be put in the form

$$dY_1 = \left(\frac{\partial Y_1}{\partial y_1}\right)_{y_2,y_3,\ldots} dy_1 + \left(\frac{\partial Y_1}{\partial y_2}\right)_{y_1,y_3,\ldots} dy_2,$$

from which

$$\left(\frac{\partial Y_1}{\partial y_1}\right)_{Y_2,y_3,\ldots} = \left(\frac{\partial Y_1}{\partial y_1}\right)_{y_2,y_3,\ldots} + \left(\frac{\partial Y_1}{\partial y_2}\right)_{y_1,y_3,\ldots} \left(\frac{\partial y_2}{\partial y_1}\right)_{Y_2,y_3,\ldots}.$$

But since, by one of the reciprocal relations pertaining to Eq. (20-10),

$$\left(\frac{\partial Y_1}{\partial y_2}\right)_{y_1,y_3,\ldots} = \left(\frac{\partial Y_2}{\partial y_1}\right)_{y_2,y_3,\ldots}.$$

the last equation becomes

$$\left(\frac{\partial Y_1}{\partial y_1}\right)_{Y_2,y_3,\ldots} = \left(\frac{\partial Y_1}{\partial y_1}\right)_{y_2,y_3,\ldots} + \left(\frac{\partial Y_2}{\partial y_1}\right)_{y_2,y_3,\ldots} \left(\frac{\partial y_2}{\partial y_1}\right)_{Y_2,y_3,\ldots}. \quad (20\text{-}11)$$

From the cycle relation between Y_2, y_1, and y_2, at constant $y_3, \ldots$, we also have

$$\left(\frac{\partial Y_2}{\partial y_1}\right)_{y_2,y_3,\ldots} = -\left(\frac{\partial Y_2}{\partial y_2}\right)_{y_1,y_3,\ldots} \left(\frac{\partial y_2}{\partial y_1}\right)_{Y_2,y_3,\ldots}.$$

If we substitute this equality in Eq. (20-11), we obtain

$$\left(\frac{\partial Y_1}{\partial y_1}\right)_{Y_2,y_3,\ldots} = \left(\frac{\partial Y_1}{\partial y_1}\right)_{y_2,y_3,\ldots} - \left(\frac{\partial Y_2}{\partial y_2}\right)_{y_1,y_3,\ldots} \left[\left(\frac{\partial y_2}{\partial y_1}\right)_{Y_2,y_3,\ldots}\right]^2. \quad (20\text{-}12)$$

But since we are considering stable equilibrium, $\mathbf{E}(y_1, y_2, \ldots)$ must be convex; hence

$$\left(\frac{\partial^2 \mathbf{E}}{\partial y_2^2}\right)_{y_1,y_3,\ldots} = \left(\frac{\partial Y_2}{\partial y_2}\right)_{y_1,y_3,\ldots} > 0,$$

so that

$$\left(\frac{\partial Y_1}{\partial y_1}\right)_{Y_2,y_3,\ldots} < \left(\frac{\partial Y_1}{\partial y_1}\right)_{y_2,y_3,\ldots}. \quad (20\text{-}13)$$

The reader will note that with the identifications $Y_1 = -P$, $y_1 = \mathbf{V}$; $Y_2 = -\mathbb{A}$, $y_2 = \xi$; $y_3 = \mathbf{S}$, we recover the inequality (20-8). Of course, it does not matter whether we indicate that $\mathbb{A}$, instead of $-\mathbb{A}$, is kept constant in the partial differential in the left-hand side.

(c) The result (20-13), and the obvious "inverse" analogous to the one incorporated in Eq. (20-9), is still not quite as general as we would like. What if instead of one or more of the y_i $(i = 3, 4, \ldots)$, we want to keep the conjugate Y_i constant? Let us illustrate the procedure by keeping Y_3, rather then

y_3, unchanged. Most of the proof is very similar and is left as an exercise for the reader. In lieu of Eq. (20-12), we readily obtain

$$\left(\frac{\partial Y_1}{\partial y_1}\right)_{Y_2,Y_3,y_4,\ldots} = \left(\frac{\partial Y_1}{\partial y_1}\right)_{y_2,Y_3,y_4,\ldots} - \left(\frac{\partial Y_2}{\partial y_2}\right)_{y_1,Y_3,y_4,\ldots}\left[\left(\frac{\partial y_2}{\partial y_1}\right)_{Y_2,Y_3,y_4,\ldots}\right]^2. \tag{20-14}$$

Exercise 20-1

Derive Eq. (20-14).

The vital question concerns the sign of $(\partial Y_2/\partial y_2)_{y_1,Y_3,y_4,\ldots}$. Let us indicate, in outline only, how one shows that this term, just like its counterpart in section 20.3(b), is still positive. It is easiest to consider the new state function

$$\mathbf{K} \equiv \mathbf{E} - Y_3 y_3. \tag{20-15}$$

Of course, **K** is not *necessarily* "new"; if $Y_3 = -P$ and $y_3 = \mathbf{V}$, then $\mathbf{K} \equiv \mathbf{H}$, the enthalpy, while if $Y_3 = T$ and $y_3 = \mathbf{S}$, $\mathbf{K} \equiv \mathbf{A}$, the Helmholtz free energy. From Eq. (20-15), using Eq. (20-10),*

$$d\mathbf{K} = Y_1\,dy_1 + Y_2\,dy_2 - y_3\,dY_3 + Y_4\,dy_4 + \ldots.$$

It is not difficult to prove that there is a minimum theorem for **K**, so that for stable equilibrium points the $\mathbf{K}(y_1, y_2, y_4, \ldots)$ surface, associated with constant Y_3, is convex.

Exercise 20-2

Prove this contention regarding the convexity of **K**. *Hint*: See section 15.3(a), where we derived other minimum theorems; use $dQ/T \leq d\mathbf{S}$ as a starting point.

Therefore,

$$\left(\frac{\partial^2 \mathbf{K}}{\partial y_2^2}\right)_{y_1,Y_3,y_4,\ldots} = \left(\frac{\partial Y_2}{\partial y_2}\right)_{y_1,Y_3,y_4,\ldots} > 0, \tag{20-16}$$

which is the desired inequality.

(d) We can now "generalize" the result (20-13) and its "inverse" to read

$$\left(\frac{\partial Y_1}{\partial y_1}\right)_{Y_2,\text{etc.}} < \left(\frac{\partial Y_1}{\partial y_1}\right)_{y_2,\text{etc.}} \tag{20-17a}$$

and

$$\left(\frac{\partial y_1}{\partial Y_1}\right)_{Y_2,\text{etc.}} > \left(\frac{\partial y_1}{\partial Y_1}\right)_{y_2,\text{etc.}}. \tag{20-17b}$$

* This is a "Legendre transformation"; see footnote, p. 85.

Following Bijvoet, we call (Y_1, y_1) the *action variables*, (Y_2, y_2) the *reaction variables*, and the remaining parameter couples the *other variables*. The word "etc." appearing with each of the partial differentials indicates that for the validity of Eqs. (20-17), it is irrelevant which of the *other* (intensive or extensive) variables are kept constant.

We already pointed out how the example of the adiabatic volume increase fits into the general formalism (20-17a). The isobaric temperature rise, our other introductory example, was associated with the inequality (20-1), which arises from

$$(T\times)\left(\frac{\partial \mathbf{S}}{\partial T}\right)_{\mathbb{A},\ \text{etc.}} > (T\times)\left(\frac{\partial \mathbf{S}}{\partial T}\right)_{\xi,\ \text{etc.}}, \tag{20-18}$$

a special case of the general equation (20-17b). It is apparent that in most examples of chemical interest, the reaction variables are always the (negative) affinity, $(-)\,\mathbb{A}$, and the degree of advancement (progress variable), ξ. In these cases, Eqs. (20-17) take on a slightly less abstract form which may be put into words as follows:

The change of an $\genfrac{}{}{0pt}{}{\textit{intensive}}{\textit{extensive}}$ *variable caused by*

changing the conjugate $\genfrac{}{}{0pt}{}{\textit{extensive}}{\textit{intensive}}$ *variable, is*

$\genfrac{}{}{0pt}{}{\textit{smaller}}{\textit{larger}}$ *if chemical equilibrium is maintained than*

if no reaction could take place in the system. (III^a_b)

Again, the Ehrenfest dichotomy is quite apparent. We should remember, however, that all these "dual formulations" *essentially* derive from a single mathematical expression (the "inverse" is hardly different in a fundamental sense).

(e) Finally, let us consider what happens when we add a component to a reaction mixture. As usual, we write the reaction in the form

$$\sum_k \nu_k k = 0. \tag{20-19}$$

We shall denote the *only* substance added externally by j. Then

$$dn_j = d_e n_j + d_i n_j, \tag{20-20}$$

in which $d_e n_j$ refers to the amount introduced in this fashion and $d_i n_j$ to changes due to process (20-19) occurring *within* the system. (Generalizations, in the sense that more than one reaction component is added externally, lead to a more complicated formalism without adding anything enlightening.)

We are dealing with an open system, for which $d\mathbf{E}$ is given by

$$d\mathbf{E} = T\,d\mathbf{S} - p\,d\mathbf{V} - \mathbb{A}\,d\xi + \mu_j d_e n_j. \tag{20-21}$$

The term $\mu_j d_i n_j$ is absorbed in $-\mathbb{A}\,d\xi$. By (IIIa) or Eq. (20-17a),

$$\left(\frac{\partial \mu_j}{\partial_e n_j}\right)_{\mathbb{A},\text{ etc.}} < \left(\frac{\partial \mu_j}{\partial_e n_j}\right)_{\xi,\text{etc.}}. \tag{20-22}$$

This implies that the "primary" increase in μ_j, caused by the external addition, is *opposed* by the equilibrium shift. In other words: *The induced chemical reaction has to lower* μ_j. At this point, the most important point to emphasize is that this conclusion by itself does not allow us to draw *any* conclusion as to the *direction* of the equilibrium shift, i.e., the sign of $\delta\xi$. To proceed, we first must specify clearly the auxiliary conditions embodied in the subscripts *etc.* in Eq. (20-22). Let us select a gas reaction at constant temperature, T, and constant *total* pressure, P. Moreover, in order to be able to treat a specific example numerically, we choose the model already introduced in section 19.2, the ideal mixture of ideal gases (see also Chapter 23). For this model we have

$$\mu_j = \mu_j^{\ominus}(T) + RT \ln |p_j| \tag{20-23a}$$

or

$$\mu_j = \mu_j^{\ominus}(T) + RT \ln \frac{n}{n} |P|, \tag{20-23b}$$

in which n is the total number of moles in the mixture. Obviously, *for* μ_j *to go down at constant* T, p_j *has to decrease.* Thus the important term is

$$\left(\frac{\partial p_j}{\partial \xi}\right)_{P,T} = \frac{P}{n^2}\left(n\frac{\partial n_j}{\partial \xi} - n_j\frac{\partial n}{\partial \xi}\right),$$

or, by Eqs. (8-46) and (8-43b),

$$\left(\frac{\partial p_j}{\partial \xi}\right)_{P,T} = \frac{P}{n^2}(n\nu_j - n_j\Delta\nu). \tag{20-24}$$

Things become simple *only* for reactions with $\Delta\nu = 0$, such as $H_2 + Cl_2 = 2HCl$. In this case $(\partial p_j/\partial\xi)_{P,T} = (P/n)(\nu_j)$. Adding a *reactant* (i.e., a component on the left) implies that ν_j is *negative*. For p_j to decrease, ξ must *increase*, i.e., the equilibrium shifts to the right. If the added j were a *product* (ν_j is *positive*), ξ would have to *decrease*. In both cases the equilibrium is displaced in such a direction as to use up some of the added component. Most chemists would say that this is in accordance with the Le Chatelier principle in form (II); we have "relieved" the "stress."

The situation is much more complicated if $\Delta\nu \neq 0$. Let us, once more, return to the example of the ammonia synthesis:

$$N_2 + 3H_2 = 2NH_3.$$

For this reaction Eq. (20-24) becomes

$$\left(\frac{\partial p_j}{\partial \xi}\right)_{P,T} = \frac{P}{n^2}(n\nu_j + 2n_j). \tag{20-25}$$

Let $j \equiv N_2$:

$$\left(\frac{\partial p_{N_2}}{\partial \xi}\right)_{P,T} = \frac{P}{n^2}(-n + 2n_{N_2})$$

or

$$\left(\frac{\partial p_{N_2}}{\partial \xi}\right)_{P,T} = \frac{P}{n}\left(2\frac{n_{N_2}}{n} - 1\right). \tag{20-26a}$$

Introducing $X_{N_2} \equiv n_{N_2}/n$,

$$\left(\frac{\partial p_{N_2}}{\partial \xi}\right)_{P,T} = \frac{P}{n}(2X_{N_2} - 1). \tag{20-26b}$$

Remember that the sign of $\delta\xi$ must be such as to reduce p_{N_2}, hence μ_{N_2}. It is obvious that the equilibrium could shift either way, depending on the numerical value of X_{N_2}: *If* $X_{N_2} > \frac{1}{2}$ (if the original mixture contains more than 50 mole % N_2), $\partial p_{N_2}/\partial\xi$ is positive, hence in order to reduce p_{N_2} the equilibrium must shift to the left. In turn this shows that *the addition of* N_2 *causes some* NH_3 *to decompose under the formation of even more* N_2. Qualitatively, this can be understood without appeal to complex mathematics: An equilibrium shift to the right, at constant T and *total* P, would cause such a large volume *contraction* that p_{N_2} would have increased notwithstanding a decrease in n_{N_2}. If we had not assumed a simple model system, namely, the ideal mixture of ideal gases, all partial pressures would have to be replaced by fugacities. This would render a *numerical* analysis much more difficult; the end result would be that the number $\frac{1}{2}$, for the "critical" mole fraction, would change, but the *general* pattern (the sign of $\delta\xi$ depends on the initial value of X_{N_2}) would be unaltered.

20.4 Conclusions and Final Comments

We shall not enter into a discussion as to whether statement (III) should still be called the Le Chatelier principle. It certainly has come a long way from the original, (I). Compared to the common form (II), it has lost all *apparent* simplicity as well as all anthropomorphic (metaphysical) flavor and, with these, most popular appeal. Perhaps it is worthwhile to mention that Ehrenfest's attention was first drawn to this subject by a student, V. R. Bursian, who found that the loosely formulated Le Chatelier principle could lead to a result of the wrong sign.* It is fortunate that we can frequently avoid such pitfalls, as well as the need to use the complex statement (III), by adapting some *basic* thermodynamic equations, as they pertain to equilibrium and stability, to *specific* types of problems. In fact, we already did this in Chapter 19. In particular, we refer to the application of van't Hoff's principle of mobile equilibrium and equations such as (19-33) (see also problem 19-1). These examples can easily be augmented; e.g., another useful equation is

$$K_p = \mathfrak{Q}_n^e \left[\frac{RT}{(1 \text{ atm})\mathbf{V}} \right]^{\Delta v}, \tag{20-27}$$

in which $\mathfrak{Q}_n$ represents the reaction quotient in terms of mole numbers.

Exercise 20-3

Derive Eq. (20-27) from the appropriate equations given in Chapter 19.

Equation (20-27) allows us to answer the question: What happens if, *at constant* T, we change $\mathbf{V}$? To see this, let us first take Δv positive, i.e., consider a dissociation reaction such as $PCl_5 = PCl_3 + Cl_2$. In this case an increase in $\mathbf{V}$ decreases $[RT/(1 \text{ atm})\mathbf{V}]^{\Delta v}$; hence $\mathfrak{Q}_n^e$ must increase for K_p to remain the same (K_p is a function of T only!). The increase in $\mathfrak{Q}_n^e$ means an equilibrium shift to the right. *Ipso facto*, if Δv is negative, as in the ammonia synthesis, an increase in $\mathbf{V}$ would cause an equilibrium shift to the left. We see that at constant T, *an increase in* $\mathbf{V}$ *results in an equilibrium shift in the direction of the largest number of molecules in the stoichiometric equation*, a familiar rule from general chemistry. Readers should, by now, be aware that K_p is only a good thermodynamic equilibrium constant for reactions in an ideal mixture of ideal gases (see section 19.2), and that the validity of Eq. (20-27) is also confined to reactions in such model systems. Thus in this example, as well as in several earlier ones, the conclusions reached have only qualitative value. Precise, quantitative predictions of equilibrium shifts may force us to abandon these simple models. This leads to serious complications (compare sections 19.3 and 19.4, as well as part II of problem 19-1), no matter what formulations and "principles," to be based solidly on the Second Law of Thermodynamics, one wishes to use as a starting point.

---

* Martin J. Klein, *Paul Ehrenfest*, North-Holland, Amsterdam/American Elsevier, New York, 1970, Vol. 1, p. 157.

REFERENCES

1. Most of the contents of this chapter were based on an article by the author: "The Principle of Le Chatelier and Braun," *J. Chem. Educ.* **34**, 375 (1957). Section 20.3 is a considerably expanded version of the treatment given in that paper.
2. This chapter owes much to Chapter V of F. E. C. Scheffer's *Toepassingen van de thermodynamika op chemische processen*, Waltman, Delft, The Netherlands, 1945.* My indebtedness to Scheffer has already been mentioned in the preface to this book.
3. Good treatments in English can be found, e.g., in P. S. Epstein's *Textbook of Thermodynamics* (Wiley, 1939), and in the text by Prigogine and Defay, already referred to on several occasions in this book (see, e.g., reference 3 of Chapter 8). See also H. B. Callen, *Thermodynamics* (Wiley, New York, 1963, p. 139), who presents a somewhat different point of view, and relates the "principle" to dynamic considerations.

---

* A second edition of this book appeared in 1958, edited by G. A. M. Diepen. Chapter V of the first edition was omitted, since, in Diepen's opinion, it "did not fit into the simple nature of the book."

Chapter 21

The Phase Rule and Related Topics

21.1 Introduction; Some Basic Concepts

The Phase Rule (J. W. Gibbs, 1876) is a formula which gives us the *variance*, $\mathbf{v}$, of a system, that is the number of (macroscopic) *intensive* variables that can be changed independently, in terms of the *number of phases*, $\mathbf{p}$, and the *number of components*, $\mathbf{c}$:

$$\mathbf{v} = \mathbf{c} + 2 - \mathbf{p}. \qquad (21\text{-}1)$$

The variance has also been called the *number of degrees of freedom* and is frequently denoted by $\mathbf{f}$ or $\mathbf{F}$. The number of phases has already been discussed in section 4.1; this concept will be elaborated upon as various examples are given. The most difficult notion is that of the number of components, sometimes referred to as the number of *independent* components, to stress that, in general, it is *not* equal to the number of different chemical species present. We have opted to arrive initially at $\mathbf{c}$ by means of an apparent "detour," due to Wind.* In our opinion this indirect approach greatly enhances the understanding of the meaning of the Phase Rule, and of the possible exceptions to its validity. Throughout this chapter we shall impose the same restrictions as enumerated at the beginning of Chapter 18 (no external fields, surface effects, etc.). Under these circumstances the *intensive* variables of interest are P, T, and the chemical potentials. In section 21.4 we discuss briefly some less known theorems, due to Bakhuis-Roozeboom and Duhem, which aim at a complete specification of a closed system in terms of *both intensive and extensive* variables.

21.2 A Prototype Derivation and Some Modifications

(a) With reference to the procedures discussed in section 18.1, we shall only follow the quickest route here, that is, use the $\delta_{P,T}\mathbf{G} = 0$ criterion. Hence we already stipulate that all phases have the same P and T. Let there

---

* C. H. Wind, *Z. Physik. Chem.* **31**, 390 (1899); see also reference 1 at the end of this chapter.

be **s** different *chemical substances*, each occurring in each of the **p** phases. The latter condition is rarely realized in practice, but we shall accept it for the time being, and consider other situations later. By definition

$$\mathbf{v} = \textit{number of intensive variables} - \textit{number of conditions between them.} \tag{21-2}$$

The variables are P, T, and **ps** chemical potentials. Within each phase there is a (generalized) Gibbs–Duhem equation:

$$S^\alpha\, dT - V^\alpha\, dP + \sum_{k=1}^{\mathbf{s}} X_k^\alpha\, d\mu_k^\alpha = 0 \qquad (\alpha = 1, \ldots, \mathbf{p}). \tag{8-29b}$$

S^α and V^α are the mean molar entropy and volume of phase α. X_k^α and μ_k^α denote the mole fraction and chemical potential, respectively, of substance k in this phase. In addition to these **p** *intraphase* relationships there are the $\mathbf{s}(\mathbf{p} - 1)$ independent *interphase* equiiibrium conditions:

$$\mu_k^\alpha = \mu_k^\beta \qquad \begin{cases} \alpha, \beta = 1, \ldots, \mathbf{p} & (\alpha < \beta) \\ k = 1, \ldots, \mathbf{s} \end{cases}. \tag{21-3}$$

Exercise 21-1

Prove Eq. (21-3) from the fundamental equilibrium condition $\delta_{P,T}\mathbf{G} = 0$.

Therefore, by Eq. (21-2): $\mathbf{v} = (2 + \mathbf{ps}) - [\mathbf{p} + \mathbf{s}(\mathbf{p} - 1)]$, or

$$\mathbf{v} = \mathbf{s} + 2 - \mathbf{p}. \tag{21-4}$$

This looks suspiciously like Eq. (21-1), but the two are identical only if $\mathbf{s} = \mathbf{c}$, which is the exception rather than the rule.

In order to establish the proper connection between Eqs. (21-4) and (21-1), we shall consider several examples. Since, in any experimental situation, it is easier to visualize changes in composition than changes in chemical potentials, we shall frequently look upon P, T and the X_k^α as the relevant intensive variables. This is allowed because the **s** quantities μ_k^α are uniquely determined by P, T, and the composition of phase α.

As *example A* (Fig. 29), consider a mixture of benzene and toluene in equilibrium with its vapor. Here $\mathbf{s} = \mathbf{p} = 2$, hence $\mathbf{v} = 2$; the system is said to be *bivariant*. This means that if *we fix any two* of the four intensive variables (P, T, X_{benzene}^{g}, $X_{\text{benzene}}^{\ell}$), *the other two are* completely determined, i.e., *out of our control*, in accordance with experiment. For example, if we consider the boiling phenomena *at constant P*, then *selecting any T* completely determines the composition of *both* the liquid *and* the vapor, in equilibrium with each other. We hope that the reader knows how to read this information off any *T,X* diagram (at constant *P*) for a binary mixture, as given in all general chemistry texts (compare Fig. 35).

(b) The restriction imposed so far, that all substances must occur in all phases, is easily removed. For every particular substance missing from any

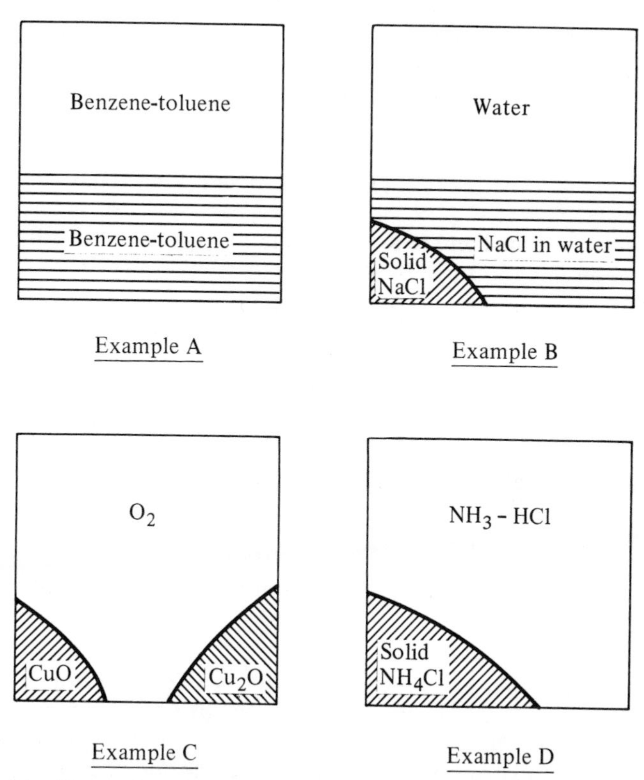

Fig. 29 Illustrating the use of various forms of the Phase Rule. (In example C, the diagram does not imply that CuO and Cu_2O actually are located in physically separated regions. This way of pictorial representation is only intended to remind us that these two solids do *not* form mixed crystals.)

one phase, there is one less chemical potential, but also one less condition of the type Eq. (21-3). In obtaining the difference (21-2), these will cancel out. As our illustrative *example* B (Fig. 29), we take a saturated solution of NaCl in water, in equilibrium with solid NaCl and water vapor. Again $\mathbf{s} = 2$ (NaCl and H_2O), but only the liquid phase will contain both substances. In this case $\mathbf{v} = \mathbf{s} + 2 - \mathbf{p} = 2 + 2 - 3 = 1$, hence $\mathbf{v} = 1$; the system is said to be *monovariant*. There are only three relevant variables: P, T, and, for example, the molality of NaCl. If we fix T, then P and the composition of the saturated solution are completely determined, again in accordance with experiment.

(c) Let us next consider a trickier *example*, C (Fig. 29): O_2 gas is in equilibrium with a mixture of CuO and Cu_2O. It is important to know that these two solids do not form mixed crystals, so that they constitute two phases. The composition of all phases is fixed and the three μ's are completely determined by P and T. In this example $\mathbf{s} = \mathbf{p} = 3$; hence by Eq. (21-4), $\mathbf{v} = 2$.

This result would imply that we can vary P $(=P_{O_2})$ and T at will. *Experimentally we know this to be incorrect*: At any given T, P_{O_2} is completely determined; the system is *monovariant* only! What went wrong? Very simply, we have to consider the reaction

$$2Cu_2O + O_2 = 4CuO$$

at equilibrium, so that there is an extra relation between the corresponding chemical potentials [cf. Eq. (19-19)]:

$$2\mu^s_{Cu_2O} + \mu^g_{O_2} = 4\mu^s_{CuO}, \tag{21-5}$$

for which we made no allowance in our derivation under subsection (a) above. In this *particular* example, Eq. (21-5) is the *only* relation between the chemical potentials; there are *none* of the type (21-3). In *general, both* types can occur, and our initial "Phase Rule" (21-4) has to be modified to read:

$$\mathbf{v} = \mathbf{s} + 2 - \mathbf{p} - \mathbf{n}, \tag{21-6}$$

where $\mathbf{n}$ is the number of *additional* conditions due to *independent* chemical reactions.

Parenthetically it should be stressed that for any particular system, $\mathbf{v}$ can always be derived "directly" (i.e., without using any "rule") by returning to its "definition," Eq. (21-2). Sometimes this is even the *simplest* procedure! In example C, there are just five relevant variables (P, T, and three μ's), and four relations [one of the type $d\mu^\alpha = -S^\alpha\, dT + V^\alpha\, dP$ for each of the three phases, plus Eq. (21-5)]; hence $\mathbf{v} = 5 - 4 = 1$.

(d) Finally, in example D (Fig. 29) we consider solid NH_4Cl in equilibrium with a vapor that consists solely of NH_3 and HCl. *We stipulate that the system has been prepared by bringing solid* NH_4Cl *in an evacuated space.* Moreover, $\mathbf{n} = 1$ due to the chemical equilibrium condition:

$$\mu^s_{NH_4Cl} = \mu^g_{NH_3} + \mu^g_{HCl}. \tag{21-7}$$

If we could apply Eq. (21-6) without further modification, we would have

$$\mathbf{v} = 3 + 2 - 2 - 1 = 2,$$

but this is incorrect! Experimentally, we know that the system is monovariant; for example, if we choose T, then P is fixed. The discrepancy is due to the way in which the system was prepared: Since we *must* have $X^g_{NH_3} = X^g_{HCl} = \frac{1}{2}$, we have already used up the degree of freedom which otherwise would be associated with the binary gas phase. Hence a further modification of our "rule" is required; we must write

$$\mathbf{v} = \mathbf{s} + 2 - \mathbf{p} - \mathbf{n} - \mathbf{m}, \tag{21-8}$$

in which **m** shall be referred to as the number of extra relations due to "special conditions." There is an inherent degree of vagueness in this notion, although in the present example we are obviously dealing with a *stoichiometric constraint.* We reiterate that both **n** and **m** arise because specific relations hold, which were not accounted for in the more *general* derivation.

21.3 Further Discussion; Relation between the Wind and the Gibbs Form of the Phase Rule; Some Special Problems

(a) Equation (21-8) essentially represents the Phase Rule in the form given by Wind.* *It is always valid.* If we *define* the *number of components* (or *independent components*) by

$$\mathbf{c} \equiv \mathbf{s} - \mathbf{n} - \mathbf{m}, \qquad (21\text{-}9)$$

we arrive at the Gibbs Phase Rule:

$$\mathbf{v} = \mathbf{c} + 2 - \mathbf{p}. \qquad (21\text{-}1)$$

As a combination of Eqs. (21-8) *and* (21-9), *Eq.* (21-1) *must also be universally valid*, and obviously there is no advantage in dealing with the *symbol* **c**. But Eq. (21-1) takes on an added interest, because Gibbs paved the way to arrive at the concept **c** more directly:

> *The number of components,* **c**, *is the smallest number of substances necessary* (≡ *the number of substances necessary and sufficient*) *to express the composition of all phases separately.* (21-10)

We prefer "to express the composition of" in lieu of the frequent alternatives "to build up" or "to construct," since the process concerned is best looked upon as purely algebraic. Gibbs himself already observed that the *choice of* components is not necessarily unambiguous, but the *number of* components, **c**, is unique. Thus, in example C (Fig. 29), one might choose

Cu and O_2 or Cu and O
CuO and O_2 or CuO and O
Cu_2O and O_2 or Cu_2O and O
CuO and Cu_2O
Cu and CuO
Cu and Cu_2O

---

* See footnote, p. 253. The notation is changed, and our formulation explicitly distinguishes the "extra relations" due to chemical reactions and those attributable to "special conditions."

Note that a prototype derivation as given in section 20.2(a) makes more sense in terms of chemical species than in terms of possible Gibbsian components; the latter may not even be present as such!

Two important questions remain:

1. Are the two Phase Rules truly equivalent? More precisely formulated: Are the definitions for **c**, Eqs. (21-9) and (21-10), *always* identical?
2. If the answer to question 1 is yes, which of the two is the easiest one to use?

(b) Unfortunately, the first question *cannot* be answered by an *unqualified* yes, but only by: *Nearly* always! The equivalence cannot be proven in a rigorous mathematical way, but it can usually be inferred. Returning to example C, $\mathbf{c} = \mathbf{s} - \mathbf{n} - \mathbf{m} = 3 - 1 - 0 = 2$; the existence of the equilibrium reaction allows us to express the chemical potential of one of the substances in terms of those of the two others. In example D, the knowledge that $X^g_{NH_3} = X^g_{HCl} = \frac{1}{2}$ allows us to express the *composition* of the gas phase as simply NH_4Cl. Thus $\mathbf{c} = 3 - 1 - 1 = 1$; in this case we *must* choose NH_4Cl as our component, neither of the other two substances will qualify as such. The only way our two Phase Rules can lead to *different* results arises when *some* condition, contained in **m** of Eq. (21-9), is not "absorbed" in the Gibbs definition (21-10). We shall discuss such a situation in subsection (d) below. In anticipation we can state that whereas Eq. (21-8) is *always* true, there are exceptions to the validity of Eq. (21-1), *if* **c** *is defined by Eq.* (21-10). In such exceptional cases, Eq. (21-1) has to be replaced by

$$\mathbf{v} = \mathbf{c} + 2 - \mathbf{p} - \phi, \tag{21-11}$$

where ϕ is the number of the unusual extra conditions *unaccounted for in Eq.* (21-10). Before giving the promised illustration, let us first try to answer the *second* question posed at the end of subsection (a), *assuming* that the two rules are equivalent, hence restricting ourselves to the case $\phi = 0$.

(c) In short, the answer concerned is: It all depends! Usually, in the Wind formulation, it is relatively easy to specify **s**, but it may be more difficult to identify all the conditions that enter into **n** and **m**. In the Gibbs formulation, the entire problem reduces to the determination of **c**, by means of Eq. (21-10). If this causes no problems, this route is the fastest. Thus, in example C, it is readily seen that the composition of all phases (individually) can be expressed in terms of Cu and O_2; hence $\mathbf{c} = 2$. In example D both phases have the composition NH_4Cl, etc. As **c** gets larger, the Gibbs approach could easily become the more difficult of the two. We should also reemphasize that in some cases (see the examples above) the "direct" derivation of **v** by means of Eq. (21-2) is quite straightforward and brief. The author always recommends to his students: "In case of doubt, try at least two, and preferably all three, methods."

Exercise 21-2

A cylinder is filled with H_2S gas. Consider the following situations:

(a) All dissociations can be disregarded.

(b) The temperature is high enough for some H_2S to dissociate into hydrogen and gaseous sulfur.

(c) Between 300 and 800°C the pressure is high enough for some sulfur, formed upon dissociation, to liquefy. None of the gases present dissolves in this liquid.

In each of these cases, obtain the variance, **v**:

(i) By "direct derivation," Eq. (21-2).

(ii) By the use of the Wind Phase Rule, Eq. (21-8).

(iii) By the use of the Gibbs Phase Rule, Eq. (21-1), with **c** obtained through the definition (21-10).

Would the molecular form of the gaseous and liquid sulfur (e.g., S_8, S_2, or a mixture thereof) influence your answer?

(d) As example E we select the ammonium bicarbonate system, as illustrated in Fig. 30. It was studied extensively by the author's mentor in chemical phase theory, Scheffer.* We start by placing NH_4HCO_3 into an evacuated space and raise the temperature to a level where both a totally dissociated vapor and a liquid phase are present. Since the solubilities of NH_3, CO_2, and H_2O in liquid NH_4HCO_3 are *not* the same, the net composition of neither the gas phase nor the liquid phase can be expressed as

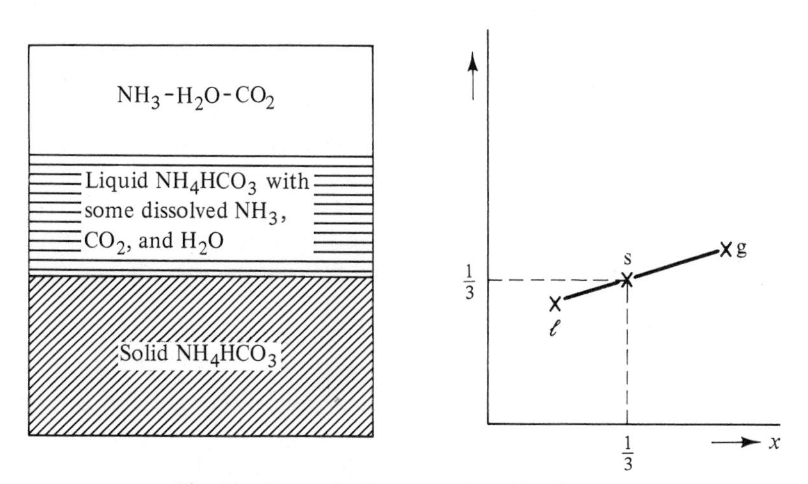

Fig. 30 Example E: ammonium bicarbonate.

---

* F. E. C. Scheffer, *Chem. Weekblad.* **20**, 224 (1923); see also J. Zernike, *Recl. Trav. Chim.* **70**, 711 (1951), and **73**, 95 (1954).

NH_4HCO_3. Any way we look at it, this requires $\mathbf{c} = 3$; the *system is ternary.* Obviously, $\mathbf{p} = 3$, so that the Gibbs Phase Rule, Eq. (21-1), gives us

$$\mathbf{v} = 3 + 2 - 3 = 2.$$

This would suggest, for example, that at given T we can choose the pressure of the equilibrium vapor at will. Scheffer's experiments showed this to be incorrect: At fixed T, all other intensive variables are completely determined, hence evidently $\mathbf{v} = 1$; *empirically the system is monovariant.* In this (exceptional) case the original Gibbs Phase Rule gives the wrong result; we have to use it in the modified form (21-11) with $\phi = 1$. Where does the unusual extra condition come from? Call $X_{NH_3} \equiv x$ and $X_{CO_2} \equiv y$, so that $X_{H_2O} = 1 - x - y$. For the solid phase all three mole fractions are $\frac{1}{3}$. We cannot say offhand what the composition of the liquid and the vapor will be, but if n^ℓ and n^g represent the total number of moles in these two phases, respectively, then we must have

$$\underset{(1)}{x^g n^g + x^\ell n^\ell = y^g n^g + y^\ell n^\ell} \underset{(2)}{=} (1 - x^g - y^g)n^g + (1 - x^\ell - y^\ell)n^\ell. \tag{21-12}$$

This expresses algebraically the fact that ℓ and g together contain the same amount of NH_3, CO_2, and H_2O. We can eliminate n^ℓ and n^g with the aid of the *two* independent identities embodied in Eq. (21-12); e.g., solve the identity (1) for n^g and substitute the result in the identity (2). The reader readily verifies that this yields

$$\frac{y^g - \frac{1}{3}}{x^g - \frac{1}{3}} = \frac{y^\ell - \frac{1}{3}}{x^\ell - \frac{1}{3}}, \tag{21-13}$$

the desired condition (see also Fig. 30). There is no problem in obtaining the correct variance by means of the Phase Rule in the modified Wind form, Eq. (21-8). The crucial point is that *the extra condition* (21-13) *enters the latter through* **m**, *while it is not incorporated in the definition of* **c**, Eq. (21-10). In turn, the last circumstance is due to Eq. (21-13) representing an *interphase condition,* while the definition (21-10), due to the last word ("separately"), which *must* be included, is concerned only with *intraphase relationships.*

Exercise 21-3

(a) Treat the ammonium bicarbonate system by means of Eq. (21-8).

(b) Obtain the variance for this system by "direct derivation," using Eq. (21-2).

(c) Show the equivalence of Eqs. (21-12) to Eq. (21-13).

(e) There is no "magic" about the Phase Rule, in whatever form one prefers. True, in the overwhelming number of practical applications it gives the correct result, but we should be alert at all times that in some instances special difficulties may arise. To mention briefly one more example, of a very different nature from the one discussed in subsection (d), such problems present themselves in the application of the Phase Rule to certain systems containing optical enantiomers.* The latter possess all the same intensive properties, except for their optical activities, which are equal in magnitude but opposite in sign. Scott pointed out that in this situation the Phase Rule may predict *too low* a variance. He suggested a modified rule, which we shall not reproduce. Wheeler showed that the anomalies concerned are best understood if we go back to a basic derivation such as the one we gave in subsection (a) above. The crucial point is that some of the Gibbs–Duhem equations of the type (8-29b) become linearly dependent as a result of the special symmetries, and the associated identical thermodynamic properties, of the enantiomers. This reduces the number of independent conditions to be entered on the right-hand side of Eq. (21-2), and increases the variance, $\mathbf{v}$, accordingly. For details we refer to the papers by Scott and Wheeler.

(f) As a postscript, we shall discuss the problem of the counting of critical phases. The simplest case is that of a one-component system. We know from experiment, as well as from the usual derivation via an equation of state,[†] that there is a nonvariant critical *point*, Cr; its pressure *and* its temperature are completely determined. To interpret this on the basis of the Phase Rule, we evidently must take $\mathbf{p} = 3$:

$$\mathbf{v} = \mathbf{c} + 2 - \mathbf{p} = 3 - \mathbf{p} = 0.$$

One could rationalize this by accepting as our three phases the gas, the liquid, and the "fluid" (see Fig. 31).

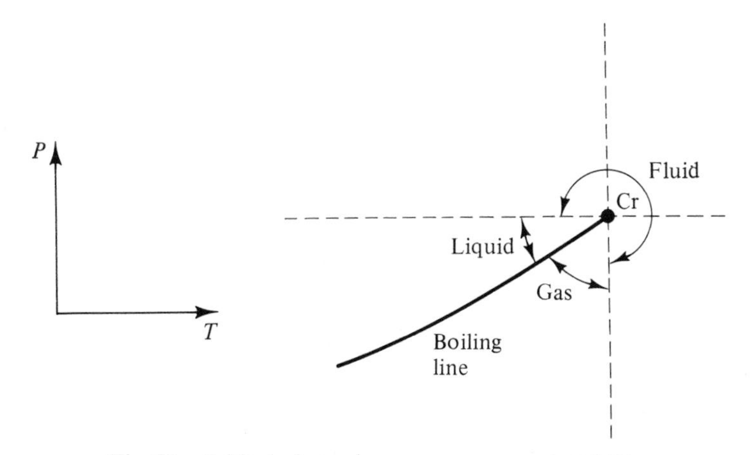

Fig. 31 Critical phases in a one-component system.

* Robert L. Scott, *J. Chem. Soc. Faraday Trans. II*, **73**, 356 (1977); John C. Wheeler, *J. Chem. Phys.*, **73**, 5771 (1980).

† This derivation can be found in any physical chemistry text, usually with special reference to the van der Waals model. See also Exercise 27-1.

This formal scheme is entirely consistent with the existence of a critical *line* in binary systems (see, e.g., line Cr_A–Cr_B in Fig. 33 at the beginning of Chapter 22). In this case $\mathbf{c} = 2$, so that the acceptance of $\mathbf{p} = 3$ leads to $\mathbf{v} = 1$, in accordance with experiment; the location of the critical point varies with the composition of the mixture.

However, anyone who has ever seen a demonstration of the critical phenomenon will balk at the idea of comparing this situation to an "ordinary" three-phase equilibrium in the manner that, e.g., solid, liquid, and gas coexist at the triple point.

To elaborate briefly upon this difference, we can introduce a finite (and, in practice, often a considerable) amount of heat into a triple-point mixture without causing any of the three phases to disappear. Consequently, the mixture will stay at the same (pressure and) temperature and constitutes an excellent thermostat (*triple-point thermostat*). On the other hand, since at the critical point the heat of vaporization is zero, an infinitesimal amount of heat may cause the entire liquid phase to disappear, after which P and T can change. Also, a triple-point mixture can normally exist within a wide range of mean molar volume (the molar volumes of the three individual phases being, in general, quite different!), but a critical state is realizable only at a particular V_{Cr} (the molar volume of the critical phases being the same), characteristic for the substance under consideration. This is not the place to discuss any further details about the singular nature of the critical point.

Zernike,* therefore, likes to speak of the existence of three *virtual phases* and shows† that $\mathbf{p}'$ phases becoming critically identical constitute $2\mathbf{p}' - 1$ such virtual phases. However, Zernike admits that what we are really doing here is the following:‡ We first define a particular concept (in our example, a phase, as in section 4.1); then a law is formulated embodying this concept (the Phase Rule). Eventually, however, the law evolves as the *definition* of the new concept. Indeed, what Zernike denotes as the number of *virtual* phases is nothing but the number of phases (at the critical point) as *defined* by the Phase Rule, in the form (21-1).§

There are two alternative ways of looking at this problem, in both of which we use the Phase Rule in the form (21-11), with $\phi > 0$. To illustrate these two distinct points of view, we return to the simple case of the liquid–gas critical point in a one-component system. In the first method, the critical condition is considered to be a *two-phase equilibrium* with a *single additional constraint*. The latter can be written, simply, as $V^\ell = V^g$ or, if one prefers

---

* J. Zernike; see p. 40 of reference 1 at the end of the chapter.

† J. Zernike, *Recl. Trav. Chim.*, **68**, 585 (1949).

‡ Ibid. He stresses that, according to Poincaré, this represents a familiar "tendency in science."

§ Compare the evolution of the First Law, as the *definition* of **E**, in section 9.2.

(cf. section 21.2), as $(\partial\mu^{\ell}/\partial P)_T = (\partial\mu^{g}/\partial P)_T$. [The equality $S^{\ell} = S^{g}$ does *not* give us an *independent* extra condition, owing to the validity of the Clapeyron equation (18-11a), and the fact that dP/dT at the critical point is neither zero nor infinite.] According to this way of reasoning, $\mathbf{v} = \mathbf{c} + 2 - \mathbf{p} - \phi = 1 + 2 - 2 - 1 = 0$. In the second method, one views the critical condition as representing a *single phase*, subjected to *two special constraints*, $(\partial P/\partial V)_T = 0$ and $(\partial^2 P/\partial V^2)_T = 0$. In this scheme $\mathbf{p} = 1$ and $\phi = 2$, with the same end result: $\mathbf{v} = 0$.

Both procedures, as given in the preceding paragraph, can be generalized for $\mathbf{c} > 1$ in a straightforward manner.* An extension to tricritical and multicritical phenomena is much more complex. There probably is, as yet, no totally satisfactory comprehensive treatment of all such kinds of phase equilibria as are known, or expected, to exist. The interested reader is referred to a paper by Bartis† and a review article by Griffith.‡

21.4 Complete Specification of a System; the Work of Bakhuis-Roozeboom and Duhem

(a) The Phase Rule tells us how many of the intensive variables P, T and, e.g., the mole fractions X_k^{α} ($k = 1, \ldots, \mathbf{c}$; $\alpha = 1, \ldots, \mathbf{p}$) are necessary and sufficient to characterize unambiguously the physicochemical *state* of a system at equilibrium, but it is not concerned with the relative amounts of the various phases. Toward the end of the nineteenth century, H. W. Bakhuis-Roozeboom and, independently, P. Duhem investigated what it takes to specify completely a (closed) system at equilibrium, in terms of both intensive and extensive variables. The rules and theorems they obtained never acquired the same popularity as the Phase Rule, probably because these did not lend themselves to so many practical applications. In this section we shall first delineate more precisely what type of questions can be asked, and subsequently give a very brief outline of how one has found some of the answers concerned. It may be worthwhile to reiterate that throughout this section, as in the preceding ones, the restrictions mentioned briefly at the end of the introduction, and listed in detail at the beginning of Chapter 18, are still imposed.

(b) For simplicity we shall assume that we can apply the Phase Rule in the form (21-1), with $\mathbf{c}$ given by the Gibbs definition (21-10). To put this

---

* The author is indebted to John C. Wheeler for enlightening correspondence on this general topic.

† James T. Bartis, *J. Chem. Phys.* **59**, 5423 (1973).

‡ Robert B. Griffith, *Phys. Rev.* **12**, 345 (1975).

another way, we shall disregard cases with $\phi \neq 0$ [cf. the discussion in connection with Eq. (21-11)]. The result of the application of the Phase Rule, our knowledge of the variance, **v**, is of vital importance in the ensuing analysis

Let n_k^α denote the number of moles of component k in phase α, n_k the total number of moles of component k (in all phases combined), and n^α the total number of moles in phase α (of all substances combined). Since

$$n_k^\alpha = X_k^\alpha n^\alpha \qquad (k = 1, \ldots, \mathbf{c};\ \alpha = 1, \ldots, \mathbf{p}), \tag{21-14}$$

we can write

$$n_k = \sum_{\alpha=1}^{\mathbf{p}} n_k^\alpha = \sum_{\alpha=1}^{\mathbf{p}} X_k^\alpha n^\alpha \qquad (k = 1, \ldots, \mathbf{c}). \tag{21-15}$$

Let us ask ourselves the following question: Given the physicochemical state of the system (hence also all the X_k^α), and **c** independent quantities n_k, how many of the **p** quantities n^α can be chosen at will? Before we attempt to answer this question we note that a knowledge of all n^α gives us, in principle, also all *extensive* properties **J**. For

$$\mathbf{J} = \sum_{\alpha=1}^{\mathbf{p}} n^\alpha J^\alpha, \tag{21-16}$$

in which the mean molar quantities J^α are *intensive*, they are "densities" in the Griffith–Wheeler sense (see section 4.2), completely determined by the state of the system. Conversely, a knowledge of **p** different **J** (e.g., **E**, **V**, and **S** for $\mathbf{p} = 3$) can give us all the n^α. Evidently, any mixture of a sufficient number of quantities **J** and n^α can be used to specify the system completely.

(c) First we consider *nonvariant systems* ($\mathbf{v} = 0$); all intensive variables are completely determined. By the Phase Rule, $\mathbf{c} = \mathbf{p} - 2$; hence the **c** amounts n_k will give us $(\mathbf{p} - 2)$ equations (21-15) between the **p** unknown n^α. This means that we can still select two n^α, or two **J**, or one of each, to determine the system completely, not only with regard to its *state*, but also insofar as the *amounts* of all phases are concerned.

As a simple example, suppose that we have n_{H_2O} moles of H_2O at its triple point. If we select, e.g., $n_{H_2O}^\ell$ and **V**, then $n_{H_2O}^s$ and $n_{H_2O}^g$ follow from the equations

$$n_{H_2O} = n_{H_2O}^s + n_{H_2O}^\ell + n_{H_2O}^g$$

and

$$\mathbf{V} = n_{H_2O}^s V_{H_2O}^s + n_{H_2O}^\ell V_{H_2O}^\ell + n_{H_2O}^g V_{H_2O}^g.$$

(d) Next we turn our attention to *monovariant systems* ($\mathbf{v} = 1$). Let us fix one intensive variable, T say, and once more all other such variables are known. In this case $\mathbf{c} = \mathbf{p} - 1$, so that the **c** quantities n_k give us $(\mathbf{p} - 1)$

independent relations (21-15) between the **p** amounts n^α. Hence we can still specify one of the latter, or one extensive property **J**, to determine the system completely.

As a simple example, consider any two-phase equilibrium in a one-component system, e.g., water in equilibrium with its vapor, at a given T. As always, we consider n_{H_2O} given, so that a specification of, e.g., **V** will give us $n^\ell_{H_2O}$ and $n^g_{H_2O}$ from

$$\mathbf{V} = n^\ell_{H_2O} V^\ell_{H_2O} + n^g_{H_2O} V^g_{H_2O}$$

and

$$n_{H_2O} = n^\ell_{H_2O} + n^g_{H_2O}.$$

Thus a monovariant system, with known n_k, has the interesting property that by choosing **V** and T, the total number of moles of each phase is determined. If we change **V**, at constant T, the pressure and the composition of all the phases will be unchanged, only the relative amounts of the phases will alter.

Exercise 21-4

Check the derivation and implications of this statement for the chemical equilibrium:

$$CaCO_3(s) = CaO(s) + CO_2(g)$$

at constant T [cf. section 19.4(b)].

(e) Finally, we treat *systems that are bivariant or polyvariant*; $\mathbf{v} = 2 + a$ ($a = 0, 1, 2, \ldots$).* Again, let us assume that we first select **v** intensive properties, so that the state of the system is completely determined. In this case $\mathbf{c} = \mathbf{p} + a$, so that the **c** known quantities n_k will give us ($\mathbf{p} + a$) independent relations (21-15) between the **p** amounts n^α. Of these relations, **p** suffice to determine all the n^α. If $a \neq 0$, the remaining equations must reduce the "effective variance" of the system. Obviously, these matters become quite complicated. Let us illustrate the relevant features for the technically important binary system ethanol–water, at a constant pressure of 1 atm (Fig. 32). It is convenient to denote X_{H_2O} simply by x, so that X_{eth} becomes $(1 - x)$.

1. First consider the two-phase equilibrium at T_2, given by the "tie line" $g_2\ell_2$. [The net composition of our mixture will have to lie between x^g_2 and x^ℓ_2; see section 18.4(c) on the *lever rule*.] The Phase Rule yields $\mathbf{v} = \mathbf{c} + 2 - \mathbf{p} = 2 + 2 - 2 = 2$, but we have already disposed of these two degrees of freedom by selecting $P = 1$ atm and $T = T_2$.

---

* Of course, the maximum **v** (corresponding to the minimum **p**, namely one) equals $\mathbf{c} + 1$. Hence in a binary system $\mathbf{v}_{max} = 3$, in a ternary system $\mathbf{v}_{max} = 4$, and so on.

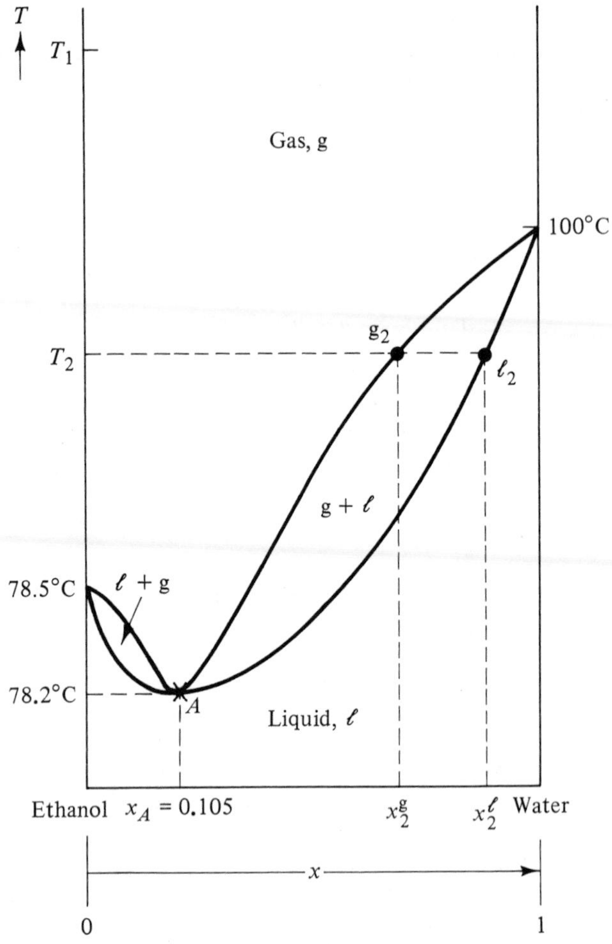

Fig. 32 Boiling diagram for ethanol-water at a constant pressure of 1 atm. ($X_{H_2O} \equiv x$; $X_{eth} \equiv 1 - x$; figure not drawn to scale).

This confirms (see Fig. 32) that the composition of the coexisting phases, as given by x_2^g and x_2^ℓ, is completely determined. Moreover, since n_{H_2O} and n_{eth} are assumed known, n^ℓ and n^g are completely determined by the two equations (21-15):

$$n_{H_2O} = x_2^\ell n^\ell + x_2^g n^g$$

and

$$n_{eth} = (1 - x_2^\ell)n^\ell + (1 - x_2^g)n_2^g.$$

All extensive properties are known as well, e.g.,

$$\mathbf{V} = n^\ell V_2^\ell + n^g V_2^g.$$

2. Next we shall consider the situation in which $a \neq 0$. To this purpose we study a gas at a temperature T_1, above the normal boiling point of water (Fig. 32). The Phase Rule gives $\mathbf{v} = \mathbf{c} + 2 - \mathbf{p} = 2 + 2 - 1 = 3$. We have already disposed of two degrees of freedom, by choosing $P = 1$ atm and $T = T_1$, so that there is only a single one left. If we were interested only in the application of the Phase Rule as such, this would simply give us the trivial result that we can select the composition of the gas, i.e., x_1^g, at will. But within the present framework, we must consider $n_{eth} = n_{eth}^g$ and $n_{H_2O} = n_{H_2O}^g$ as *given*, so that $x_1^g \equiv X_{H_2O}^g\,[\equiv n_{H_2O}/(n_{H_2O} + n_{eth})]$ is also completely determined. Hence the specification of P and T, *in addition to* n_{eth} and n_{H_2O}, has made the system *effectively* nonvariant.

In this entire section we have used the Gibbs Phase Rule, in the form (21-1). If we had used the Wind version (21-8), we could have interpreted our knowledge of n_{eth}^g and $n_{H_2O}^g$ in example (2) as a *special condition*, not entirely unrelated to that in example D of section 21.2. In fact, we have fixed the composition of the gas, so that $\mathbf{m} = 1$ and $\mathbf{v} = \mathbf{s} + 2 - \mathbf{p} - \mathbf{n} - \mathbf{m} = 2 + 2 - 1 - 0 - 1 = 2$. We immediately obtain the result that the system is, in effect, bivariant. Obviously, a choice of P and T for a known quantity of a binary gas, *with known composition*, completely determines anything there is to know about such a system.

(f) If we review all the systems, treated in subsections (c) through (e), we note that they have an interesting feature in common: *Two* variables always suffice for their specification. This is *Duhem's Theorem* (1899):

> *The equilibrium state of a closed system, for which the mole numbers of all components are given, is completely determined by two independent variables* (*irrespective of the number of components, the number of phases, or the number of chemical reactions that can occur*).

Exercise 21-5

Combine the techniques used in section 21.2 with some of those in section 21.4, and obtain Duhem's Theorem for a general situation, by subtracting the *total* number of conditions from the *total* number of variables (intensive *plus* extensive).

In demonstrating the validity of this theorem in the specific examples given above, we always selected $\mathbf{v}$ intensive variables, so that there remained $(2 - \mathbf{v})$ extensive variables to be chosen. Usually, we have more flexibility. To see this, consider the simple example of n moles of a single-component gas. This system is not only completely determined by P and T, but alternatively by P and $\mathbf{V}$, or by $\mathbf{V}$ and T. We end this section by an interesting example, in which such a choice is *not* immaterial.

To this purpose we return to the ethanol–water system and consider the *azeotropic* point, A (Fig. 32). For this *constant-boiling mixture* $x_A^\ell = x_A^g \equiv x_A$, and the two relevant equations (21-15) read

$$n_{H_2O} = x_A(n_A^\ell + n_A^g)$$

and

$$n_{eth} = (1 - x_A)(n_A^\ell + n_A^g).$$

These equations are indeterminate; they give no unique solution for n_A^ℓ and n_A^g. Thus, with given n_{H_2O} and n_{eth}, at given P and T, we can still have an infinite number of equilibrium states. Such an azeotropic point represents a particular case of what Duhem has called an *indifferent state.** *Apparently*, in this unusual situation, Duhem's Theorem fails. However if, in lieu of P and T, we select $\mathbf{V}$ and T as independent variables, there is no anomaly whatsoever.

Exercise 21-6

Verify the contents of the last sentence.

REFERENCES

1. The "indirect approach" to the phase rule, along the lines suggested by Wind, can also be found in J. Zernike's *Chemical Phase Theory*, Æ. E. Kluwer, Deventer, The Netherlands, 1955, Chapter I.
2. The material given in section 21.4 was discussed, in part, by G. E. Uhlenbeck in his 1967 summer lectures at the University of Colorado.
3. For Duhem's theorem and its implications, see also I. Prigogine and R. Defay, reference 3 of our Chapter 8, in particular Chapter XIII and part of Chapter XXIX.

---

* P. Duhem, *J. Phys. Chem.*, **2**, 31 (1898).

Chapter 22

Some Aspects of Phase Equilibria in Binary Systems

22.1 Introduction; the P,T,x Diagram

For a binary system A–B, the Phase Rule gives us

$$\mathbf{v} = 4 - \mathbf{p}.$$

Since $\mathbf{p}_{min} = 1$, $\mathbf{v}_{max} = 3$. Hence the comprehensive properties of such a system must be represented in three dimensions, e.g., in the appropriate P,T,x diagram ($x \equiv X_B = 1 - X_A$). Figure 33 depicts such a three-dimensional arrangement, for a system with the following characteristics:

1. Melting and boiling points (hence also the vapor pressures) of A and B are of comparable magnitudes (in the sense that we exclude a system such as $NaCl$–H_2O).
2. A and B are completely miscible in the liquid phase but completely immiscible as solids (i.e., they form no solid solutions).
3. A and B form no compounds, nor do they undergo dissociation, in any phase.
4. The boiling points (and vapor pressures) of the binary mixtures vary monotonically between those of A and B; there are no maxima or minima in the liquid–gas surfaces, hence no azeotropes [compare section 21.2(f) with Fig. 32 and also section 22.4(b)].

Even with all these restrictions and stipulations, the P,T,x diagram (Fig. 33) is very complicated, as uninitiated readers will readily admit. Its properties have been studied exhaustively by Bakhuis-Roozeboom and his pupils, notably Kuenen, Büchner, Schreinemakers, and Aten, with special emphasis on the systematic drawing of two-dimensional projections of—and cuts through—this three-dimensional model. A detailed discussion of these matters falls beyond the scope of this book. Nevertheless, we shall draw attention to some of the less complex features, displayed in Fig. 33, which will give the

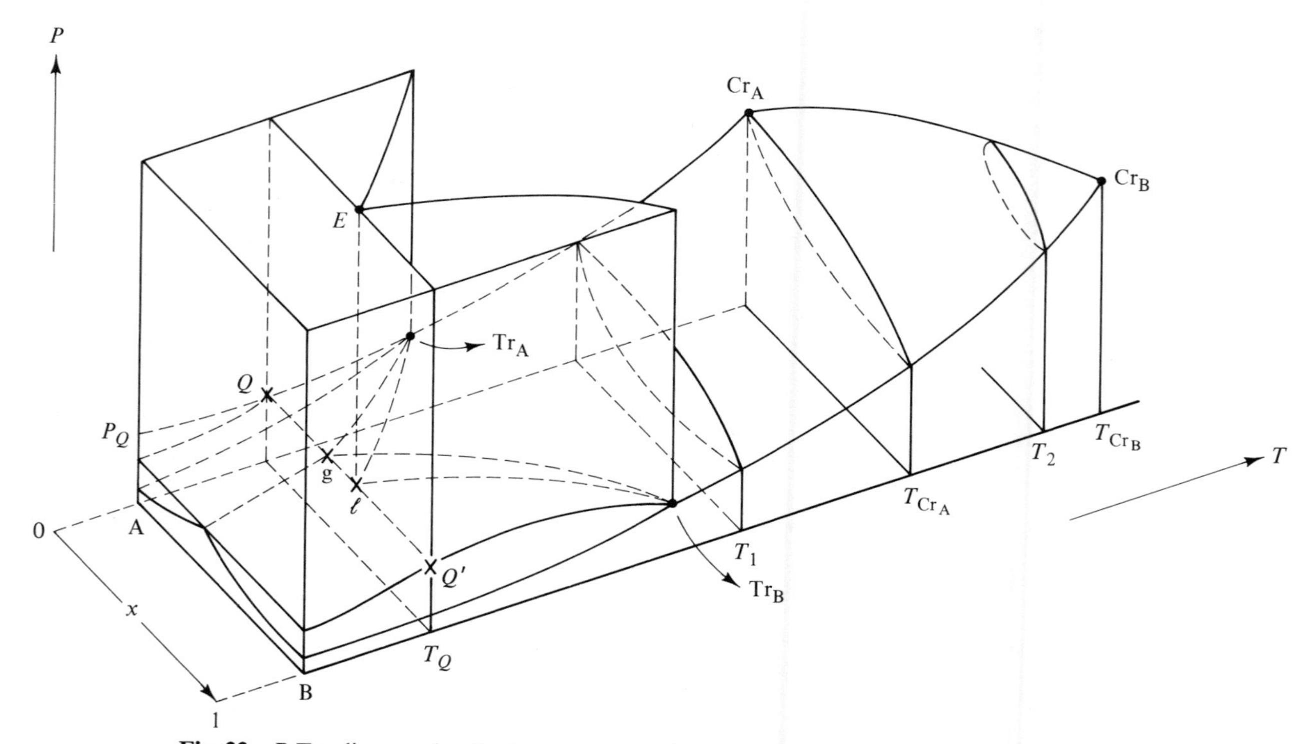

Fig. 33 P,T,x diagram for the binary system A–B, as specified in the text, showing certain familiar T,x and P,x cuts within a single three-dimensional framework. For the purposes of our discussion it is not necessary to understand the intricacies of various features in the "interior" of the diagram.

readers at least a start in learning how to disentangle the maze of surfaces and curves involved. Not all of the latter need be identified for our purpose.

First, in the two "side planes," one readily recognizes the familiar P,T diagrams of the pure components (compare Fig. 20 in Chapter 18). Tr_A and Tr_B are the two triple points, and Cr_A and Cr_B the respective critical points. Cr_A–Cr_B is the *critical line* referred to in section 21.2(f). Only at a pressure P_Q *and* a temperature T_Q can four phases coexist in equilibrium, namely, solid A (represented by point Q) and solid B (represented by point Q'), as well as a gas and a liquid of a particular composition (represented by the points g and ℓ, respectively). This nonvariant state (since $\mathbf{p} = 4$, $\mathbf{v} = 0$!) is routinely referred to as the *quadruple point* of the system, although it appears as a single *point* only in a P,T projection.

The T,x cut at high pressure, containing a *eutectic point*, E, reflects an ordinary melting diagram, as drawn in more detail in Fig. 36, and discussed in the accompanying text of section 22.5(a). The P,x cut at T_1 illustrates the familiar binary vapor-pressure relationships. A corresponding T,x cut at constant pressure, not included in Fig. 33, is drawn in Fig. 35 (bottom part). Note that in our example, since A has the *highest* vapor pressure at a given T, it must have the *lowest* boiling temperature of the two components at a particular P. For future reference, we draw attention to the P,x cut at T_2, between the critical temperatures of A and B. It embodies the possibility of *retrograde condensation* phenomena, to be discussed in section 27.3 (see, in particular, Fig. 47b and the accompanying analysis).

Following in the footsteps of the pioneering research of J. W. Gibbs, the thermodynamics of phase equilibria was developed by a number of scientists, in particular J. D. van der Waals. In the rest of this chapter we discuss some of the more general features of this work. In Chapter 27 we shall return to this subject, in a discussion of van der Waals's treatment of binary liquid–gas equilibria, by means of an extension of his well-known equation of state to two-component systems. No attempt will be made to maintain a rigid separation between the general thermodynamic approach and the special results obtained by the introduction of such models.

A characteristic of the relevant work of van der Waals is his preference to use, as much as possible, the Helmholtz free energy, **A** (cf. section 18.4 and Chapters 17 and 27). Today, we usually take P and T as independent variables, hence prefer to couch the discussion in terms of the Gibbs free energy, **G**.* This we shall do in the remaining sections of this chapter. As to notation, A and B will continue to identify the two components. Coexisting phases will be denoted, in general, by superscripts **1** and **2** (no further references to three- or four-phase equilibria will be encountered).

* Compare section 15.2(c).

This chapter is very much open-ended. Of course, the topics selected reflect, in part, the author's preference. However, it also seemed advisable to confine the application of some of the general thermodynamic results to *those* phase equilibria with which most readers can be expected to have at least *some* familiarity. Unusual, very complex phase diagrams have been avoided. For further details, interested readers should consult the references at the end of the chapter.

22.2 The G,x Curve at Constant P and T; Coexistence

(a) For 1 *mole* of a single phase we have

$$G = (1 - x)\mu_A + x\mu_B, \tag{22-1}$$

in which $x \equiv X_B$, as usual. For this case Eq. (19-6) gives

$$dG = -S\,dT + V\,dP + (\mu_B - \mu_A)\,dx. \tag{22-2}$$

Hence

$$\left(\frac{\partial G}{\partial x}\right)_{P,T} = \mu_B - \mu_A. \tag{22-3}$$

The first and third of these equations can be solved to yield μ_A and μ_B:

$$\mu_A = G - x\left(\frac{\partial G}{\partial x}\right)_{P,T} \tag{22-4a}$$

and

$$\mu_B = G + (1 - x)\left(\frac{\partial G}{\partial x}\right)_{P,T}. \tag{22-4b}$$

A typical G,x curve is drawn in Fig. 34. As x increases, μ_B increases and μ_A decreases. Hence we see that $(\mu_B - \mu_A)$ increases monotonically with x. Therefore, Eq. (21-3) implies that $(\partial^2 G/\partial x^2)_{P,T}$ is positive over the entire range. Simple geometrical considerations show that in the point U,

$$-tg\alpha = \left(\frac{\partial G}{\partial x}\right)_{P,T}, \quad \text{hence} \quad RS = \left[-x\left(\frac{\partial G}{\partial x}\right)_{P,T}\right]_{\text{in } U}.$$

Thus we see that *for a mixture represented by U, μ_A is the distance from the* lower left corner of the G,x diagram to the point R. The latter point is the intersection of the tangent in U with the A axis. Similarly, μ_B is the distance

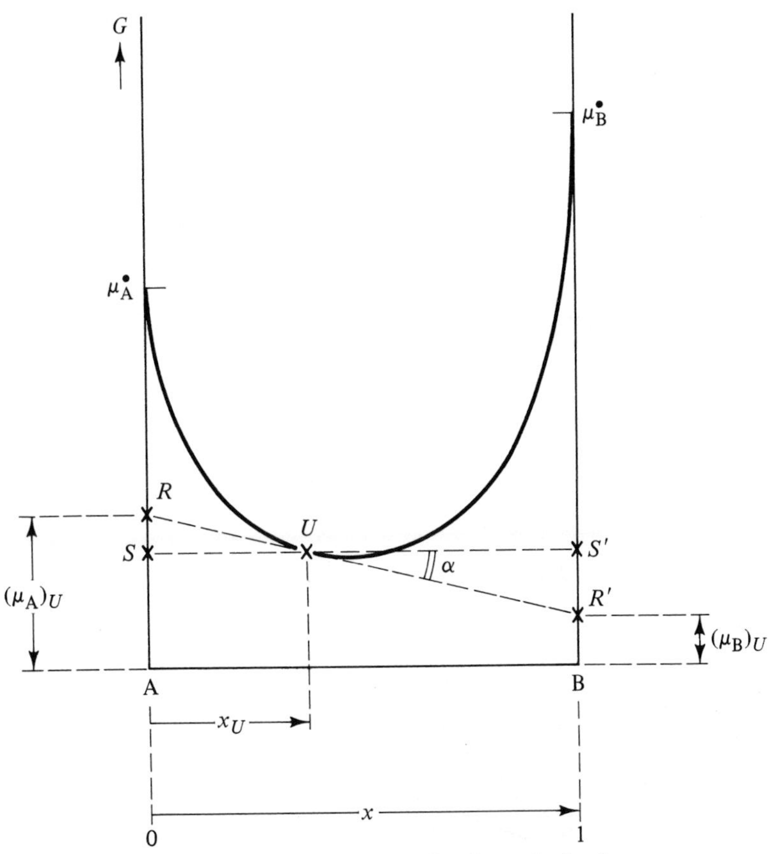

Fig. 34 G,x diagram at constant P and T for a single phase.

Obviously, $\lim_{x\to 0} \mu_A = \mu_A^\bullet$, and $\lim_{x\to 1} \mu_B = \mu_B^\bullet$. The shape of the G,x curve also suggests that $\lim_{x\to 0} \mu_B = \lim_{x\to 1} \mu_A = -\infty$ so that, by (22-3),

$$\lim_{x\to 0} (\partial G/\partial x)_{P,T} = -\infty \quad \text{and} \quad \lim_{x\to 1} (\partial G/\partial x)_{P,T} = +\infty.$$

Analytically, this limiting behavior of μ_A and μ_B can be deduced from their expressions in an infinitely dilute solution, to be given in Chapter 26. A related discussion for the A,x curve [Fig. 44(b)] will be given in section 27.1.

from the lower right corner of the diagram to the point R', the intersection of this tangent line with the B axis.

(b) When two phases, **1** and **2**, coexist we must have the thermodynamic equilibrium conditions

$$\mu_A^1 = \mu_A^2 \quad \text{and} \quad \mu_B^1 = \mu_B^2. \tag{22-5}$$

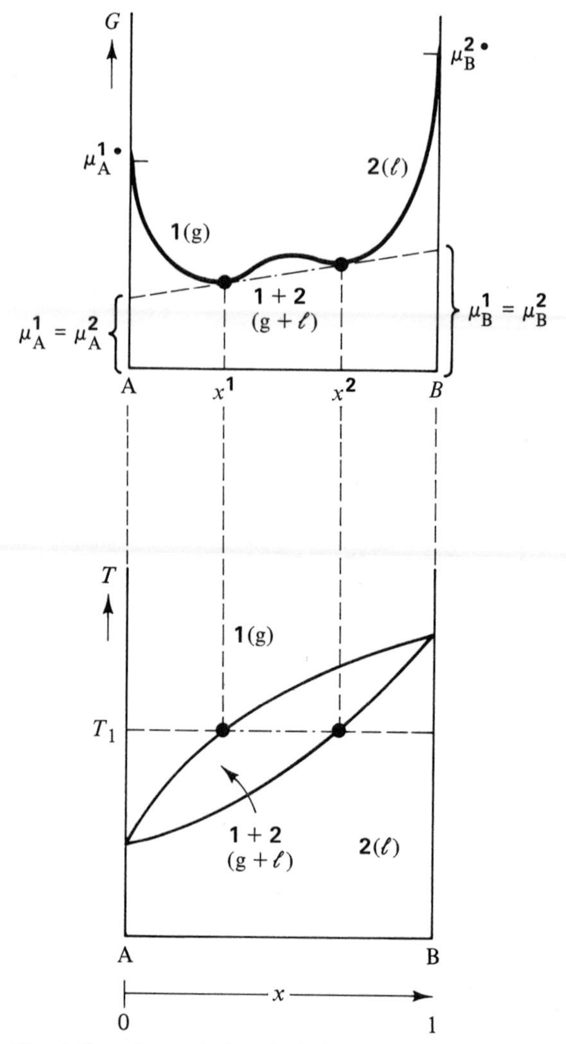

Fig. 35 Coexisting phases, **1**(g) and **2**(ℓ) in (top) a G,x diagram at a particular P_1,T_1, and (bottom) the related T,x diagram at P_1. (See footnote below.)

In general, the coexisting phases will have *different* compositions; Fig. 35 is drawn for the special case **1** ≡ g and **2** ≡ ℓ.* The upper part gives the G,x curve at some chosen P_1, T_1, the lower part the corresponding more familiar

* We have assumed a metastable loop between the stable G^ℓ and G^g parts, as assumed in a van der Waals type model. Whether this loop is real or not need not concern us in this context. Certainly, for other coexisting phases (besides ℓ and g), such a metastable continuation is not present in general; the G^1 and G^2 curves proceed "inward" independently.

binary boiling diagram at P_1. The line of stable coexistence, the dashed-dotted line in Fig. 35, a simple *tie line* in the T,x diagram, is obtained in the G,x diagram by a *double tangent construction*, reminiscent of a similar construction for a two-phase equilibrium in a single-component system. The reader is referred to section 18.4(d) and Fig. 25, in which we used the A,V plane instead, as suggested by van der Waals.

22.3 The van der Waals Equation for Two-Phase Blades

(a) We first remind the readers that in a *one-component system* every two-phase equilibrium is represented by a *single line* in the P,T plane (Fig. 20), satisfying the Clapeyron equation

$$\frac{dP}{dT} = \frac{\Delta S}{\Delta V}. \qquad \text{(18-11a)}$$

The derivation, and applications, of this famous relation have been discussed extensively in sections 18.1 and 18.2, respectively.

In a *binary system* every two-phase equilibrium is always associated with *a pair of surfaces* ("*blades*") in the P,T,x space. For example, the coexisting phases ℓ and g give rise to an upper pressure ℓ-blade and a lower pressure g-blade. In Fig. 33 they both stretch from the gℓ portion on the "quadruple point line" QgℓQ′, all the way to the critical line $\mathrm{Cr_A}$–$\mathrm{Cr_B}$, and are bordered in the $x = 0$ and $x = 1$ planes by the boiling lines of A and B, respectively. As a second example, for the coexisting phases s_A (pure crystalline A) and ℓ, there is a curved ℓ-surface, which extends from the eutectic line $E\ell$ to the melting line of A, and a flat s_A surface, which is simply the appropriate part of the P,T plane at $x = 0$. In general, for the coexisting phases **1** and **2**, we must obtain *two different* equations, in lieu of a *single* Clapeyron equation, one relating dP, dT, and $dx^{\mathbf{1}}$, and the other one relating dP, dT, and $dx^{\mathbf{2}}$. Following van der Waals, we shall derive a single expression, which is symmetrical in **1** and **2**. This will give us either blade for any desired two-phase equilibrium, by making the proper identifications.

(b) From the first of Eqs. (22-5), for a small displacement

$$d\mu_A^{\mathbf{1}} = d\mu_A^{\mathbf{2}}. \qquad \text{(22-6)}$$

By Eq. (22-4a),

$$d\mu_A^\alpha = dG^\alpha - \left(\frac{\partial G^\alpha}{\partial x^\alpha}\right)_{P,T} dx^\alpha - x^\alpha d\left(\frac{\partial G^\alpha}{\partial x^\alpha}\right)_{P,T}. \qquad (\alpha = \mathbf{1}, \mathbf{2}) \qquad \text{(22-7)}$$

Since G is a function of P, T, and x^α,

$$dG^\alpha = \left(\frac{\partial G^\alpha}{\partial T}\right)_{P,x^\alpha} dT + \left(\frac{\partial G^\alpha}{\partial P}\right)_{T,x^\alpha} dP + \left(\frac{\partial G^\alpha}{\partial x^\alpha}\right)_{P,T} dx^\alpha$$

or

$$dG^\alpha = -S^\alpha\,dT + V^\alpha\,dP + \left(\frac{\partial G^\alpha}{\partial x^\alpha}\right)_{P,T} dx^\alpha \qquad (\alpha = \mathbf{1}, \mathbf{2}). \tag{22-8}$$

But by Eqs. (22-3) and (22-5),

$$\left(\frac{\partial G^{\mathbf{1}}}{\partial x^{\mathbf{1}}}\right)_{P,T} = \left(\frac{\partial G^{\mathbf{2}}}{\partial x^{\mathbf{2}}}\right)_{P,T}, \tag{22-9}$$

reflecting the double tangent construction discussed in the preceding section (see Fig. 35, top). Substitute Eqs. (22-8) and (22-9) in Eq. (22-7) and equate the results for $\alpha = \mathbf{1}$ and $\mathbf{2}$, according to Eq. (22-6):

$$(V^{\mathbf{2}} - V^{\mathbf{1}})\,dP = (S^{\mathbf{2}} - S^{\mathbf{1}})\,dT + (x^{\mathbf{2}} - x^{\mathbf{1}})\,d\left(\frac{\partial G^{\mathbf{1}}}{\partial x^{\mathbf{1}}}\right)_{P,T}. \tag{22-10}$$

Obviously, the last term makes this expression different from the Clapeyron equation. We can expand this term as follows:

$$d\left(\frac{\partial G^{\mathbf{1}}}{\partial x^{\mathbf{1}}}\right) = \frac{\partial^2 G^{\mathbf{1}}}{\partial x^{\mathbf{1}}\,\partial T}\,dT + \frac{\partial^2 G^{\mathbf{1}}}{\partial x^{\mathbf{1}}\,\partial P}\,dP + \left(\frac{\partial^2 G^{\mathbf{1}}}{\partial (x^{\mathbf{1}})^2}\right)_{P,T} dx^{\mathbf{1}}. \tag{22-11}$$

This allows us to rewrite Eq. (22-10) in the convenient form:

$$V^{\mathbf{21}}\,dP = S^{\mathbf{21}}\,dT + (x^{\mathbf{2}} - x^{\mathbf{1}})\left(\frac{\partial^2 G^{\mathbf{1}}}{\partial (x^{\mathbf{1}})^2}\right)_{P,T} dx^{\mathbf{1}}, \tag{22-12}$$

in which

$$V^{\mathbf{21}} \equiv (V^{\mathbf{2}} - V^{\mathbf{1}}) - (x^{\mathbf{2}} - x^{\mathbf{1}})\frac{\partial^2 G^{\mathbf{1}}}{\partial x^{\mathbf{2}}\,\partial P} = (V^{\mathbf{2}} - V^{\mathbf{1}}) - (x^{\mathbf{2}} - x^{\mathbf{1}})\left(\frac{\partial V^{\mathbf{1}}}{\partial x^{\mathbf{1}}}\right)_{P,T} \tag{22-13a}$$

and

$$S^{\mathbf{21}} \equiv (S^{\mathbf{2}} - S^{\mathbf{1}}) + (x^{\mathbf{2}} - x^{\mathbf{1}})\frac{\partial^2 G^{\mathbf{1}}}{\partial x^{\mathbf{1}}\,\partial T} = (S^{\mathbf{2}} - S^{\mathbf{1}}) - (x^{\mathbf{2}} - x^{\mathbf{1}})\left(\frac{\partial S^{\mathbf{1}}}{\partial x^{\mathbf{1}}}\right)_{P,T} \tag{22-13b}$$

Equation (22-12) represents the desired result, first obtained by van der Waals.

To give an example, in studying an ℓ,g equilibrium the identification $\mathbf{1}$ = g and $\mathbf{2} = \ell$ gives us the equation for the gas blade, while $\mathbf{1} = \ell$ and $\mathbf{2}$ = g yields the equation for the liquid blade. Since the relations we have obtained are of quite general validity, they can be used to draw *some* general conclusions. However, most detailed applications require a knowledge the two $G(P,T,x^\alpha)$. This, in turn, usually involves the adaptation of a suitable model.

(c) Frequently, in practice, we consider a two-phase equilibrium in a binary system at constant P or at constant T. This means that we focus our attention on a T,x diagram (such as the one drawn in the bottom part of Fig. 35), or a P,x diagram (several of which are drawn in Fig. 33), respectively. In these diagrams two curves* separate the areas of homogeneous **1** and homogeneous **2** from the area of coexisting phases, **1** + **2**. The properties of these curves are readily obtained from the van der Waals equation (22-12). First, *at constant temperature,*

$$\left(\frac{\partial P}{\partial x^1}\right)_T = \frac{(x^2 - x^1)}{V^{21}}\left(\frac{\partial^2 G^1}{\partial (x^1)^2}\right)_{P,T} \tag{22-14a}$$

and

$$\left(\frac{\partial P}{\partial x^2}\right)_T = \frac{(x^1 - x^2)}{V^{12}}\left(\frac{\partial^2 G^2}{\partial (x^2)^2}\right)_{P,T}. \tag{22-14b}$$

If we divide the first of these equations by the second, we obtain

$$\left(\frac{\partial x^2}{\partial x^1}\right)_T = -\frac{V^{12}}{V^{21}}\frac{(\partial^2 G^1/\partial (x^1)^2)_{P,T}}{(\partial^2 G^2/\partial (x^2)^2)_{P,T}}. \tag{22-15}$$

Similarly, at constant pressure,

$$\left(\frac{\partial T}{\partial x^1}\right)_P = -\frac{x^2 - x^1}{S^{21}}\left(\frac{\partial^2 G^1}{\partial (x^1)^2}\right)_{P,T}, \tag{22-16a}$$

$$\left(\frac{\partial T}{\partial x^2}\right)_P = -\frac{x^1 - x^2}{S^{12}}\left(\frac{\partial^2 G^2}{\partial (x^2)^2}\right)_{P,T}, \tag{22-16b}$$

and

$$\left(\frac{\partial x^2}{\partial x^1}\right)_P = -\frac{S^{12}}{S^{21}}\frac{(\partial^2 G^1/\partial (x^1)^2)_{P,T}}{(\partial^2 G^2/\partial (x^2)^2)_{P,T}}. \tag{22-17}$$

(d) An example will illustrate the use of these equations. (For additional applications, see section 22.4.) Figure 36 presents a simple melting diagram at constant pressure, as we already observed in connection with Fig. 33.

It is "simple" in the sense that all the assumptions outlined in section 22.1(a) still hold. In particular, there are no compounds or solid solutions (mixed crystals); only the pure components A and B can crystallize out. The homogeneous liquid area occupies the upper part of the diagram. (At higher temperature boiling *may* occur, depending on the pressure chosen. This is disregarded in Fig. 36, since it is of no interest within the present context.) The readers should readily recognize three two-phase areas, $s_A + s_B$, $s_A + \ell$, and $\ell + s_B$. In each of these we have drawn a single tie line (dashed-dotted line) to indicate the two coexisting phases (heavy dots) at a particular temperature. Any point on such a tie line represents a heterogeneous

* One continuous loop if we go beyond the critical point of one of the components (see also Fig. 47b in Chapter 27).

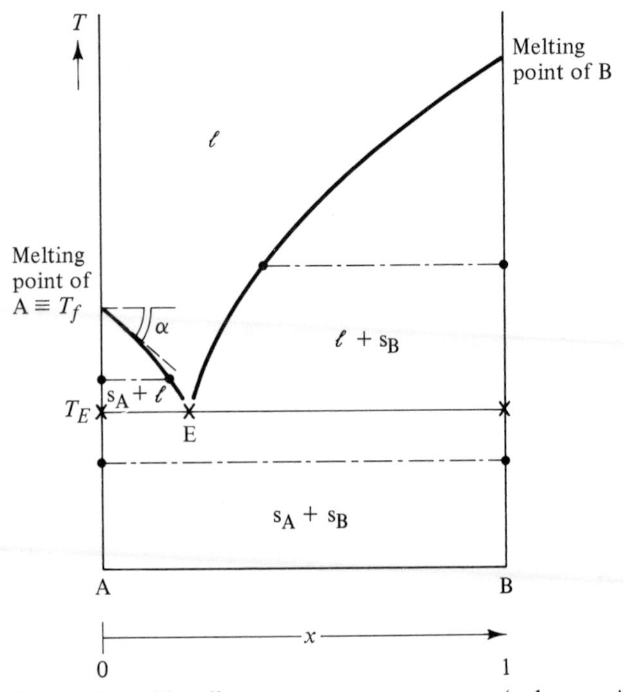

Fig. 36 Simple melting diagram, at constant pressure (only pure A and pure B can crystallize out).

mixture of these two phases, in the proportions given by the lever rule of section 18.4(c). At T_E we have the possibility of a three-phase equilibrium, s_A-eutectic melt ℓ_E – s_B, as indicated by three crosses.

Exercise 22-1

Apply the Phase Rule to obtain the variance in the areas identified in the small print above. In particular, check whether we have any freedom in selecting T_E and the composition of the eutectic melt.

In Fig. 36 we have indicated the melting point (freezing point) of pure A by T_f. If we consider A as a solvent and B as a single solute, the curve T_f–E represents the *freezing-point lowering* of the solvent as more and more solute is dissolved. The usual interest focuses on the limiting case of this lowering as $x^\ell \to 0$ (*ideally—or infinitely—dilute solution*). The relevant *conventional* thermodynamic analysis will be given in section 26.3(c). In the present context we wish to point out that the essential result is already embodied in the van der Waals equation (22-12), applied at constant pressure. From Eq. (22-16a), with $\mathbf{1} \equiv \ell$ and $\mathbf{2} \equiv s_A \equiv s$, we have immediately ($x^s = 0$!)

$$\left(\frac{\partial T}{\partial x^\ell}\right)_{P,T\to T_f} = \frac{x^\ell}{S^{s\ell}}\left(\frac{\partial^2 G^\ell}{\partial (x^\ell)^2}\right)_{P,T\to T_f}, \tag{22-18}$$

in which $S^{s\ell}$ is given by Eq. (22-13b):

$$S^{s\ell} = S^{\bullet s} - S^{\ell} - x^{\ell}\left(\frac{\partial^2 G^{\ell}}{\partial x^{\ell}\,\partial T}\right)_{T\to T_f}. \tag{22-19}$$

To proceed we must anticipate a single result for the model at hand (see Chapter 26):

$$G^{\ell} = f(P, T) + RT[(1 - x^{\ell}) \ln (1 - x^{\ell}) + x^{\ell} \ln x^{\ell}]. \tag{22-20}$$

The readers should verify that this functional form of G^{ℓ} leads to

$$(S^{s\ell})_{T\to T_f} = S^{\bullet s} - S^{\bullet\ell} + [\text{terms of order } (x^{\ell})^2];$$

hence for small x^{ℓ},

$$(S^{s\ell})_{T\to T_f} = \frac{-\Delta H^{\bullet}_{\text{fus}}}{T_f} \tag{22-21a}$$

and

$$\left(\frac{\partial^2 G^{\ell}}{\partial (x^{\ell})^2}\right)_{P,T\to T_f} = \frac{RT_f}{x^{\ell}}. \tag{22-21b}$$

In Eq. (22-21a), $\Delta H^{\bullet}_{\text{fus}}$ is the heat of melting 1 mole of pure A at P, T_f. Upon substitution of Eqs. (22-21) in Eq. (22-18), we obtain the desired relation,

$$\left(\frac{\partial T}{\partial x^{\ell}}\right)_{P,T\to T_f} = -\frac{RT_f^2}{\Delta H^{\bullet}_{\text{fus}}}. \tag{22-22}$$

Exercise 22-2

Accepting Eq. (22-20), go through the details of the derivation of Eq. (22-22) from Eq. (22-18).

There are many alternative ways to write this final result, first obtained by van't Hoff late in the nineteenth century. We shall give some of the alternative forms and discuss their implications after we have given the more conventional derivation in section 26.3(a).

22.4 The Gibbs–Konovalow Theorems and Related Topics; the Duhem–Margules Equation(s)

(a) The Gibbs–Konovalow theorems are an immediate consequence of the general validity of Eqs. (22-14) and (22-16), coupled with the fact that for all phases α in stable equilibrium, $(\partial^2 G^{\alpha}/\partial(x^{\alpha})^2)_{P,T}$ has to be positive [see Table 15.1 in section 15.3(c) and Fig. 34 in the second section of the present chapter]. It should be evident that this necessitates a one-to-one correspondence between the vanishing of $(x^1 - x^2)$ and an extremum property of $P(x^{\alpha})$

at constant T, as well as of $T(x^\alpha)$ at constant P, for $\alpha = \mathbf{1}$ *and* $\mathbf{2}$. This is expressed by the two Gibbs–Konovalow theorems, which can be combined conveniently to read:

If, for a binary system, in a series of two-phase equilibria at constant $\begin{array}{l}\textit{temperature}\\ \textit{pressure}\end{array}$*, the composition of the two phases becomes the same, the* $\begin{array}{l}\textit{pressure}\\ \textit{temperature}\end{array}$ *must pass through an extremum; conversely, at a point where the* $\begin{array}{l}\textit{pressure}\\ \textit{temperature}\end{array}$ *goes through an extremum, the two coexisting phases must have the same composition.*

In the rest of this subsection we shall investigate the presence of maxima and minima,* an example of which has already been encountered in our discussion of the system ethanol–water in section 21.2(f). We refer in particular to the azeotropic mixture, A, in Fig. 32. First we want to obtain the *sign* of $(\partial^2 P/\partial(x^{\mathbf{1}})^2)_T$ and of $(\partial^2 P/\partial(x^{\mathbf{2}})^2)_T$. Keeping in mind our special interest for the case $x^{\mathbf{1}} = x^{\mathbf{2}}$, we obtain, by differentiation of Eq. (22-14a) with respect to $x^{\mathbf{1}}$ at constant T,

$$\left(\frac{\partial^2 P}{\partial(x^{\mathbf{1}})^2}\right)_T = (x^{\mathbf{2}} - x^{\mathbf{1}})\begin{pmatrix}\text{two terms not to}\\ \text{be written out}\end{pmatrix} + \left[\left(\frac{\partial x^{\mathbf{2}}}{\partial x^{\mathbf{1}}}\right)_T - 1\right]\frac{(\partial^2 G^{\mathbf{1}}/\partial(x^{\mathbf{1}})^2)_{P,T}}{V^{21}}$$

For $x^{\mathbf{1}} = x^{\mathbf{2}}$ this simplifies to

$$\left(\frac{\partial^2 P}{\partial(x^{\mathbf{1}})^2}\right)_T = \left[\left(\frac{\partial x^{\mathbf{2}}}{\partial x^{\mathbf{1}}}\right)_T - 1\right]\frac{(\partial^2 G^{\mathbf{1}}/\partial(x^{\mathbf{1}})^2)_{P,T}}{V^{\mathbf{2}} - V^{\mathbf{1}}}. \tag{22-23}$$

At $x^{\mathbf{1}} = x^{\mathbf{2}}$, $(\partial x^{\mathbf{2}}/\partial x^{\mathbf{1}})_T$ becomes, by Eqs. (22-15) and (22-13a),

$$\left[\left(\frac{\partial x^{\mathbf{2}}}{\partial x^{\mathbf{1}}}\right)_T\right]_{x^{\mathbf{1}}=x^{\mathbf{2}}} = \frac{(\partial^2 G^{\mathbf{1}}/\partial(x^{\mathbf{1}})^2)_{P,T}}{(\partial^2 G^{\mathbf{2}}/\partial(x^{\mathbf{2}})^2)_{P,T}}, \tag{22-24}$$

so that Eq. (22-23) becomes

$$\left(\frac{\partial^2 P}{\partial(x^{\mathbf{1}})^2}\right)_T = \underbrace{\frac{(\partial^2 G^{\mathbf{1}}/\partial(x^{\mathbf{1}})^2)_{P,T}}{(\partial^2 G^{\mathbf{2}}/\partial(x^{\mathbf{2}})^2)_{P,T}}}_{\text{(I)}}\underbrace{\frac{1}{V^{\mathbf{2}} - V^{\mathbf{1}}}}_{\text{(II)}}\underbrace{\left[\left(\frac{\partial^2 G^{\mathbf{1}}}{\partial(x^{\mathbf{1}})^2}\right)_{P,T} - \left(\frac{\partial^2 G^{\mathbf{2}}}{\partial(x^{\mathbf{2}})^2}\right)_{P,T}\right]}_{\text{(III)}}. \tag{22-25}$$

---

* The occurrence of points of inflection cannot be excluded. The associated phase equilibria are of a more complex nature and will not be discussed in this book.

As mentioned at the beginning of this section, term (I) is always positive. Offhand, we cannot say anything about term (II), although its sign is usually evident once **1** and **2** have been identified with specific phases. The sign of (III) is not readily accessible. Thus at this stage it is hard to say whether Eq. (22-25) represents a maximum or a minimum. Nevertheless, we can draw some interesting conclusions. We first note that if we interchange the phases, **1** and **2**, terms (II) and (III) *both* change sign. Hence $(\partial^2 P/\partial(x^{\mathbf{2}})^2)_T$ *has the same sign as* $(\partial^2 P/\partial(x^{\mathbf{1}})^2)_T$. Moreover, in entirely similar fashion as we obtained Eq. (22-25), we can derive

$$\left(\frac{\partial^2 T}{\partial(x^{\mathbf{1}})^2}\right)_P = \underbrace{\frac{(\partial^2 G^{\mathbf{1}}/\partial(x^{\mathbf{1}})^2)_{P,T}}{(\partial^2 G^{\mathbf{2}}/\partial(x^{\mathbf{2}})^2)_{P,T}}}_{\text{(I)}} \underbrace{\frac{1}{S^{\mathbf{2}} - S^{\mathbf{1}}}}_{\text{(IV)}} \underbrace{\left[\left(\frac{\partial^2 G^{\mathbf{2}}}{\partial(x^{\mathbf{2}})^2}\right)_{P,T} - \left(\frac{\partial^2 G^{\mathbf{1}}}{\partial(x^{\mathbf{1}})^2}\right)_{P,T}\right]}_{-\text{(III)}}, \tag{22-26}$$

and verify that $(\partial^2 T/\partial(x^{\mathbf{2}})^2)_P$ *has the same sign as* $(\partial^2 T/\partial(x^{\mathbf{1}})^2)_P$.

Exercise 22-3

Derive Eq. (22-26).

Let us now return to the example of an *azeotropic mixture* (Fig. 32), by making the identification $\mathbf{1} \equiv \ell$ and $\mathbf{2} \equiv$ g. In that case not only term (I), but also (II) and (IV), are positive (unless we have a simultaneous occurrence of a *critical* phenomenon), so that the expression (III) completely determines the sign of $(\partial^2 P/\partial(x^\alpha)^2)_T$ and $(\partial^2 T/\partial(x^\alpha)^2)_P$ for $\alpha = \ell$ *and* g. A comparison of the expressions (22-25) and (22-26) clearly shows that *these two signs are opposite.* Hence *if we have a double minimum in the T,x plane at constant P* (as is indeed the case for the system ethanol–water, illustrated in Fig. 32), *the corresponding P,x diagram at constant T must exhibit a double maximum at the same composition.* Conversely, a *maximum*-boiling azeotrope (as is present, for example, in the system acetone–chloroform, CH_3COCH_3–$CHCl_3$) is associated with a *minimum* vapor pressure.* Note that if we vary P, T, and x in such a way that our *mixture remains azeotropic*, we move along a *maximum or minimum vapor-pressure curve* associated with the liquid, gas blades in P,T,x space. Along this curve, the fundamental van der Waals equation (22-12) reduces to

$$\frac{dP}{dT} = \frac{S^g - S^\ell}{V^g - V^\ell}, \tag{22-27}$$

---

* The question as to *why* certain binary systems have a maximum or a minimum in their vapor pressures, hence show azeotropic behavior, can be answered only on a corpuscular level. We return to this point briefly in section 25.1.

formally identical with the ordinary Clapeyron equation [compare Eq. (18-11a)]. In such contexts an azeotrope is sometimes called a *pseudo-one-component* system. Equation (22-27) gives the slope along the projection of the maximum (or minimum) vapor pressure curve onto the P,T plane.

(b) In section 8.2 we introduced the *Gibbs–Duhem equation* for a single phase and pointed out that there are actually several different relations in the literature which are designated as such. An extension to two-phase equilibria, known as the *Duhem–Margules equation* suffers from similar ambiguities. In some books it is associated with a relation of very general validity, and elsewhere it refers to an equation that can be obtained only if we introduce approximations, such as the adoption of specific models.

From the outset we shall restrict our discussion to liquid–gas equilibria. As before, the two components are called A and B, and x is a shorthand for X_B (so that $1 - x = X_A$). For 1 mole of the binary liquid, the (generalized) Gibbs–Duhem equation, in the form Eq. (8-29b), gives us

$$(1 - x^\ell)(d\mu_A^\ell + \bar{S}_A^\ell\, dT) + x^\ell(d\mu_B^\ell + \bar{S}_B^\ell\, dT) - V^\ell\, dP = 0, \qquad (22\text{-}28)$$

in which V^ℓ is the *mean* molar volume of the liquid. The conditions for heterogeneous equilibrium are

$$\mu_i^\ell = \mu_i^g \qquad (i = \text{A, B}), \qquad (21\text{-}3)$$

or, by Eq. (19-54),

$$\mu_i^\ell = \mu_i^{g\ominus}(T) + RT \ln |f_i^g| \qquad (i = \text{A, B}), \qquad (22\text{-}29)$$

in which we use the shorthand

$$|f_i^g| \equiv \frac{f_i^g \text{ (in atm)}}{1 \text{ atm}}.$$

Upon substitution of Eq. (22-29) into Eq. (22-28), after dividing through by RT, we obtain

$$(1 - x^\ell)d \ln |f_A^g| + x^\ell\, d \ln |f_B^g| + \frac{1}{RT}\{(1 - x^\ell)[d\mu_A^{g\ominus}(T) + \bar{S}_A^\ell\, dT]$$

$$+ x^\ell[d\mu_B^{g\ominus}(T) + \bar{S}_B^\ell\, dT]\} - \frac{V^\ell}{RT}\, dP = 0. \qquad (22\text{-}30)$$

The term in braces can be simplified considerably. Since $\mu_i^{g\ominus}$ is a function of the temperature only,

$$d\mu_i^{g\ominus} = \frac{d\mu_i^{g\ominus}}{dT}\, dT = -\bar{S}_i^{g\ominus}\, dT. \qquad (22\text{-}31)$$

Call

$$T(\bar{S}_i^{g\ominus} - \bar{S}_i^\ell) \equiv \mathbf{L}_i. \qquad (22\text{-}32)$$

This is the heat needed to evaporate 1 mole of i out of an infinite amount of the liquid into a gas phase with unit fugacity at the pressure and temperature of interest. Introduce the additional shorthand

$$\mathbf{L} \equiv (1 - x^{\ell})\mathbf{L}_{\mathrm{A}} + x^{\ell}\mathbf{L}_{\mathrm{B}}. \tag{22-33}$$

With Eqs. (22-31) through (22-33), Eq. (22-30) becomes

$$(1 - x^{\ell})d \ln |f_{\mathrm{A}}^{\mathrm{g}}| + x^{\ell}\, d \ln |f_{\mathrm{B}}^{\mathrm{g}}| - \frac{\mathbf{L}}{RT^2}\, dT - \frac{V^{\ell}}{RT}\, dP = 0. \tag{22-34}$$

This is what *some* authors call the Duhem–Margules equation.* It is representative of the typical equations in phenomenological thermodynamics in that it is of very general validity, but relatively useless because it contains several quantities that are not readily accessible.

Of course, we can reduce Eq. (22-34) to isobaric or isothermal form. Frequently, the latter receives the most attention. At constant T, we obtain

$$(1 - x^{\ell})d \ln |f_{\mathrm{A}}^{\mathrm{g}}| + x^{\ell}\, d \ln |f_{\mathrm{B}}^{\mathrm{g}}| - \frac{V^{\ell}}{RT}\, dP = 0. \tag{22-35}$$

Since V^{ℓ} is usually known, Eq. (22-35) gives us a relation between $f_{\mathrm{A}}^{\mathrm{g}}$ and $f_{\mathrm{B}}^{\mathrm{g}}$. Strictly speaking, we cannot put *both* dP *and* dT in Eq. (22-34) equal to zero. For, by the Phase Rule (compare Example A in section 21.1), $\mathbf{v} = 2$. Hence at fixed P and T, the system becomes nonvariant; x^{ℓ} and x^{g}, as well as $f_{\mathrm{A}}^{\mathrm{g}}$ and $f_{\mathrm{B}}^{\mathrm{g}}$, are entirely determined. Nevertheless, one frequently finds the relation

$$(1 - x^{\ell})d \ln |f_{\mathrm{A}}^{\mathrm{g}}| + x^{\ell}\, d \ln |f_{\mathrm{B}}^{\mathrm{g}}| = 0. \tag{22-36}$$

This Duhem–Margules equation should be looked upon as an *approximate* form of Eq. (22-35), justifiable only if, as is often the case, $PV^{\ell} \ll RT$. A final approximation stipulates that the equilibrium vapor is an ideal mixture of ideal gases, so that

$$d \ln |f_i^{\mathrm{g}}| = d \ln |p_i| = \left(\frac{\partial \ln |p_i|}{\partial x^{\ell}}\right)_T dx^{\ell} \qquad (i = \mathrm{A, B}).$$

In this case, Eq. (22-36) becomes

$$(1 - x^{\ell})\left(\frac{\partial \ln |p_{\mathrm{A}}|}{\partial x^{\ell}}\right)_T + x^{\ell}\left(\frac{\partial \ln |p_{\mathrm{B}}|}{\partial x^{\ell}}\right)_T = 0. \tag{22-37}$$

Several authors refer only to *this* relation as the Duhem–Margules equation. We have certainly come a long way from the exact Eq. (22-34)! Note that throughout the entire analysis, we have retained the composition *of the*

---

* This is certainly incorrect historically, since Duhem and Margules published the relevant papers before the turn of the century, when fugacities were not yet introduced.

liquid as an independent variable. Equation (22-37) shows that, at given temperature, the shape of the $p_A(x^\ell)$ curve completely determines that of the $p_B(x^\ell)$ curve. Of course, the approximation $f_i^g \approx p_i$ can also be introduced into Eqs. (22-34) and (22-35), yielding yet two more forms of occasional utility.

Exercise 22-4

Remembering that $p_i \equiv X_i^g P$ $(i = \text{A, B})$, use Eq. (22-37) to confirm that for an azeotropic mixture,

$$\left(\frac{\partial P}{\partial x^\ell}\right)_T = \left(\frac{\partial P}{\partial x^g}\right)_T = 0,$$

a result already obtained in subsection (a).

REFERENCES

Owing to the relative neglect this subject has suffered in recent decades, the references to this chapter are couched in a somewhat more elaborate descriptive framework than is our custom.

By now, the readers should be convinced that a study of the thermodynamics of heterogeneous equilibria has much more meaning if one has some familiarity with phase diagrams. Unfortunately, the drawing and reading of these diagrams has become somewhat of a lost art, and a stepchild in the teaching of physical chemistry. It may be true that such diagrams do not "explain" anything, but they are as essential to many in chemical industry, metallurgy, glass and porcelain technology, and geology, as maps are to geographers. The best teaching aids concerned are, in the author's opinion, not just "classics" but "collector's items," books that often are only accessible through the older and better scientific libraries. In this category one must certainly place the sequence of monographs, *Die heterogenen Gleichgewichte vom Standpunkte der Phasenregel*, started by H. W. Bakhuis-Roozeboom around 1900 and "finished" (?) after his death by his pupils Büchner, Aten, and Schreinemakers. The most famous summary in English is given by Alexander Findley in *The Phase Rule and Its Applications*, first published in 1903 and revised and brought up-to-date in many subsequent editions. The 9th edition was reissued by Dover, in 1951, as revised and enlarged by A. N. Campbell and N. O. Smith. A more thorough alternative is J. Zernike's *Chemical Phase Theory*, already referred to several times in this book (see, e.g., reference 1 of Chapter 21). The "classics" also include Gustav Tammann's *Lehrbuch der heterogenen Gleichgewichte* (1924) and W. Eitel's *Physikalische Chemie der Silikate* (1941), of particular interest to metallurgists and to scientists interested in glass or ceramics, respectively.

Turning our attention specifically to the *thermodynamics* of this vast field, *some* of which is included in all of the above references as well, we must first mention J. W. Gibb's *Collected Works* and P. Duhem's "Traité Élémentaire de Mécanique Chimique" (1899). For much of the material in the present chapter, the necessary background is provided by J. D. van der Waals and Ph. Kohnstamm's *Lehrbuch der Thermodynamik* (1908). The most comprehensive discussion in more recent publications can be found in the textbook by Prigogine and Defay (1954) frequently quoted in this book (see, e.g., reference 3 of Chapter 8). Excellent chapters can also be found in, among others, the books *Chemical Thermodynamics* by J. G. Kirkwood and I. Oppenheim (1961) and *Chemical Thermodynamics* by M. L. McGlashan (Academic Press, London, 1979). The latter may represent the only one of these texts which is still in print at the time *this* book goes to press.

The author has observed a certain renaissance of this subject in the current research literature. For example, there are several recent papers by John G. Wheeler *et al.*, some of which have been mentioned already in earlier chapters [see, in particular, sections 18.3(b) and 21.3(e)].

Part IV

Model Systems

("Special" Thermodynamics)

Chapter 23

Ideal Mixtures of Ideal Gases

23.1 Introduction: the Single-Component Ideal Gas Revisited

(a) The single-component ideal gas was introduced in Chapter 2, and discussed further in a variety of contexts throughout the first three parts of this book [see, e.g., (parts of) sections 3.3, 10.3, 12.2, 14.4 and 19.3]. If we *define* this model (phenomenologically) by its equation of state,

$$P\mathbf{V} = nRT, \tag{23-1}$$

we can *derive* all the important thermodynamic functions, by using only the First and Second Laws, remembering that **E**, **H**, **S**, **A**, and **G** are *extensive* properties. The results can be summarized as follows:

$$\mathbf{E} = nE(T), \tag{23-2}$$

$$\mathbf{H} = nH(T), \tag{23-3}$$

$$\mathbf{S} = n\left[S^*(T) + R\ln\frac{\mathbf{V}}{n}\right], \tag{23-4}$$

$$\mathbf{A} = n\left[A^*(T) - RT\ln\frac{\mathbf{V}}{n}\right], \tag{23-5}$$

and

$$\mu = \frac{\mathbf{G}}{n} = \frac{\mathbf{A} + P\mathbf{V}}{n} = \mu^{\ominus}(T) + RT\ln|P|. \tag{23-6}$$

All logarithms refer to dimensionless quantities; thus $S^*(T)$ must contain a term $-R\ln(\mathbf{V}^*/n)$ (compare section 14.4), $\ln|P|$ stands for $\ln(P \text{ in atm}/1 \text{ atm})$, and so on.

Exercise 23-1

Derive Eqs. (23-2) through (23-6) from Eq. (23-1).

The explicit form of the various temperature functions (usually determined up to an arbitrary constant only) need not interest us, but for future reference, we list the following important relations between them:

$$H = E + RT, \tag{23-7}$$

$$T\frac{dS^*}{dT} = \frac{dE}{dT}, \tag{23-8}$$

$$A^* = E - TS^*, \tag{23-9}$$

and

$$\mu^\ominus = H - TS^\dagger, \tag{23-10}$$

in which

$$S^\dagger(T) \equiv S^*(T) + R \ln RT. \tag{23-11}$$

Exercise 23-2

Verify Eqs. (23-7) through (23-10).

(b) The present chapter is concerned with an *ideal mixture of ideal gases.* This model has already been used in several places throughout the third part of this book, notably in section 19.2. On a corpuscular level it consists of mass points which do not interact with each other; all inter- as well as intraspecies forces are neglected. In this context the author recommends a rereading of section 19.2(a) and reminds readers that we do not use the simple designation *ideal gas mixture* to avoid confusion with other models (see Chapter 24). All that we needed in section 19.2 was to anticipate the famous equation for the chemical potential:

$$\mu_k = \mu_k^\ominus(T) + RT \ln \frac{p_k}{p_k^\ominus}, \tag{23-12}$$

in which the partial pressure of k in this mixture, p_k, is *defined* by*

$$p_k \equiv X_k P. \tag{23-13}$$

In many introductory textbooks, Eq. (23-12) is simply given, or at best is justified by analogy with the expression (23-6) for a *single-component* ideal gas. A few authors postulate Eq. (23-12) as the (phenomenological) *definition* of an ideal mixture of ideal gases. Readers who diligently derived Eqs. (23-2) through (23-6) from the equation of state (23-1) (see Exercise 23-1) might expect that the corresponding familiar equation of state,

$$P = \frac{RT}{\mathbf{V}} \sum_{k=1}^{r} n_k, \tag{23-14}$$

* For a critical discussion of the partial pressure concept, see section 23.4.

is an equally powerful definition for the ideal mixture of ideal gases and that *it alone* should suffice for an analogous derivation of the various thermodynamic functions, including μ_k as given by Eq. (23-12). Unfortunately, this is not the case: Eq. (23-14) has to be *supplemented* by the requirement

$$\mathbf{E} = \sum_{k=1}^{r} n_k E_k^{\bullet}(T), \tag{23-15}$$

in which $E_k^{\bullet}(T)$ is the energy for 1 mole of pure k at the same T. The two-pronged definition (23-14) and (23-15) indeed suffices to derive all other properties of this model, but this procedure is still quite complicated (see section 23.5).

The very specific additivity relationship (23-15) *cannot* be derived from the equation of state (23-14) and the Second Law. All that we can obtain in this fashion is that $\mathbf{E} = nE(T)$, where n is the *total* number of moles in our mixture and $E(T)$ its *mean* molar internal energy.

In the next section we shall introduce, and analyze, a much more powerful "single-statement definition," which, in a sense, goes back to Gibbs.*

23.2 The Preferred Phenomenological Definition and Its Implications

(a) We define an ideal mixture of ideal gases by the statement

$$\mathbf{A}(\mathbf{V}, T) = \sum_{k=1}^{r} n_k \left[A_k^*(T) - RT \ln \frac{\mathbf{V}}{n_k} \right], \tag{23-16}$$

in which A_k^* has the same value as for 1 mole of pure k, taken to be ideal, at this temperature [cf. Eq. (23-5)]. We shall first show how this definition allows us to derive all relevant thermodynamic properties in quite straightforward fashion, and postpone an analysis of its "meaning" until subsection (c) below.

1. From

$$P = -\left(\frac{\partial \mathbf{A}}{\partial \mathbf{V}}\right)_{T,\ \text{all}\ n_k} \tag{23-17}$$

we immediately obtain the equation of state (23-14).

---

* It can also be found in Prigogine and Defay, reference 3 of Chapter 8, p. 124, and it appears as the direct result of a statistical treatment in E. A. Guggenheim's *Thermodynamics*, 3rd ed., North-Holland, Amsterdam, 1957, p. 222.

2. From

$$\mathbf{S} = -\left(\frac{\partial \mathbf{A}}{\partial T}\right)_{\mathbf{V},\,\text{all } n_k} \tag{23-18}$$

Eq. (23-16) gives us initially

$$\mathbf{S} = \sum_{k=1}^{r} n_k\left(-\frac{dA_k^*}{dT} + R\ln\frac{\mathbf{V}}{n_k}\right). \tag{23-19}$$

But by Eqs. (23-8) and (23-9),

$$\frac{dA_k^*}{dT} = -S_k^*,$$

so that Eq. (23-19) becomes

$$\mathbf{S} = \sum_{k=1}^{r} n_k\left(S_k^* + R\ln\frac{\mathbf{V}}{n_k}\right). \tag{23-20}$$

Comparison with Eq. (23-4) gives the very important result

$$\mathbf{S}(\mathbf{V}, T) = \sum_{k=1}^{r} n_k S_k^{\bullet}(\mathbf{V}, T). \tag{23-21}$$

Since $n_k S_k^{\bullet}$ is the entropy of n_k moles of pure k (considered ideal) at the same $\mathbf{V}$ and T, Eq. (23-21) expresses the *additivity of* $\mathbf{S}(\mathbf{V}, T)$.

3. From $\mathbf{E} = \mathbf{A} + T\mathbf{S}$, by Eqs. (23-16) and (23-19),

$$\mathbf{E} = \sum_{k=1}^{r} n_k\left(A_k^* - T\frac{dA_k^*}{dT}\right) \tag{23-22}$$

$$= \sum_{k=1}^{r} n_k(E_k^{\bullet} - TS_k^* + TS_k^*).$$

Hence

$$\mathbf{E}(T) = \sum_{k=1}^{r} n_k E_k^{\bullet}(T). \tag{23-23}$$

Similarly,

$$\mathbf{H}(T) = \sum_{k=1}^{r} n_k H_k^{\bullet}(T), \tag{23-24}$$

as the reader may verify. These two equations demonstrate the *additivity of* $\mathbf{E}(T)$ *and* $\mathbf{H}(T)$. Apparently, $\bar{E}_k = E_k^{\bullet}(T)$ and $\bar{H}_k = H_k^{\bullet}(T)$ for this model.

4. From

$$\mu_k = \left(\frac{\partial \mathbf{A}}{\partial n_k}\right)_{\mathbf{V}, T, n_{j \neq k}}, \tag{23-25}$$

Eq. (23-16) gives us initially

$$\mu_k = A_k^* - RT \ln \frac{\mathbf{V}}{n_k} + RT. \tag{23-26}$$

By means of Eqs. (23-9) through (23-11) and (23-14), this is readily transformed into the familiar

$$\mu_k = \mu_k^{\ominus}(T) + RT \ln X_k|P|. \tag{23-27}$$

In Eq. (23-27),

$$\mu_k^{\ominus}(T) = A_k^* + RT - RT \ln RT, \tag{23-28}$$

and Eq. (23-13) may be used to introduce the partial pressure p_k (see also section 23.4).

(b) It is left to the reader to obtain initially

$$\mathbf{S}(P, T) = \sum_{k=1}^{r} n_k(S_k^* + R \ln RT - R \ln |P| - R \ln X_k) \tag{23-29}$$

and convert this to

$$\mathbf{S}(P, T) = \sum_{k=1}^{r} n_k S_k^{\bullet}(P, T) - R \sum_{k=1}^{r} n_k \ln X_k. \tag{23-30}$$

This shows that $\mathbf{S}(P, T)$ *is not additive*; the last term gives us the "entropy of mixing" and it (or its equivalent, $-R \sum_{k=1}^{r} X_k \ln X_k$, for *1 mole* of mixture) is also known as the "Gibbs paradox" term. We shall comment briefly on this designation when we encounter a similar expression for an ideal solution (see section 25.3).

Exercise 23-3

Derive Eqs. (23-29) and (23-30).

In the experience of the author, many students are puzzled by the fact that $\mathbf{S}(\mathbf{V}, T)$ *is additive whereas* $\mathbf{S}(P, T)$ *is not*. This mystery should disappear as soon as we realize that *the mixing processes involved are quite different*. Let us illustrate this for the case that $\frac{1}{2}$ mole of A and $\frac{1}{2}$ mole of B are combined to yield 1 mole of mixture (Fig. 37). The entropy increase upon mixing *at constant P* is just given by the Gibbs paradox term:

$$-R(X_A \ln X_A + X_B \ln X_B) = -R(\tfrac{1}{2} \ln \tfrac{1}{2} + \tfrac{1}{2} \ln \tfrac{1}{2}) = +R \ln 2.$$

The mixing *at constant* $\mathbf{V}$ can be considered to occur *in two stages* ($\mathbf{S}$ is a state function!); *first mix at constant pressure* and *then compress from* $2\mathbf{V}$ *back to* $\mathbf{V}$.

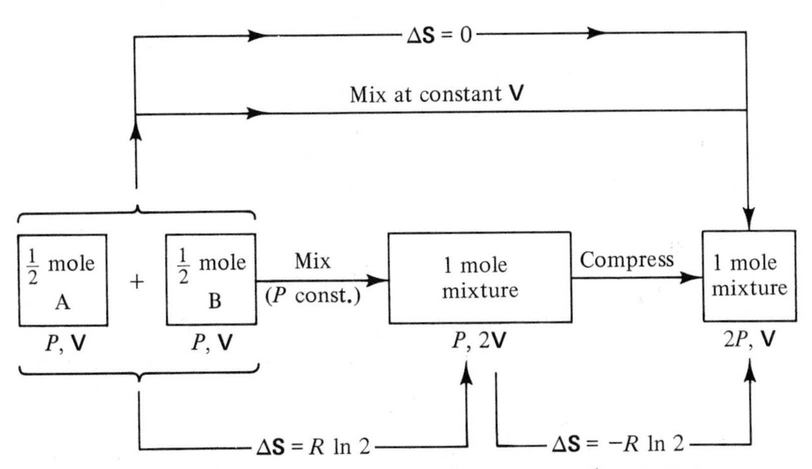

Fig. 37 Entropy change upon mixing $\frac{1}{2}$ mole A and $\frac{1}{2}$ mole B (at constant T) to give 1 mole mixture.

The entropy change in the last process is

$$R \ln \frac{\mathbf{V}_{\text{final}}}{\mathbf{V}_{\text{initial}}} = R \ln \frac{\mathbf{V}}{2\mathbf{V}} = R \ln \tfrac{1}{2} = -R \ln 2,$$

which just cancels out the entropy of mixing at constant pressure.

(c) We now take a closer look at the "meaning" of the definition (23-16), which, at first glance, appears to be so "horribly abstract." It should be clear that there are two important aspects:

1. $\mathbf{A}(\mathbf{V}, T)$ is additive in the components.
2. Each n_k moles of k, if by itself at the same $\mathbf{V}$ and T of the mixture, exhibits ideal behavior.

On a corpuscular level, the second aspect implies* that there are no *intraspecies* forces. As to the first aspect, phenomenologically it tells us that upon mixing these species, the Helmholtz free energy is unchanged. According to the discussion in section 15.2(b), this means, in turn, that the total work done in the reversible mixing of these components is zero. As one immediately realizes, this strongly suggests that there are no *interspecies* forces either. We return to this point in the next section. In the meantime we conclude that our preferred phenomenological definition (23-16) is not so mysterious

---

* In addition to all gas molecules being mass points. *Intraspecies* forces refer to (intermolecular) interactions between molecules of the same species, k. *Interspecies* forces refer to such interactions between different species $k, l, \ldots$.

after all; its two parts relate to the absence of inter- and intraspecies forces, respectively.

23.3 The Reversible Mixing of Gases with Particular Reference to Some Models of Interest

(a) The easiest way to visualize the mixing of two (or more) gases is to put them in adjacent compartments and open up a stopcock between these [Fig. 38(a)]. Unfortunately, this is a "very" *irreversible* process. The conceptualization of *reversible* mixing already worried some of the most famous scientists of the nineteenth (to early twentieth) century. Two entirely different types of thought experiments have been suggested; one method utilizes external forces, most frequently the force of gravity, and the other employs suitable semipermeable membranes. The first of these methods goes back,

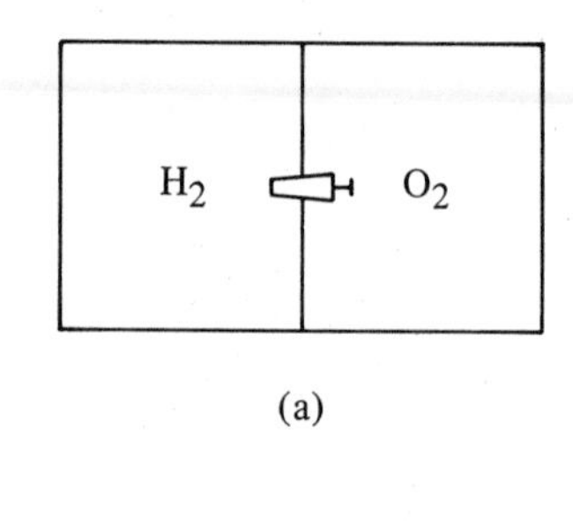

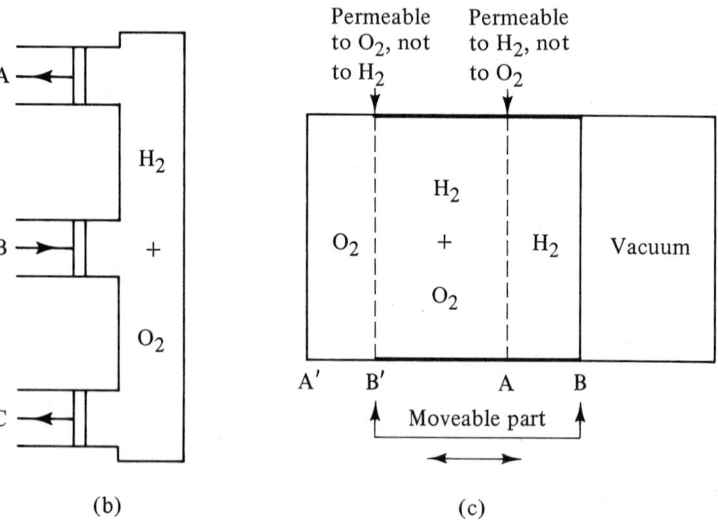

Fig. 38 Mixing of H_2 and O_2. (a) Irreversible; (b) reversible, utilizing the force of gravity; (c) reversible, utilizing semipermeable membranes. (See text.)

at least, to 1875, when it was described by Lord Rayleigh* (Fig. 38b). It has also been advocated by Lorentz.† One considers a long vertical column (Fig. 38b) with a mixture of H_2 and O_2, say. *If there are no interspecies forces*, then at equilibrium each gas is distributed as if it were by itself. As a result, the top of the column is richer in H_2, the bottom richer in O_2.‡ By means of small horizontal side tubes [Fig. 38(b) is not drawn to scale], equipped with slow-moving pistons, one can reversibly *remove* a small amount of *these* two mixtures and *add* some gas *in the middle* so that the overall composition in the main column (as given by n_{H_2} and n_{O_2}) is unchanged. In a qualitative sense it is readily seen that the net effect is the reversible demixing of H_2 and O_2, but the detailed justification of this method and the accompanying mathematical analysis is quite complex and will not be reproduced here. The most general expression, obtained by Lord Rayleigh, gives the work for the isothermal mixing of n_1 moles of 1 at $\mathbf{V}_1$ with n_2 moles of 2 at $\mathbf{V}_2$ into a final volume $\mathbf{V}$. It takes the form

$$W = RT\left(n_1 \ln \frac{\mathbf{V}}{\mathbf{V}_1} + n_2 \ln \frac{\mathbf{V}}{\mathbf{V}_2}\right). \tag{23-31}$$

For the special case (of particular importance to us in the present context) $\mathbf{V}_1 = \mathbf{V}_2 = \mathbf{V}$, we obviously obtain

$$W = 0. \tag{23-32}$$

(b) The alternative method is often attributed to Planck, who gave a very elaborate discussion in his textbook,§ but it goes back at least to L. Boltzmann, who already described it in 1878.‖ It has survived the course of time better than the Rayleigh method, and can still be found in some current textbooks.¶ It is quite simple; no complex mathematical analysis is required. In Fig. 38(c) the isothermal mixing of H_2 and O_2 is again taken as an example. In a hollow cylinder, the partition A consists of a *fixed* semi-permeable surface ("membrane"), permeable to H_2 but impermeable to O_2. The two partitions B and B', the former totally impermeable, the latter per-meable to O_2 only, can be moved *jointly* in such a way that the distance BB' equals AA' at all times. *If there are no interspecies forces*, a *slow* motion of this type leaves a uniform distribution of O_2 molecules between A and

* *Phil. Mag.*, XLIX, 311 (1875).

† H. A. Lorentz, *Thermodynamica*, E. J. Brill, Leiden, 1929, Chapter V.

‡ See also Problem 23-3.

§ Max Planck, *Thermodynamics*, A. Ogg, trans., Dover, New York, 1926, p. 219. The first German edition appeared in 1897.

‖ L. Boltzmann, *Wiener Sitzungsberichte*, 1878, II, p. 754.

¶ For example, J. Waser, *Basic Chemical Thermodynamics*, W. A. Benjamin, New York, 1966, section 8.5. See also H. A. Buchdahl, reference 4 of Chapter 1, p. 170.

A' and of H_2 molecules between B and B'. Thus no work is associated with this process, and the result (23-32) is recovered.

Exercise 23-4

If you first obtain Eq. (23-32) by the second method, how would you subsequently derive Eq. (23-31)?

(c) Two final comments are in order. In the first place, *some* of the semipermeable membranes, required in the method discussed in subsection (b), may actually not exist. For example, while Sir William Ramsay already knew in 1894 that at 300°C Pd foil is permeable to H_2 but not to O_2, the author still does not know of a membrane permeable to O_2 but not to H_2. This has created a controversy. On the one hand, some scientists could not care less; material scientists may not be clever enough as yet, but this should not affect thermodynamics! On the other hand, this point of view has strong opponents, who wish to emphasize the experimental basis of phenomenological thermodynamics at all times; *thought* experiments should not be considered if the corresponding *real* experiments cannot be carried out! Either point of view has its merits, and preferences are based primarily on one's convictions regarding the "nature" of thermodynamics.

Second, it is important to note that *both* methods lead to the result (23-32) in the absence of *interspecies* forces. Nothing has to be stipulated about *intraspecies* forces, although they too would be absent if we are dealing with an ideal mixture of ideal gases.

23.4 A Further Digression: Dalton's Law and the Partial Pressure Concept

(a) A critical discussion of these topics logically follows the preceding section(s). In many treatments the validity of Dalton's Law is associated exclusively with the ideal mixtures of ideal gases. This was not Dalton's intention at all; his own words follow:*

> *The elastic or repulsive power of each particle is confined to those of its own kind; and consequently the force of such a fluid, retained in a given vessel, or gravitating, is the same in a separate as in a mixed state, depending upon its proper density and temperature.*

---

* John Dalton, *Mem. Manchester Lit. Phil. Soc.* (2), **1**, 244–258 (1805). Also published as Alembic Club Reprint 2, by Oliver & Boyd, Edinburgh, 1923.

In modern language the stipulation is clearly that there shall be *intraspecies* forces only! In 1876, Gibbs paraphrased this very simply:*

> *It is in this sense that we should understand the law of Dalton, that every gas is as a vacuum to every other gas . . . .*

This obviously pertains to a model, which also played a central part in the preceding section in connection with the derivation of Eq. (23-32) for reversible mixing. Henceforth we shall refer to it as a *Dalton gas mixture*; the ideal mixture of ideal gases evidently constitutes a special case.

(b) Gibbs may not have been the first one to put Dalton's law in the familiar form

$$P = \sum_{k=1}^{r} p_k, \tag{23-33}$$

but he certainly *was* the first author who clearly realized that the *partial pressures*, p_k, *have to be defined*, and that this concept is by no means trivial. A century later we can point to at least five different definitions in the literature. Of these, the following three are found most prominently:

(I) $\quad p_k \equiv X_k P \qquad (k = 1, \ldots, r).$ (23-34)

This definition is preferred by Guggenheim, Prigogine and Defay, Lewis and Randall, and many, many others. Paul, who gives an exceptionally comprehensive discussion of these issues,† suggests that we call (I) the *formal partial pressure*. This definition is *operational* in any sense of the word but except for certain model systems, it represents not much more than a convenient *shorthand*. Parenthetically, we note that if we accept the p_k in this fashion, Eq. (23-33) becomes a mathematical identity without true scientific content. Consequently, one would have to rephrase Dalton's Law, which is no trivial matter, unless we are willing to confine its applicability to ideal mixtures of ideal gases. Thus Guggenheim calls

$$p_k = n_k \frac{RT}{\mathbf{V}} \qquad (k = 1, \ldots, r) \tag{23-35}$$

"Dalton's Law of partial pressures." This certainly is a distortion of history.

(II) $\quad$ *p_k is the pressure the same amount of k would exert if it alone would occupy the same* $\mathbf{V}$ *and T.*

---

* See the Scientific Papers of J. Willard Gibbs, *Thermodynamics*, Dover, New York, 1961, Vol. I, p. 157.

† Martin A. Paul, *Principles of Chemical Thermodynamics*, McGraw-Hill, New York, 1951, Chapter 7.

We find this definition in Moore's *Physical Chemistry.** The corresponding p_k have been called *Dalton partial pressures* or *Dalton–Gibbs partial pressures,* since they give the Gibbs equation (23-33) a scientific content, in accordance with Dalton's intentions, as discussed in subsection (a). This definition, too, is obviously operational.

(III) *p_k is the pressure exerted by pure k in equilibrium with the gas mixture through a heat-conducting membrane, permeable to k alone.*

This definition is adopted by Kirkwood and Oppenheim in their advanced text, which we have mentioned several times in this book. It is operational only in Bridgman's "paper and pencil" sense, *unless* we can manufacture all the necessary membranes *and* establish all equilibria involved experimentally. Sir William Ramsay succeeded in measuring the partial pressure of hydrogen along these lines [see also section 23.3(c)], but in general this type of experiment is rare because:

1. The relevant membranes are not available [see section 23.3(c)].
2. Equilibrium is established very slowly.
3. They are confined to a low-pressure range (less than 2 atm in Ramsay's case), which is not very interesting in this context, since in that case we approach the ideal mixture of ideal gases, and much easier measurements would give essentially the same numerical results (see below).

(c) There is no reason whatever to assume (although actual experimental data are scarce) that in real gas mixtures, in particular at high pressures, the three p_k, (I), (II), and (III), are the same. As mentioned above, definition (I) may always be adopted as a mere convenient *shorthand*, but *as soon as interspecies forces become operative, the entire concept of a partial pressure associated with a particular constituent loses its physical meaning.* In such situations its use should be severely limited.

If there are no *inter*species forces (regardless of whether *intra*species interactions are present or not), i.e., for a "Dalton gas mixture," definition (II) makes the most sense to us. The validity of Dalton's Law (23-33) now appears as an immediate consequence of the additivity of the Helmholtz free energy, $\mathbf{A}(\mathbf{V}, T)$. For if

$$\mathrm{A}(\mathbf{V}, T) = \sum_{k=1}^{r} n_k A_k^{\bullet}, \tag{23-36}$$

then since

$$P = -\left(\frac{\partial \mathbf{A}}{\partial \mathbf{V}}\right)_{T,\ \mathrm{all}\ n_k}. \tag{23-37}$$

* 4th ed., Prentice-Hall, Englewood Cliffs, N.J., 1972, p. 28.

we have

$$P = \sum_{k=1}^{r} -\left(\frac{\partial n_k A_k^\bullet}{\partial \mathbf{V}}\right)_{T,\,\text{all } n_k}. \tag{23-38}$$

But if we adopt definition (II), the sum on the right-hand side of Eq. (23-38) is just $\sum_{k=1}^{r} p_k$.

(d) Equation (23-36) is just the first part of our definition of an ideal mixture of ideal gases [see section 23.2(c)], but for *that* model there are no *intraspecies* forces *either. Only for an ideal mixture of ideal gases do measurements based on the three definitions for* p_k *(and others not mentioned here) all necessarily lead to the same numerical values* [see also remark 3 in subsection (b)]. This contention is readily proven. For example, let us accept definition (II). Then, by the *second* part of the definition of our model (components also exhibit ideal gas behavior),

$$p_k = n_k \frac{RT}{\mathbf{V}}. \tag{23-39}$$

Combining Eq. (23-39) with Eq. (23-14) immediately yields

$$p_k = \frac{n_k}{\sum_{k=1}^{r} n_k} P = X_k P,$$

i.e., Eq. (23-34). Moreover, this result, combined with Eq. (23-27), gives

$$\mu_k = \mu_k^\ominus(T) + RT \ln |p_k|.$$

For k in the mixture to be in equilibrium with pure k, across a semi-permeable membrane (permeable to k only), the latter must have $\mu = \mu_k$; hence, by Eq. (23-6), $P = p_k$.

Exercise 23-5

In similar fashion, show the equivalence of the three definitions for p_k in the case of an ideal mixture of ideal gases

(a) starting with definition (I);
(b) starting with definition (III).

23.5 *Alternative Phenomenological Definitions of Ideal Mixtures of Ideal Gases*

The last subsection has brought us back to the main subject of this chapter, the ideal mixture of ideal gases. A number of digressions should have enhanced our "understanding" of the preferred phenomenological definition (23-16). We are now in a better situation to discuss possible alternatives, already referred to in section 23.1(b). In particular we wish to reconsider the "popular" procedure, used by Planck, Lorentz, Uhlenbeck, and others, which starts with the two-pronged definition:

$$P = \frac{RT}{\mathbf{V}} \sum_{k=1}^{r} n_k \tag{23-14}$$

and

$$\mathbf{E} = \sum_{k=1}^{r} n_k E_k^{\bullet}(T). \tag{23-15}$$

The next step is to apply the First Law,

$$\Delta\mathbf{E} = Q - W,$$

to the *reversible mixing* experiment *at constant* $\mathbf{V}$ *and* T. By Eq. (23-15), $\Delta\mathbf{E} = 0$, and we concluded, as a result of certain thought experiments in section 23.3, that $W = 0$. Therefore, $Q = 0$ or, since the process we are considering is *reversible*, $\Delta\mathbf{S} = 0$. In other words, by means of a lengthly detour, we have recovered

$$\mathbf{S}(\mathbf{V}, T) = \sum_{k=1}^{r} n_k S_k^{\bullet}(\mathbf{V}, T). \tag{23-21}$$

Without this last equation, Eqs. (23-14) and (23-15) are powerless; *with* it we readily recover the remaining equations of relevance. For since we now have P, $\mathbf{V}$, $\mathbf{E}$, and $\mathbf{S}$, we also have $\mathbf{A}$ and $\mathbf{G}$, hence also μ. Readers are suggested to complete the process themselves.

Exercise 23-6

Using the defining equations (23-14) and (23-15) and the consequence (23-21), derive all other relevant properties of the ideal mixture of ideal gases, as summarized in section 23.2.

The author much prefers the procedure outlined throughout the earlier sections of this chapter. Readers can make their own choice.

23.6 Appendix: Ideal Gases at Absolute Zero

S. Goudsmit has been credited with the formulation of a list of questions that one can raise after any presentation on any topic in physics. The author has never seen such a list, but one of these questions is alleged to be: How do things work out at absolute zero? To give an answer within the context of the present chapter is no simple matter. This we readily see by considering the entropy of mixing at constant P and T. As we derived in section 23.2, per mole of gas mixture formed, it is given by

$$\Delta S_{\text{mix}} = -R \sum_{k=1}^{r} X_k \ln X_k,$$

independent of the nature of the components and of the values of P and T. However, according to the *Nernst Heat Theorem* of section 16.2, this entropy of mixing should vanish as $T \to 0$. This apparent paradox can be resolved only if we realize that a number of familiar phenomenological statements for the single-component ideal gas, as well as for the ideal mixture of ideal gases, are no longer valid at such low temperatures. On a corpuscular level, we can still define these gases as consisting of noninteracting mass points, which can be treated individually either by means of classical mechanics or by means of the quantum mechanical *particle in a box* model. But near absolute zero, the macroscopic system can no longer be treated by the use of Boltzmann statistics. It becomes a *quantum fluid* which has to satisfy Fermi–Dirac or Bose–Einstein statistics instead (which one of these depends on the nature of the particles concerned). Consequently, the equation of state is no longer given by Eq. (23-1) or (23-14), and $\mathbf{E}$ is no longer a function of T only (independent of $\mathbf{V}$). There are several other differences, and the result is that much of the preceding discussion in this chapter would have to be modified rather drastically. Obviously, a detailed understanding of these matters would require some knowledge of both quantum mechanics and quantum statistics. Fortunately, such details are of no great concern in the application of thermodynamics to most chemical processes, but they do play a role in the understanding of the basis for the Third Law. In the language of section 16.3, the entropy of mixing of ideal gases vanishes at absolute zero because both the initial and final state of mixing are nondegenerate, i.e., have $\Omega = 1$.

Perhaps the following comments will shed some light on this difficult topic, from a slightly different angle. With reference to section 23.2, we remind the readers that *at constant volume*, ΔS_{mix} is *always* zero, even at high temperatures. If we consider the mixing *at constant pressure* in two stages (see Fig. 37), namely, the mixing at constant volume followed by an expansion to restore the desired pressure, we clearly see that the entire entropy change is due to the second process. The unique feature at absolute zero is that the entropy change accompanying the expansion is zero as well. The vital difference concerned is already apparent if we just consider the reversible isothermal expansion of one mole of a *single-component* ideal gas from V to $2V$. At high temperatures $\Delta S = R \ln 2$, $\Delta E = 0$, and $W = RT \ln 2$. At absolute zero $\Delta S = 0$, ΔE is negative (the increase in volume lowers the energy levels of the "particle in a box") and $W = -\Delta E$.

*Exercise 23-7**

Take 1 mole of a single-component ideal gas through the following reversible cycle:

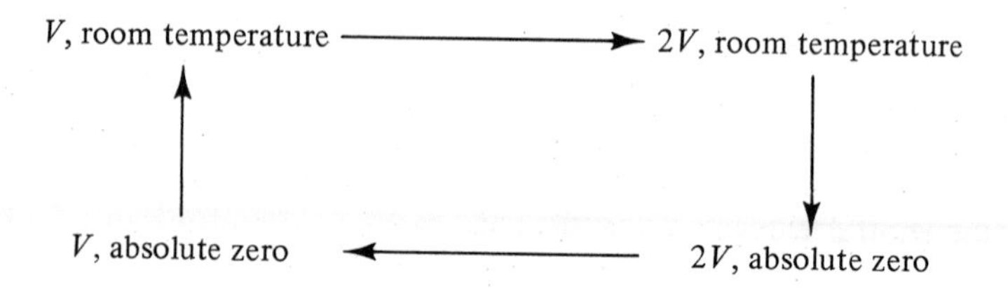

(a) From the fact that ΔS for the cycle must be zero, what can you conclude about the values of C_v at V relative to those at $2V$ near absolute zero? Compare the situation at room temperature, as governed by the equations given in section 14.1(c).

(b) (Optional for those readers who are familiar with the quantum mechanical treatment of the particle in a box—*not* an exercise in phenomenological thermodynamics.) Can you understand this difference in C_v values?

PROBLEMS

Problem 23-1

Compute ΔS_{mix} when X_A moles of the ideal gas A (at pressure P and temperature T) are mixed with X_B moles of the ideal gas B (at the same P and T) to yield 1 mole of an ideal mixture of ideal gases (at this pressure, P, and temperature, T), with X_A ranging from 0 to 1 at intervals of 0.1. Plot a graph of ΔS_{mix} versus X_A.

Problem 23-2

A mixing experiment is carried out by means of an adiabatically enclosed apparatus of the general type illustrated in Fig. 38a. One compartment has a volume of 1 ℓ and contains helium gas at 500 K, 1 atm pressure. The other compartment has a volume of 1.5 ℓ and is filled with neon gas at 400 K, 0.5 atm pressure. The tap between the two compartments is opened and the gases are allowed to mix. After equilibrium is reached:

(a) What is the final P and T?

(b) What is the entropy change accompanying the process?

[Consider helium and neon ideal gases ($C_v = \frac{3}{2}R$) and assume that the resulting system is an ideal mixture of ideal gases.]

---

* The author is indebted to his colleague S. J. Strickler for suggesting this exercise.

Problem 23-3

In Problem 17-1 we introduced the "barometric formula" for the variation of the pressure of a single-component ideal gas with altitude.

Assume that the air at sea level, with a barometric pressure of 1 atm, is an ideal mixture of 78 mole % of ideal nitrogen, 21 mole % of ideal oxygen, and 1 mole % of ideal argon. Adapt the barometric formula to compute the partial pressure, as well as the mole percent, of each of these three constituents at altitudes of 1 km and 5 km above sea level, assuming a uniform temperature of 20°C throughout. Compare the (total) barometric pressure at 1 km above sea level with that computed in Problem 17-1(c).

Chapter 24

Ideal Mixing in the Gas Phase

24.1 The Phenomenological Definition of Ideal Mixing and Its Corpuscular Implications; Amagat's Law

(a) In 1923, G. N. Lewis and M. Randall *defined* an ideal gas mixture by requiring that at *total* pressure P, the fugacities are

$$f_k = X_k f_k^\bullet(P, T). \tag{24-1a}$$

The reader is reminded that the closed-dot superscript refers to the pure substance k. Using Eq. (19-54), one readily obtains the alternative in terms of chemical potentials:*

$$\mu_k = \mu_k^\bullet(P, T) + RT \ln X_k. \tag{24-1b}$$

It should be understood that for the mixture to be ideal, *these equations must hold for all components* k $(= 1, \ldots, r)$ *at all compositions*, i.e., *for all possible values of a set of* $(r - 1)$ *independent* X_k.

Within a qualitative or at best semiquantitative framework, it is customarily stated that Eqs. (24-1) stipulate the *absence of preferential forces*; that is, all intra- as well as interspecies forces should be the same. Of course, this is trivially true for the ideal mixture of ideal gases, discussed in Chapter 23, since all such forces are simply *zero*. The usefulness of the present model lies in its (approximate) applicability to certain mixtures at such high pressures that these forces are certainly no longer negligible. In fact, the model extends to the liquid state; the corresponding *ideal solutions* will be discussed in Chapter 25, to which we also refer for some additional properties of ideal mixtures (in the Lewis–Randall sense) in general.

The precise corpuscular basis of the model is still somewhat controversial. To pursue this matter it is useful, as well as instructive, to establish first the

* In several textbooks one follows the original methodology of Lewis and Randall by carrying out most mathematical manipulations in terms of fugacities. In the author's opinion, working with chemical potentials is often easier and more "direct," at least for theoretical discussions. In practice, calculations may focus on fugacities and fugacity coefficients (see section 24.2).

link between its phenomenological definition and Amagat's Law (also known as Leduc's Law).

(b) While Dalton's Law (of partial pressures) is widely known, if sometimes misunderstood (see section 23.4), Amagat's Law (of additive volumes) is less frequently mentioned.* This is most regrettable, since in many instances the latter is much more useful than the former. In 1880, E. H. Amagat carried out experiments at pressures up to 3000 atm (which had to be considered truly remarkable at that time) and for several mixtures established the approximate validity of

$$\mathbf{V} = \sum_{k=1}^{r} n_k (V_k^\bullet)_P, \tag{24-2}$$

in which $(V_k^\bullet)_P$ is the volume of 1 mole of pure k at the *total* pressure, P, of the mixture under consideration. (A fixed temperature, T, is implied throughout.) Since, by Eq. (8-12), we *always* have

$$\mathbf{V} = \sum_{k=1}^{r} n_k \bar{V}_k, \tag{24-3}$$

Amagat's Law will be valid if

$$\bar{V}_k = (V_k^\bullet)_P \qquad \text{for all } k = 1, \ldots, r. \tag{24-4}$$

It is readily seen that Eq. (24-4) is an immediate consequence of the Lewis–Randall definition of ideal mixing. All we have to do is differentiate Eq. (24-1b) with respect to P at constant T and composition, to yield

$$\left(\frac{\partial \mu_k}{\partial P}\right)_{T,\,\text{composition}} = \left(\frac{\partial(\mu_k^\bullet)P}{\partial P}\right)_T. \tag{24-5}$$

By Eq. (19-57) the left-hand side just equals $\bar{V}_k$, and the right-hand side represents $(V_k^\bullet)_P$, so that we indeed recover Eq. (24-4).

(c) The equivalence, just demonstrated, is noteworthy because we can derive Amagat's Law from a *truncated virial expansion*:

$$\mathbf{V} = \frac{nRT}{P} + \frac{1}{n} \sum_{j,k=1}^{r} n_j n_k B_{jk}(T), \tag{24-6}$$

in which, as usual,

$$n \equiv \sum_{k=1}^{r} n_k. \tag{24-7}$$

If the right-hand side of Eq. (24-6) contained only the *first* term, we would recover Eq. (23-14), the equation of state for an ideal mixture of ideal gases. The *second* term takes care of (inter- as well as intraspecies) pair interactions;

---

* A fairly elaborate discussion is given by Paul; see footnote, p. 297.

$B_{jk}(T)$ is the *second virial coefficient*, arising from the interaction between j and k particles. The neglect of third- and higher-order contributions in the expansion implies that P is low enough for the effect of three- and many-body encounters to be negligible.

By definition, for any specific component i $(= 1, \ldots, r)$,

$$\bar{V}_i \equiv \left(\frac{\partial \mathbf{V}}{\partial n_i}\right)_{P,T,n_{j\neq i}}. \tag{24-8}$$

Substituting $\mathbf{V}$ from Eq. (24-6) yields, after considerable but straightforward algebra,

$$\bar{V}_i = \frac{RT}{P} + \sum_{j,\,k=1}^{r} X_j X_k (2B_{ij} - B_{jk}). \tag{24-9}$$

Exercise 24-1

Derive Eq. (24-9).

For pure i ($X_i = 1$, all $X_{j\neq i} = 0$) at pressure P equal to the total pressure of the mixture,

$$(V_i^{\bullet})_P = \frac{RT}{P} + B_{ii}. \tag{24-10}$$

The only approximation made in obtaining the last two equations is the *truncation* of the virial expansion. Without further simplifications we do not yet obtain the desired result (24-4).

However, if we *assume* that at the temperature of interest,

$$B_{jk} = \tfrac{1}{2}(B_{jj} + B_{kk}) \qquad \text{for all } j, k = 1, \ldots, r, \tag{24-11}$$

the result (24-9) becomes

$$\bar{V}_i = \frac{RT}{P} + B_{ii}. \tag{24-12}$$

Exercise 24-2

Derive Eq. (24-12) from Eqs. (24-9) and (24-11).

Comparison of Eqs. (24-10) and (24-12) yields Eq. (24-4), the necessary and sufficient condition for the validity of Amagat's Law, and by the analysis in subsection (b), the Lewis–Randall ideal mixing conditions.

The concept of *no preferential forces*, in the present framework, corresponds to

$$B_{jj} = B_{kk} = B_{jk} \qquad \text{for all } j, k = 1, \ldots, r. \tag{24-13}$$

This, indeed, would justify the crucial equation (24-11), and everything would fall into place. Of course, for mixtures containing very few, preferably only two, components, the latter equation could also be true "by accident." But the unexplained fact remains that the approximate validity of Amagat's Law and the Lewis–Randall rule extends to situations in which there is no reason whatever for *either* Eq. (24-13) *or* (24-11) to be valid. For example, it is found experimentally* that for mixtures of N_2 and H_2 at 0°C, Amagat's Law and the Lewis–Randall rule hold to good approximation up to pressures of at least 1000 atm. In this case Eq. (24-13) is certainly not true, and the validity of Eq. (24-11) would be fortuitous. In addition, at these pressures, the use of a truncated two-term virial expansion, such as Eq. (24-6), has to be considered very questionable. Hence an element of mystery remains!

(d) Finally, it is sometimes useful to express the Lewis and Randall rule in terms of the fugacity *coefficients*, defined by

$$\gamma_k \equiv \frac{f_k}{X_k P}. \tag{19-55}$$

With Eq. (24-1a),

$$\gamma_k = \frac{X_k(f_k^\bullet)_P}{X_k P} = \frac{(f_k^\bullet)_P}{P}. \tag{24-14}$$

By Eq. (19-50), this simply reads

$$\gamma_k = (\gamma_k^\bullet)_P; \tag{24-15}$$

the fugacity coefficient of component k in an ideal mixture is equal to that of pure k at the *total* pressure of the gas. Thus we have access to three different ways to describe the model of interest: in terms of chemical potentials, in terms of fugacities, and in terms of fugacity coefficients. Of course, we can also work in terms of relative fugacities or *activities*, defined by the general equation

$$a_k \equiv \frac{f_k}{f_k^\ominus}. \tag{19-64}$$

From the discussion in section 19.4(a) it follows that as long as we are confining our attention to gaseous systems, all $f_k^\ominus$ are 1 atm, and the activities are just the numerical values of the fugacities, $|f_k|$, when the latter are expressed in units of atmospheres. In lieu of Eq. (19-55) we can write $\gamma_k = |f_k|/X_k|P| = a_k/X_k|P|$, whence some authors refer to these γ_k as *activity coefficients.*

---

* Martin A. Paul; see footnote, p. 297, pp. 312–313.

24.2 Applications to Chemical Reactions in Gases at High Pressures

In section 19.3(b) we showed that for a chemical reaction in a real gas mixture the correct thermodynamic equilibrium constant, K_f, a function of the temperature only, can be written as follows:

$$K_f = \text{“}\mathfrak{Q}_\gamma^{e}\text{”} \frac{\mathfrak{Q}_p^e}{(1\ \text{atm})^{\Delta\nu}}. \tag{19-61}$$

If the reaction mixture had been an ideal mixture of ideal gases, the correct equilibrium constant would have been

$$K_p = \frac{\mathfrak{Q}_p^e}{(1\ \text{atm})^{\Delta\nu}}, \tag{19-29}$$

so that "$\mathfrak{Q}_\gamma^e$", defined as $\prod_k (\gamma_k^e)^{\nu_k}$, gives us the deviation from this model. As we pointed out before, we place $\mathfrak{Q}_\gamma^e$ between quotation marks because, notwithstanding its *formal* appearance, it can hardly be considered a true reaction quotient.

The only widely publicized method to arrive at numerical values of the γ_k, in order to compute "$\mathfrak{Q}_\gamma^e$", uses the following two approximations in succession:

1. First we adopt the Lewis–Randall model of ideal mixing introduced in section 24.1, hence assume the validity of Eq. (24-15). This reduces the problem to that of obtaining the $(\gamma_k^\bullet)_P$, the fugacities of the pure reactants and products at the *total* pressure of the equilibrium mixture.
2. Next we estimate these $\gamma_k^\bullet$ from "universal" $\gamma^\bullet(P_r, T_r)$ tables or graphs, as alluded to in section 19.3(a).

As an important example of the use of this procedure, we return to the technologically important ammonia synthesis, several aspects of which have already been discussed in Chapter 19 (see also Problem 19-1). Let us consider the reaction in the form

$$\tfrac{1}{2}N_2 + \tfrac{3}{2}H_2 = NH_3.$$

This equilibrium has been subjected to extensive experimental studies,* the detailed results of which, in one form or another, are reproduced in many textbooks on physical chemistry or chemical thermodynamics.† If one mixes

---

* A. L. Larson and R. L. Dodge, *J. Am. Chem. Soc.* **45**, 2918 (1923); A. L. Larson, *ibid.*, **46**, 367 (1924); L. J. Winchester and B. F. Dodge, *Am. Inst. Chem Eng. J.*, **2**, 431 (1956).

† See, e.g., W. J. Moore, *Physical Chemistry*, 4th ed., Prentice-Hall, Inc., Englewood Cliffs, N.J., 1972, pp. 304–305; N. A. Gokcen, *Thermodynamics*, Techscience, Hawthorne, Calif., 1975, p. 287.

H_2 and N_2 in the stoichiometric ratio (3:1) at 723 K, then, as the total pressure is increased from 10 atm to 600 atm, $\mathfrak{Q}_p^e$ roughly doubles its value from 0.0066 atm^{-1} to 0.0130 atm^{-1}, reflecting the desired increase in the yield of ammonia at higher pressures. Parenthetically, these data once more remind us that the equilibrium mixture hardly behaves as an ideal mixture of ideal gases. At the same temperature, over the same range of pressures, "$\mathfrak{Q}_\gamma^e$", as obtained by the approximate procedure under discussion, is roughly cut in half; at 10 atm it is still very close to unity (i.e., we are still, effectively, dealing with an ideal mixture of ideal gases), but at 600 atm it is reduced to about 0.50. The net result is that at total pressures up to 600 atm, K_f, computed as the product of the experimental $\mathfrak{Q}_p^e$ and the approximate "$\mathfrak{Q}_\gamma^e$" (divided by 1 atm^{-1}), remains roughly constant ($\approx 0.0063 \pm 0.0002$)! This is, of course, as it *should* be for K_f to be a "good" thermodynamic equilibrium constant. It ought to be mentioned, however, that at even higher pressures (admittedly no longer of great technological importance, since the required industrial equipment becomes too expensive) the approximation method breaks down; e.g., at $P = 2000$ atm, K_f computed in this fashion reaches a value of 0.046! It is safe to assume that at such high pressures, N_2, H_2, and NH_3 no longer form ideal mixtures in the Lewis–Randall sense (compare the general critical discussion in section 24.1), although it is difficult to ascertain how much the use of "universal" fugacity charts for the pure components contributes to the error. Several attempts have been made to come up with alternative models for the determination of fugacities and equilibrium constants in real gas mixtures.* To date, such attempts have only met with limited success, and much remains to be done in this field.

---

* O. Redlich and J. N. W. Kwong, *Chem. Rev.* **44**, 223 (1949); L. J. Gillespie and J. A. Beattie, *Phys. Rev.* **36**, 743 (1930); I. Prigogine and R. Defay, p. 148 of reference 3 of Chapter 8.

Chapter 25

The Ideal Solution and Some Related Models

25.1 An Introduction to Chapters 25 and 26

In these chapters we shall deal primarily with *liquid* mixtures and solutions.* However, it is important to remember that there is a close connection between the ideal solution of the present chapter and the ideal *gas* mixture of Chapter 24. Moreover, some concepts to be introduced presently can be transferred, with few or no modifications, to the realm of *solid* solutions, which will not be discussed specifically.

We shall consider two well-known models in detail, the *ideal solution* and the *ideally* or *infinitely dilute solution* (Chapter 26). Unfortunately, there is no generally accepted terminology to distinguish between these. It is particularly confusing that some authors talk about ideal behavior when referring to the *second* model, while others call the latter simply "the" dilute solution, opening up the possibility of a mix-up between this *abstract model* and a *real system* which is dilute only in the colloquial sense. There are some lesser known models, such as Hildebrand's *regular solutions* and Guggenheim's *simple solutions*, to which we shall return in the last part of this chapter.

On a corpuscular basis, the *ideally dilute* solution arises if the forces between all *solute* particles can be neglected. Of course, the *solvent* molecules do interact, and the dissolving process normally implies preferential attractions between solvent and solute particles. It should be clear that, in principle, *any* solution can be made ideally dilute if we only make the solute concentration(s) low enough. In this respect it is *this* model (not the *ideal* solution) which resembles the ideal gas; *any* gas can be made ideal if we only make its pressure low enough.

How easily the ideally dilute solution model is approached obviously depends primarily on the nature of the solute particles. At one end of the scale, we have non-

* For the use of these terms, we refer to section 8.1.

polar molecules with low polarizability. At the other extreme, it is rarely allowable to neglect the interactions between ions, in particular if they carry a high charge.

The situation is quite different for the *ideal* (liquid) solution: The various species should be as much alike as possible, which not only correlates with the absence of *preferential* forces (compare section 24.1) but also requires all particles to be of approximately the same size and shape. Thus *whether a liquid solution is ideal or not is settled primarily by the nature of the components;* the crucial features are generally independent of the composition, hardly influenced by a change in pressure, and only to a minor extent sensitive to a change in temperature [compare section 25.4(d)]. To give a few examples, benzene and toluene form a "pretty decent" ideal solution, benzene and monodeuterobenzene an even better one; chloroform and acetone will markedly deviate [due to preferential hydrogen bonding, according to $Cl_3CH \cdots O = C(CH_3)_2$], and water–sulfuric acid mixtures are nowhere near ideal. Mixtures containing ions rarely behave as ideal solutions.

The conventional statistical mechanical theories of ideal mixing* require a "random" spatial arrangement of the various species. This would indeed be achieved if there are no preferential forces and the particles have the same size and shape. The reader may note that this second feature was not mentioned in our earlier discussion of ideal *gaseous* mixtures. It is obviously of more importance in condensed phases, but there is certainly no sharp distinction in this context. As we hinted in section 24.1(c), and as will be apparent again upon reading section 25.4, the corpuscular basis for these models is by no means crystal clear, as yet. A further discussion of these matters, which probably need some revision in the light of modern statistical mechanical theories of liquids, falls beyond the scope of this book.

A few miscellaneous remarks will bring this introductory section to a close. Some textbooks confine all discussions concerned to two-component systems. This leads to the simplest possible algebraic manipulations, since all formal summations are avoided. In the case of ideal solutions, this restriction is not unrealistic as far as the practical importance of the model is concerned. It has the added advantage that various properties can be plotted conveniently in a closed graph (of the type already encountered in Chapter 22), ranging from pure A to pure B. The algebraic extension to many-component systems is straightforward in principle, and we shall occasionally opt to give our analysis immediately within such a more general framework. In the case of ideally dilute solutions, the restriction to two components implies the presence of a single solute. This is certainly *not* a desirable situation, since it eliminates the possibility of describing chemical reactions between several solutes.

---

* See, e.g., G. S. Rushbrooke, *Introduction to Statistical Mechanics*, Oxford University Press, London, 1949, Chapter X.

To analyze the properties of ideal solutions, the mole fraction appears to be the most convenient, hence most widely used, concentration unit. As to the ideally dilute solution, we shall first have to distinguish between solvent and solutes. Following the scheme outlined in section 8.1, the former will be designated by 1, the latter by i $(= 2, \ldots, r)$. If the "concentration" of the solvent has to be specified, the mole fraction, once more, is most appropriate. For the solutes one has the choice of working in terms of mole fractions ("rational" system), molalities ("practical" system), or molarities (less desirable, but not uncommon). In this context the reader is advised to review the relevant parts of section 8.1.

25.2 Ideal Solutions, Their Equilibrium Vapor, and Raoult's Law

(a) The phenomenological definition of an ideal solution is formally identical to that of an ideal gas mixture [section 24.1(a)]. As usual we have the option of expressing it in terms of fugacities or chemical potentials; thus

$$f_k^\ell = X_k^\ell f_k^{\bullet\ell}(P, T) \tag{25-1a}$$

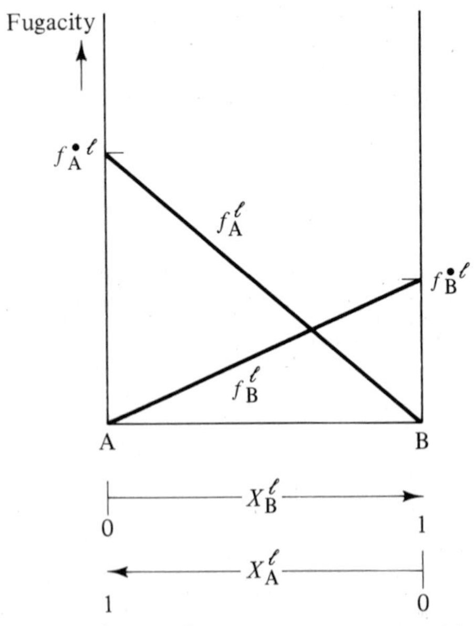

Fig. 39 Fugacities, f_A^ℓ and f_B^ℓ, for a two-component ideal solution at constant P and T. [See Eq. 25-1a).]

or

$$\mu_k^\ell = \mu_k^{\bullet\ell}(P, T) + RT \ln X_k^\ell, \tag{25-1b}$$

in which, once more, it is stipulated that these equations should hold for all components k (= 1, . . . , r), at all compositions, i.e., for all possible values of a set of $(r - 1)$ independent X_k^ℓ. Note that we have added superscripts ℓ to specify the liquid phase and to distinguish the quantities concerned from the corresponding ones in the equilibrium vapor (obviously at the same P and T), which will be identified by superscripts g. For a solution of two components, A and B, we have plotted the fugacities, according to Eq. (25-1a), in Fig. 39.

(b) If the components are volatile, we study the properties of an equilibrium vapor. It is instructive to apply the Phase Rule, which pertains to the general situation, i.e. without reference to any model in any phase. In the language of Chapter 21, in all examples of current interest $\mathbf{c} = \mathbf{s}$, the number of independent components is identical to the number of (chemical) species present. In Fig. 40 we have indicated four cases, all at the same fixed T. If necessary, the reader should review the relevant portions of Chapter 21

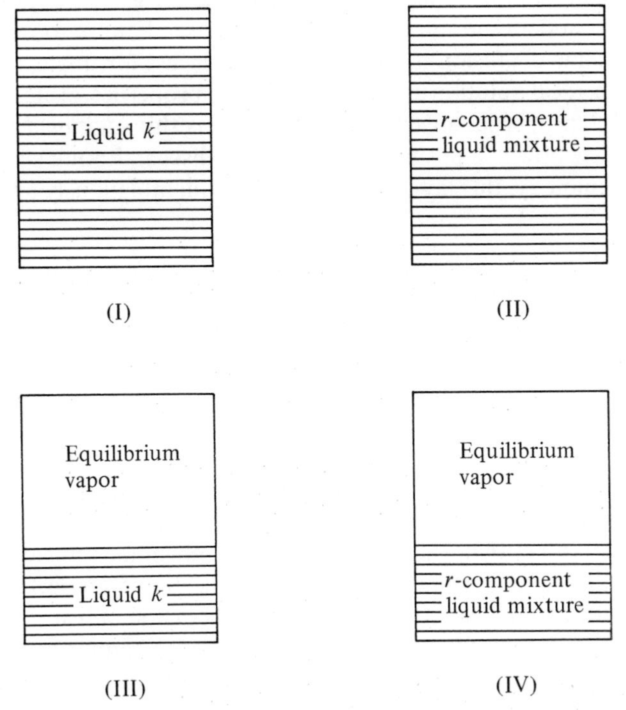

Fig. 40 Four cases to be analyzed by means of the Phase Rule, all pertaining to the same (constant) T. [See subsection 25.2(b).]

and verify that for systems I, . . . , IV, the variance $\mathbf{v} = 2, (r + 1), 1$, and r, respectively. This implies that in case I we can still choose P, but in case III it is already determined. In case II we can select P as well as the $(r - 1)$ independent composition variables. In case IV, of primary interest in this section, we can, for example, choose the composition of the liquid, but that exhausts the available degrees of freedom (remember that T has already been specified!); both the pressure P as well as the composition of the equilibrium vapor are completely determined. In some cases, some or all of the components may not be volatile under the conditions concerned. But *if* the vapor is pulled into the discussion, as is frequently the case, we lose an important degree of freedom.

(c) Let us return to the ideal solution. We are free to discuss its properties with or without reference to its equilibrium vapor. Since we know that all components have to be very much alike, it is safe to state that *either all or none* of the components are volatile. Let us assume the former; then it is quite likely that this gas phase will also be ideal in the Lewis–Randall sense, but there is no reason whatever to assume that it is an ideal mixture of ideal gases. If, as is frequently the case, this is stipulated nevertheless, we are making an *additional assumption* regarding the gas phase, independent of whatever model is adopted for the liquid phase. Failure to recognize this has led to some confusion in the literature. It is less fortunate, for example, to *define* an ideal solution in terms of certain properties of its equilibrium vapor *if* such properties are dependent on making this *additional* assumption.

(d) We are now in a position to derive, and discuss, Raoult's Law (1886).* In our terminology, this law can be expressed unambiguously and explicitly as

$$X_k^g P = X_k^\ell P_k^\bullet \qquad \text{for } k = 1, \ldots, r. \tag{25-2}$$

If we adopt the formal definition of partial pressure [Eq. (23-34)], Eq. (25-2) reads more simply

$$p_k = X_k^\ell P_k^\bullet \qquad \text{for } k = 1, \ldots, r. \tag{25-3}$$

In words: The partial pressure for any component k is equal to the mole fraction of k *in the liquid* times the vapor pressure of pure k at the same temperature.

The relation between Raoult's Law and the phenomenological definition of an ideal solution is somewhat controversial. Some go so far as to call Eq. (25-1a) Raoult's Law. This is not only a distortion of history, but it also suggests a *necessary* equivalence to Eq. (25-2), which cannot be justified. As we shall see, the derivation of Raoult's Law requires *additional* assumptions

---

* Attention should be drawn to an excellent article by A. G. Williamson, *J. Chem. Educ.* **43**, 211 (1966).

[cf. subsection (c) above], which makes it equally inelegant to use *it*, in the form (25-3), as the *definition* of an ideal solution.

(e) Let us look first at the simple equilibrium between pure liquid k and its vapor (Fig. 40, case III), for which we must have

$$\mu_k^{\bullet\ell}(P_k^\bullet, T) = \mu_k^{\bullet g}(P_k^\bullet, T). \tag{25-4}$$

Assume that the vapor is ideal; then Eq. (25-4) becomes

$$\mu_k^{\bullet\ell}(P_k^\bullet, T) = \mu_k^{\ominus g}(T) + RT \ln \frac{P_k^\bullet}{1\text{ atm}}. \tag{25-5}$$

For the liquid–vapor equilibrium of component k in a mixture (Fig. 40, case IV), we have

$$\mu_k^\ell(P, T, \text{composition}) = \mu_k^g(P, T, \text{composition}). \tag{25-6}$$

Next *assume that* ℓ *is ideal AND that* g *is an ideal mixture of ideal gases*, so that Eq. (25-6) becomes

$$\mu_k^{\bullet\ell}(P, T) + RT \ln X_k^\ell = \mu_k^{\ominus g}(T) + RT \ln \frac{p_k}{1\text{ atm}}. \tag{25-7}$$

Subtract Eq. (25-5) from Eq. (25-7) and rearrange terms to yield

$$RT \ln \frac{p_k}{P_k^\bullet} = RT \ln X_k^\ell + [\mu_k^{\bullet\ell}(P, T) - \mu_k^{\bullet\ell}(P_k^\bullet, T)]. \tag{25-8}$$

This will only reduce to Raoult's Law, Eq. (25-3), *if the "correction term" in straight brackets vanishes.* Since $(\partial\mu_k^\bullet/\partial P)_T = V_k^\bullet$, and the molar volume of a substance in the liquid phase is hardly dependent on the small pressure variations involved, the "correction term" can be written as

$$V_k^{\bullet\ell}(T)(P - P_k^\bullet). \tag{25-9}$$

In the true limiting case of the model, all components k become alike; therefore, all $P_k^\bullet$ become the same. Since we have also assumed an ideal mixture of ideal gases, Dalton's Law must hold, too; hence

$$P = \sum_{k=1}^{r} p_k = \sum_{k=1}^{r} X_k^\ell P_k^\bullet = P_k^\bullet \sum_{k=1}^{r} X_k^\ell = P_k^\bullet. \tag{25-10}$$

This shows that if we "truly" had our ideal solution in equilibrium with an ideal mixture of ideal gases, Eq. (25-9) vanishes and the strict validity of Raoult's Law is consistent with $P = P_k^\bullet$ for all k. In practice, deviations from Raoult's Law may be due to any one of, or any combination of, three factors:

1. The liquid does not behave as an ideal solution.
2. The equilibrium vapor does not behave as an ideal mixture of ideal gases.
3. The term (25-9) is not negligible.

More details about the influence of the third factor can be found in the article by Williamson (see footnote p. 314). The second point has been analyzed by Fried,* while the first, most "obvious," one usually receives due attention [see also section 25.2(g)]. To a first approximation, these factors are given in order of decreasing importance, but the respective contributions cannot always be disentangled.

(f) Next consider a two-component system for which Raoult's Law is (approximately) valid, even though $P^{\bullet}_{A} \neq P^{\bullet}_{B}$. In Fig. 41(a) we recognize the straight lines

$$p_A = X^{\ell}_{A} P^{\bullet}_{A} = (1 - X^{\ell}_{B}) P^{\bullet}_{A} \equiv (1 - x^{\ell}) P^{\bullet}_{A}$$

and

$$p_B = X^{\ell}_{B} P^{\bullet}_{B} \equiv x^{\ell} P^{\bullet}_{B}.$$

The total pressure becomes

$$P = p_A + p_B = P^{\bullet}_{B} + (1 - x^{\ell})(P^{\bullet}_{A} - P^{\bullet}_{B}). \tag{25-11}$$

This is represented by the straight line from the point giving the vapor pressure of pure A to that of pure B. One can see this purely geometrically, by noting that $ab = cd$. It was this linear behavior of the total vapor pressure (as a function of the composition of the liquid) that originally struck Raoult in the 1880s. In a P,x phase diagram, the *straight line* appears as the liquid

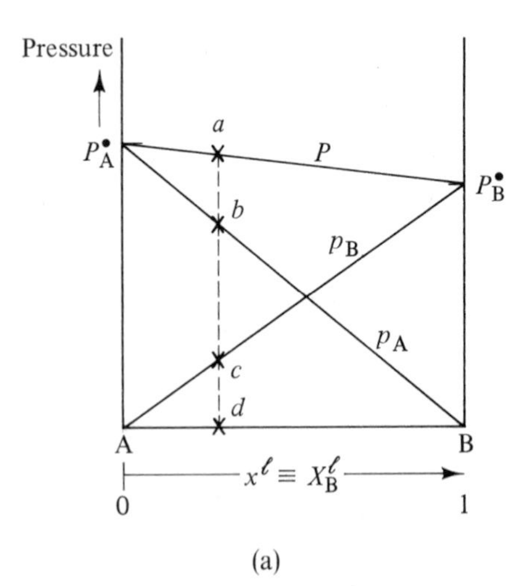

(a)

Fig. 41(a) Illustrating Raoult's Law for a two-component system (T constant).

---

* V. Fried, *J. Chem. Educ.* **45**, 720 (1968).

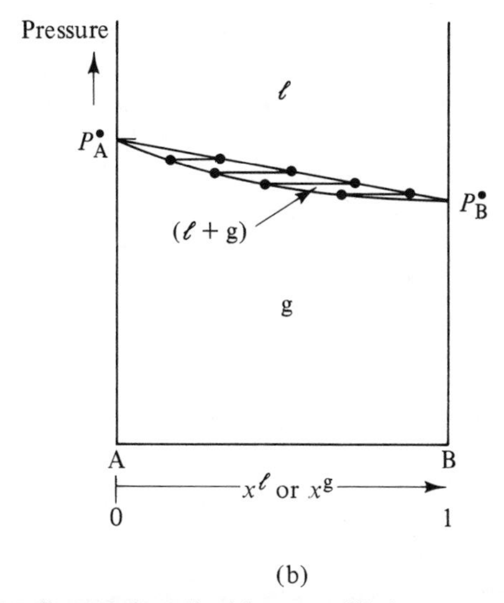

(b)

Fig. 41(b) *P*,*x* diagram for a liquid-gas equilibrium at constant *T*, when Raoult's Law is valid.

border of the coexistence area [Fig. 41(b)]. The other (gas) border is obtained by adding up

$$\frac{X_A^\ell}{P} = \frac{X_A^g}{P_A^\bullet} \quad \text{and} \quad \frac{X_B^\ell}{P} = \frac{X_B^g}{P_B^\bullet}$$

to yield

$$\frac{1}{P} = \frac{X_A^g}{P_A^\bullet} + \frac{X_B^g}{P_B^\bullet}$$

or

$$\frac{1}{P} = \frac{1 - x^g}{P_A^\bullet} + \frac{x^g}{P_B^\bullet}, \tag{25-12}$$

a *hyperbola*. In the nonideal case, the *P*,*x* cut at constant *T* shows a coexistence loop, whose ℓ and g borders have a more general curvature (compare the *P*,*x* cut, at T_1, in Fig. 33).

(g) Although strictly speaking, this falls beyond the scope of this chapter, a few remarks should be made about very strong deviations from Raoult's Law, as reflected in the appearance of maxima or minima in the *P*,*x* curves. This, in turn, leads to *azeotropic* mixtures, as we already discussed in a different context in section 22.4. At "normal" pressures, such strong deviations cannot be due to the second and third factors listed at the end of subsection (e) above. Here we are truly dealing with liquids that are far from ideal, i.e., exhibiting strong *preferential attractions* (associated with the P,x *minima*) or strong *preferential repulsions* (associated with the P,x *maxima*) between the constituents. Mixing such components together leads to very low and very high "escaping tendencies," respectively.* Strong interspecies repulsions may further lead to immiscibility or partial miscibility (i.e., separation into layers) in the liquid phase. Although models describing liquid mixtures deviating *slightly* from ideality have been put forward with some degree of success (see section 25.4), these *strongly* deviating systems are, as yet, hard to deal with quantitatively on a theoretical basis.

25.3 Ideal Solutions and the Thermodynamics of Mixing

(a) As before, let **J** be any thermodynamic (extensive) property. We frequently wish to know its change when, at constant P and T, we add together X_A moles of A and X_B moles of B, to form 1 mole of mixture. This quantity is known as the (molar or molal) **J** *of mixing*, and it is written as ΔJ_{mix}. It is given by the general expression

$$\Delta J_{mix} = X_A(\bar{J}_A - J_A^\bullet) + X_B(\bar{J}_B - J_B^\bullet). \tag{25-13}$$

The extension of this concept, and of Eq. (25-13), to many-component systems is obvious.

Suppose that for a specific **J**, in a particular mixture,

$$\bar{J}_A = J_A^\bullet \quad \text{and} \quad \bar{J}_B = J_B^\bullet$$

or, more generally,

$$\bar{J}_k = J_k^\bullet \quad \text{for all } k. \tag{25-14}$$

Then, by Eq. (25-13),

$$\Delta J_{mix} = 0, \tag{25-15}$$

which can also be expressed by the statement that in this situation, **J** *is additive in the components.* This is *not* the case, *in general*, and the magnitude of ΔJ_{mix} can give us valuable information about the inter- versus intraspecies interactions in various mixtures of interest. For this reason the thermodynamics of mixing is important to physical chemists.

(b) After this general introduction, we turn our attention to the ideal solution, the model of primary interest in this chapter. A proof, formally identical to that of Eq. (24-4), establishes the equality

$$\bar{V}_k^\ell = V_k^{\bullet\ell} \quad \text{for all } k. \tag{25-16}$$

---

* The "escaping tendency" nomenclature of G. N. Lewis [see section 19.1(c)] is particularly appropriate in this context.

Moreover, from the phenomenological definition (25-1b),

$$\left(\frac{\partial \mu_k^\ell/T}{\partial T}\right)_{P,\ \text{composition}} = \left(\frac{\partial \mu_k^{\bullet\ell}/T}{\partial T}\right)_P. \tag{25-17}$$

In Chapter 15 we obtained

$$-\frac{\mathbf{H}}{T^2} = \left(\frac{\partial \mathbf{G}/T}{\partial T}\right)_P, \tag{15-24}$$

so that a familiar theorem about partial molar quantities (see section 8.3) gives us

$$-\frac{\bar{H}_k}{T^2} = \left(\frac{\partial \mu_k/T}{\partial T}\right)_{P,\ \text{composition}} \tag{25-18}$$

Comparison of Eqs. (25-17) and (25-18) yields the result

$$\bar{H}_k^\ell = H_k^{\bullet\ell}. \tag{25-19}$$

Equations (25-16) and (25-19) show that *in an ideal solution*, $\mathbf{V}$ *and* $\mathbf{H}$ *are additive in the components.* Since $\mathbf{H} \equiv \mathbf{E} + P\mathbf{V}$, it must also be true for $\mathbf{E}$, and it is readily seen that this property extends to many related quantities, such as $\mathbf{C}_p$ and $\mathbf{C}_v$, as well. For obvious reasons, Eq. (25-19) also implies that the *heat of mixing* is *zero.* Readers should convince themselves that all these results are entirely consistent with the *corpuscular* picture of an ideal solution, based on the absence of *preferential* forces between particles of the same size and shape (see section 25.1 and the related analysis in section 24.1).

Experimentally, it is found that in solutions approaching the ideal model, several other physicochemical properties, some of which are not so easily accessible to thermodynamic analysis, are also additive in the components. Among these are the *molar refraction* and the *parachor*.

(c) Even in an ideal solution there are thermodynamic properties that are obviously not additive in the components. In this context, the entropy and the (Gibbs) free energy immediately come to mind, and it is easy to obtain explicit expressions for $\Delta S_{\text{mix}}^\ell$ and $\Delta G_{\text{mix}}^\ell$. As a starting point, we rewrite Eq. (25-13), with $\mathbf{J} \equiv \mathbf{G}$, for a liquid:

$$\Delta G_{\text{mix}}^\ell = X_{\text{A}}^\ell(\mu_{\text{A}}^\ell - \mu_{\text{A}}^{\bullet\ell}) + X_{\text{B}}^\ell(\mu_{\text{B}}^\ell - \mu_{\text{B}}^{\bullet\ell}). \tag{25-20}$$

This *defines* $\Delta G_{\text{mix}}^\ell$ for *any* binary solution. If this solution is ideal, we can substitute the defining equation (25-1b) into Eq. (25-20) to yield

$$\Delta G_{\text{mix}}^\ell = RT(X_{\text{A}}^\ell \ln X_{\text{A}}^\ell + X_{\text{B}}^\ell \ln X_{\text{B}}^\ell). \tag{25-21}$$

Since $\Delta G^{\ell}_{\text{mix}} = \Delta H^{\ell}_{\text{mix}} - T\,\Delta S^{\ell}_{\text{mix}}$, and we have already established that $\Delta H^{\ell}_{\text{mix}} = 0$ [cf. Eq. (25-19)], we immediately obtain $\Delta S^{\ell}_{\text{mix}}$ for this model:

$$\Delta S^{\ell}_{\text{mix}} = -R(X^{\ell}_{\text{A}} \ln X^{\ell}_{\text{A}} + X^{\ell}_{\text{B}} \ln X^{\ell}_{\text{B}}). \tag{25-22}$$

Exercise 25-1

Convince yourself that the *signs* of $\Delta G^{\ell}_{\text{mix}}$ and $\Delta S^{\ell}_{\text{mix}}$ are consistent with what you would *expect* for a spontaneous process under the appropriate conditions.

Note that both results, Eqs. (25-21) and (25-22), are entirely independent of the *nature* of A and B (as long as they form an ideal solution); $\Delta S^{\ell}_{\text{mix}}$ is fixed by the composition of the mixture formed, while $\Delta G^{\ell}_{\text{mix}}$ is completely determined by this composition and the temperature. The expression (25-22) is reminiscent of Eq. (23-30) for an ideal mixture of ideal gases. [See the discussion following Eq. (23-30), as well as Problem 23-1.]

As we already mentioned in Chapter 23, the expression $(X_{\text{A}} \ln X_{\text{A}} + X_{\text{B}} \ln X_{\text{B}})$ or its equivalent for a many-component mixture, $\sum_k X_k \ln X_k$, is frequently referred to as the *Gibbs paradox* term. Its assumed "paradoxical" nature is attributed to the fact that this expression makes its *full contribution* as long as A and B are distinguishable (no matter how "closely alike" they are), but *drops abruptly* to zero if A and B become identical. However, Sommerfeld* has pointed out that on account of the atomistic nature of matter, the difference between the properties of the relevant species cannot be made arbitrarily small. In other words, the *continuous* changing of these properties, by infinitesimal amounts, to a limit of zero, is a purely *mental process* that *cannot be carried out* in the laboratory. This is another instance in which the question may be raised to what extent thought processes are allowed to enter the discussion of thermodynamics [cf. section 23.3(c)]. A more extensive discussion belongs to the realm of the philosophy of science.

25.4 Some Related Models; "Simple" Solutions

(a) The ideal solution has very limited practical usefulness, because few mixtures of interest to chemists approach this model closely enough. Apparently, not many substances consist of molecules that satisfy the dual requirements of (1) exhibiting identical intra- and interspecies interactions, as well as (2) having the same size and shape, to the necessary degree of approximation. For *binary liquids*, to which we shall confine our attention in this section, it has been possible to define related models of greater utility, by a

---

* Quoted by J. Kestin, *A Course in Thermodynamics*, Blaisdell, Waltham, Mass., 1966, Vol. I, p. 578. Originally published in Vol. V of A. Sommerfeld's *Lectures on Theoretical Physics*.

limited relaxation of one or both of these requirements. The resulting new descriptions of liquid mixtures have been surrounded by a certain amount of confusion and controversy, due to at least three reasons:

1. There is no general agreement on nomenclature. This pertains, for example, to the distinction between *simple* and *regular* solutions [see subsection (d) below].
2. The defining criteria of these models are often *of necessity* somewhat vague, because there are no sharp boundaries between several of the categories of liquid mixtures involved.
3. The corpuscular basis for these models is not always completely understood.

This last weakness has carried over into the statistical mechanical theories, which should form a bridge between the corpuscular and the phenomenological descriptions. These difficult theories have sometimes relied on assumptions and approximations, which some experts in this area consider quite unrealistic. Thus we find ourselves confronted with a vast wide-open field, in which much work remains to be done. It presents a real challenge to any attempt at summarizing the most important phenomenological features, without getting sidetracked into many related discussions, which fall outside the scope of this book. For supplementary reading we have to refer to the fairly extensive list of references given at the end of the chapter.

(b) Since the models we wish to discuss were conceived in terms of their deviations from the ideal solution, it is useful to express some of their thermodynamic properties of interest in terms of (molar) *excess functions*, $\Delta J^{\mathbf{E}}_{\text{mix}}$, defined as the difference between the actual ΔJ_{mix} and this quantity for the corresponding hypothetical ideal solution:

$$\Delta J^{\mathbf{E}}_{\text{mix}} \equiv \Delta J_{\text{mix}} - \Delta J^{\text{id}}_{\text{mix}}. \tag{25-23}$$

Since our attention will henceforth be focused nearly exclusively on the *liquid*,* phase, we simplify our notation by omitting all superscripts ℓ.

From the discussion in section 25.3 it should be clear that for $\mathbf{J} = \mathbf{V}, \mathbf{H}, \mathbf{E}$, and several other properties, $\Delta J^{\mathbf{E}}_{\text{mix}}$ and ΔJ_{mix} are identical. Therefore, the newly introduced excess functions are needed primarily in conjunction with the quantities, **S**, **G**, and **A**.

Caloric measurements of ΔH_{mix} and dilatometric measurements of ΔV_{mix} are quite straightforward, at least in principle. On the other hand, the evaluation of ΔG_{mix}, hence also of $\Delta G^{\mathbf{E}}_{\text{mix}}$, usually involves a more intricate experimental procedure.† The limited amount of data obtained in this fashion has

---

* The exception is the brief reference to vapor pressures in subsection (c) below. See Eqs. (25-37).

† For an excellent survey, see McGlashan, reference 4 at the end of the chapter.

served to test the validity of the theoretical framework we shall turn to presently. Whether the difficult acquisition of further experimental data is a worthwhile venture depends to a large extent on the equally challenging continued development and refinement of these theoretical models.

(c) Following Guggenheim, we shall call a binary solution *simple* if its excess free energy of mixing is given by

$$\Delta G^{\mathbf{E}}_{\text{mix}} = x(1-x)\mathbf{w}. \tag{25-24}$$

In this phenomenological definition, x stands for X_{B} $(= 1 - X_{\mathrm{A}})$, as usual, and $\mathbf{w}$ is a characteristic parameter with the dimensions of energy per mole,* dependent on pressure and temperature but independent of the composition. In the rest of this subsection we shall summarize the most important corollaries to the definition (25-24). In subsection (d) we shall give some examples, discuss the sign as well as the magnitude of $\mathbf{w}(P, T)$ and its derivatives, and identify two special types of simple solutions. Finally, in subsection (e), we shall make a few comments about the molecular interpretation of the parameter $\mathbf{w}$.

Many of the implications of the definition (25-24) are immediately written down by observing that Eqs. (15-20) through (15-24) have to be valid if each $\mathbf{J}$ is replaced by the respective ΔJ_{mix} or $\Delta J^{\mathbf{E}}_{\text{mix}}$. That is, in the language of section 15.2(c), $\Delta G^{\mathbf{E}}_{\text{mix}}$ acts as the "characteristic excess function" of the independent variables P and T. Thus we obtain

$$\Delta V^{\mathbf{E}}_{\text{mix}} = \Delta V_{\text{mix}} = x(1-x)\left(\frac{\partial \mathbf{w}}{\partial P}\right)_T, \tag{25-25}$$

$$\Delta E^{\mathbf{E}}_{\text{mix}} = \Delta E_{\text{mix}} = x(1-x)\left[\mathbf{w} - T\left(\frac{\partial \mathbf{w}}{\partial T}\right)_P - P\left(\frac{\partial \mathbf{w}}{\partial P}\right)_T\right], \tag{25-26}$$

$$\Delta H^{\mathbf{E}}_{\text{mix}} = \Delta H_{\text{mix}} = x(1-x)\left[\mathbf{w} - T\left(\frac{\partial \mathbf{w}}{\partial T}\right)_P\right], \tag{25-27}$$

$$\Delta C^{\mathbf{E}}_{p,\text{mix}} = \Delta C_{p,\text{mix}} = -x(1-x)T\left(\frac{\partial^2 \mathbf{w}}{\partial T^2}\right)_P, \tag{25-28}$$

$$\Delta S^{\mathbf{E}}_{\text{mix}} = -x(1-x)\left(\frac{\partial \mathbf{w}}{\partial T}\right)_P, \tag{25-29}$$

and

$$\Delta A^{\mathbf{E}}_{\text{mix}} = x(1-x)\left[\mathbf{w} - P\left(\frac{\partial \mathbf{w}}{\partial P}\right)_T\right]. \tag{25-30}$$

---

* In some books $\mathbf{w}$ has dimensions of energy *per molecule*, in which case the right-hand side of Eq. (25-24) is multiplied by the Avogadro number.

Exercise 25-2

Show that the relations (15-20) through (15-24) can be used indeed as indicated, and derive Eqs. (25-25) through (25-30), not necessarily in this order, from the phenomenological definition (25-24).

If we define the *excess chemical potentials* by

$$\mu_k^{\mathbf{E}} \equiv \mu_k - \mu_k^{\mathrm{id}} \qquad (k = \mathrm{A, B}), \tag{25-31}$$

we have for our simple solution,

$$\mu_{\mathrm{A}}^{\mathbf{E}} = x^2 \mathbf{w} \tag{25-32a}$$

and

$$\mu_{\mathrm{B}}^{\mathbf{E}} = (1 - x)^2 \mathbf{w}. \tag{25-32b}$$

Note that the term x^2 ($\equiv X_{\mathrm{B}}^2$) appears with $\mu_{\mathrm{A}}^{\mathbf{E}}$, and the factor $(1 - x)^2$ ($\equiv X_{\mathrm{A}}^2$) with $\mu_{\mathrm{B}}^{\mathbf{E}}$.

Exercise 25-3

Adapt Eqs. (22-4) and derive Eqs. (25-32).

From Eqs. (25-31) and (25-32), after substituting the definition (25-1b) for μ_k^{id},

$$\mu_{\mathrm{A}} - \mu_{\mathrm{A}}^{\bullet} = RT \ln (1 - x) + x^2 \mathbf{w} \tag{25-33a}$$

and

$$\mu_{\mathrm{B}} - \mu_{\mathrm{B}}^{\bullet} = RT \ln x + (1 - x)^2 \mathbf{w}. \tag{25-33b}$$

These results are readily transformed into the associated expressions for the activities and activity coefficients. First, by using

$$\mu_k = \mu_k^{\ominus} + RT \ln a_k \tag{19-65}$$

with the identification $\mu_k^{\ominus} = \mu_k^{\bullet}$, Eqs. (25-33) yield

$$RT \ln a_{\mathrm{A}} = RT \ln (1 - x) + x^2 \mathbf{w} \tag{25-34a}$$

and

$$RT \ln a_{\mathrm{B}} = RT \ln x + (1 - x)^2 \mathbf{w}. \tag{25-34b}$$

Next, defining the activity coefficients (in terms of mole fractions), γ_k, by

$$\gamma_k \equiv \frac{a_k}{X_k}, \tag{25-35}$$

we obtain:

$$\ln \gamma_A = x^2 \frac{\mathbf{w}}{RT} \tag{25-36a}$$

and

$$\ln \gamma_B = (1 - x)^2 \frac{\mathbf{w}}{RT}. \tag{25-36b}$$

Let us remind ourselves that both sequences of equations (25-25) through (25-30) and (25-32) through (25-36) follow, without any approximations, from the phenomenological definition of a simple solution, Eq. (25-24). These sets of equations by no means exhaust all possibilities; with expressions for the μ_k, a_k, and γ_k at our disposal, it is easy, at least in principle, to obtain several other properties of interest. To give but one example, equations for the partial pressures p_k ($\equiv X_k^g P$) can be derived along the lines illustrated in section 25.2. If we still make the assumption that the equilibrium vapor can be considered an ideal mixture of ideal gases, the results are

$$\frac{p_A}{P_A^\bullet} = (1 - x)e^{x^2\mathbf{w}/RT} \tag{25-37a}$$

and

$$\frac{p_B}{P_B^\bullet} = xe^{(1-x)^2\mathbf{w}/RT}, \tag{25-37b}$$

provided that the "correction term," analogous to the expression in brackets on the right-hand side of Eq. (25-8), can be neglected. Please remember that in Eqs. (25-37), x still stands for x^ℓ.

Exercise 25-4

Derive Eqs. (25-37).

Exercise 25-5

(Optional for those with a particular interest in phase equilibria.) As we lower (in exceptional cases *raise*) the temperature of a binary liquid, a temperature may be reached, the critical mixing temperature, T_c, below which the liquid splits into two phases. General thermodynamic considerations give the critical mixing conditions in a number of related forms, e.g. (see reference 2 or 3 at the end of the chapter),

$$\left(\frac{\partial \mu_A}{\partial x}\right)_{P,T_c} = \left(\frac{\partial \mu_B}{\partial x}\right)_{P,T_c} = \left(\frac{\partial^2 \mu_A}{\partial x^2}\right)_{P,T_c} = \left(\frac{\partial^2 \mu_B}{\partial x^2}\right)_{P,T_c} = 0.$$

Show that for a simple solution, the critical mixture has $x = 0.5$, and that $T_c = \mathbf{w}/2R$. Show that for this mixture $\gamma_A = \gamma_B = 1.65$.

Of course, it would be much better to treat the gas phase in a more realistic fashion, e.g., through the use of the van der Waals equation for binary mixtures (see section 27.1) or the use of a truncated virial expansion [cf. Eq.

(24-6)]. Needless to say, such refinements render the analysis much more difficult.

(d) From the discussion in subsection (c) it should be clear that by a judicious interpretation of measurements on ΔH_{mix}, $\Delta C_{p,\text{mix}}$, ΔV_{mix}, as well as vapor pressures (sometimes carried out at more than one temperature*), we should be able to obtain numerical values of the characteristic parameter $\mathbf{w}$ and its derivatives. Such data also enable us to check the internal consistency of Eqs. (25-24) through (25-37), hence the degree of relevance of the model for real mixtures. Since most measured properties of interest are composition dependent (note that the ΔJ^{E} are parabolic functions of x), it is customary to refer published data to mixtures with $X_{\text{A}} = X_{\text{B}} = x = 0.5$. This will always be implied when we quote numerical values for any ΔJ_{mix} or $\Delta J^{\text{E}}_{\text{mix}}$ in the final part of this chapter.

The analysis of available data along the lines suggested above has met with limited success. In surveying the values of $\mathbf{w}$ and its derivatives, one is not struck by an overwhelming display of systematic behavior. Nevertheless, two regularities stand out:

1. In the vast majority of cases studied experimentally, $\mathbf{w}$ is positive. By Eqs. (25-27) and (25-36) and the sign of $(\partial \mathbf{w}/\partial T)_P$ (see below), this means that in these mixtures ΔH_{mix} and $\ln \gamma_k$ are positive as well. The positive ΔH_{mix}, in turn, suggests that A and B exhibit preferential repulsions; A and B molecules tend to "segregate," which in the extreme can lead to phase separation.† In a few instances, $\mathbf{w}$ is negative; A and B molecules tend to "intermingle." The system chloroform–acetone, mentioned in the introductory section 25.1, represents such an exceptional case; the preferential A–B attractions are due to hydrogen bonds.
2. To the best of the author's knowledge, $(\partial \mathbf{w}/\partial T)_P$ is negative in nearly all cases investigated. Since $\mathbf{w}$ is a parameter with the dimensions of energy, expressing differential interactions between neighbor molecules [see subsection (e) below], it is not illogical to assume that such differences are *somewhat* "washed out" as the temperature is increased. But within the framework of the simple solution model, by Eq. (25-29), a negative $(\partial \mathbf{w}/\partial T)_P$ is associated with a positive $\Delta S^{\text{E}}_{\text{mix}}$. This is certainly not easily interpreted by means of the customary order–disorder arguments [see section 16.3(a)].

---

* Measurements are rarely, if ever, carried out at more than one pressure (if this degree of freedom is available).

† Of course, in such an extreme situation the mixture at hand can hardly be characterized any longer as *related to* an ideal solution. Nevertheless, *critical mixing* phenomena are predicted *qualitatively* by Eq. (25-24); see Exercise 25-5.

Mixtures of CCl_4 and cyclohexane are usually considered to be prototypes of the *simple* solution model. For this system, at room temperature and atmospheric pressure, $\mathbf{w}$ is about 300 J mole^{-1}, and $(\partial \mathbf{w}/\partial T)_P$ roughly -1 J K^{-1} mole^{-1}. By way of comparison, under similar circumstances, the system benzene–toluene, traditionally considered to be a good representative of an *ideal* solution, has a $\mathbf{w}$ of about -10 J mole^{-1}, an order of magnitude smaller, although $(\partial \mathbf{w}/\partial T)_P$ is still about -0.6 J K^{-1} mole^{-1}.* At the other end of the spectrum, CS_2 and acetone are still said to form simple solutions, although $\mathbf{w}$ is quite large (≈ 4 kJ mole^{-1} at room temperature) and $(\partial \mathbf{w}/\partial T)_P$ is exceptionally large as well (≈ -5 J K^{-1} mole^{-1}). We reiterate a remark made earlier to the extent that, for practical purposes, there are no sharp boundaries between ideal and simple solutions on the one hand, or between simple and "no longer simple" solutions on the other hand.

In Hildebrand's original definition of *regular* solutions (1929) it was assumed that $\mathbf{w} \neq 0$, but $(\partial \mathbf{w}/\partial T)_P = 0$. Such a solution has $\Delta H_{\text{mix}} = x(1-x)\mathbf{w}$, but $\Delta S^{\mathbf{E}}_{\text{mix}} = 0$. Guggenheim† pointed out (1949) that it is unlikely for a regular solution to exist that is not also ideal, i.e., that it is unlikely for a mixture to have a nonvanishing parameter of differential interaction, without its entropy of mixing being affected at all. This led him to define *simple* solutions by means of equation (25-24). In Guggenheim's terminology, the name *regular* solutions is retained for hypothetical mixtures with $\Delta H^{\mathbf{E}}_{\text{mix}} = \Delta H_{\text{mix}} \neq 0$ and $\Delta S^{\mathbf{E}}_{\text{mix}} = 0$, while the name *athermal* solutions is introduced for the other nonexistent extreme case, namely, for mixtures with $\Delta H^{\mathbf{E}}_{\text{mix}} = \Delta H_{\text{mix}} = 0$ and $\Delta S^{\mathbf{E}}_{\text{mix}} \neq 0$. Prigogine and Defay‡ never use the designation *simple*, but similarly define regular and athermal solutions as limiting cases of (what they only refer to as) nonideal solutions, as follows: A solution is called regular if $|\Delta H_{\text{mix}}| \gg T|\Delta S^{\mathbf{E}}_{\text{mix}}|$, and a solution is called athermal if $|\Delta H_{\text{mix}} \ll T|\Delta S^{\mathbf{E}}_{\text{mix}}|$. An example of a regular solution, in *this* sense, is CCl_4–benzene, for which $\Delta H_{\text{mix}} \cong 110$ J mole^{-1} and $T\Delta S^{\mathbf{E}}_{\text{mix}} \approx 30$ J mole^{-1}, at 300 K. Solutions of polymers in low-molecular-weight solvents are the most important representatives of athermal mixtures. Not all authors consider these distinctions useful, since they involve further boundaries that are not necessarily sharply defined. McGlashan, in his book,§ heads the section concerned: "*Simple* ('*Regular*') *Mixtures.*" Apparently, he considers the designations "simple" and "regular" synonymous and never introduces the term "athermal." All this illustrates the disagreement on nomenclature, already referred to in subsection (a) above.

Usually little is said about $(\partial \mathbf{w}/\partial P)_T$. We shall be able to get an idea about its magnitude from experimental ΔV_{mix} data by using Eq. (25-25) (see also problem 25-2). ΔV_{mix} can be either positive (e.g., CCl_4–*tert*-butyl alcohol)

---

* Not too much significance should be attached to the negative sign of the small $\mathbf{w}$ for this system. Published data show some disagreements, and a temperature lowering could easily change this sign.

† Reference 2 at the end of the chapter.

‡ Reference 3 at the end of the chapter.

§ Reference 4 at the end of the chapter.

or negative [e.g., CCl_4–$C(CH_3)_4$], and it is not always easy to explain this on a molecular basis. The order of magnitude ranges typically from a few hundredths $cm^3 \, mole^{-1}$ to as much as about 1 $cm^3 \, mole^{-1}$. At $x = \frac{1}{2}$, the last number corresponds to $(\partial \mathbf{w}/\partial P)_T \approx 0.4 \, J \, atm^{-1} \, mole^{-1}$.

(e) The most frequently taken approach toward a corpuscular understanding of the phenomenological features of simple solutions is based on the "*quasi-crystalline lattice*" *model.** As long as all molecules are assumed to have approximately the same size and shape, we visualize each of them occupying a single lattice site with a common *coordination number* (the number of nearest neighbors), $\mathbf{z}$. (In a *true* crystal, $\mathbf{z}$ is a constant; in our *quasi*-crystalline liquid, $\mathbf{z}$ must be regarded as a statistical average). We further assume that only nearest-neighbor-pair interactions contribute to the energy. Let ε_{AA}, ε_{BB}, and ε_{AB} denote the interaction energy between two adjacent A molecules, two adjacent B molecules, and an A and a B molecule, respectively. Then ε, defined by

$$\varepsilon \equiv \varepsilon_{AB} - \tfrac{1}{2}(\varepsilon_{AA} + \varepsilon_{BB}), \tag{25-38}$$

can be considered as the energy gained on mixing per "creation" of an AB pair. Subsequently, we define a molar energy of preferential interaction (sometimes called a molar configurational energy), $\mathbf{w}$, by

$$\mathbf{w} = N_{Av}\mathbf{z}\varepsilon. \tag{25-39}$$

The hope is that this $\mathbf{w}$ adequately represents the parameter of the phenomenological theory, introduced in Eq. (25-24). Of course, $\mathbf{z}$ is a positive number; hence $\mathbf{w}$ has the sign of ε.

Since attractions dominate in the interactions, all energies on the right-hand side of Eq. (25-38) are negative. If all intermolecular interactions were to arise solely from van der Waals–London dispersion forces, it can be shown that† the mean of ε_{AA} and ε_{BB} is always more negative than ε_{AB}, so that ε would, of necessity, be positive. This could explain the predominance of positive values of $\mathbf{w}$ as observed experimentally [see subsection (d) above]. A negative $\mathbf{w}$ may indicate that there are important electrostatic forces of attraction between A and B molecules, e.g., those associated with hydrogen bonds in the system chloroform–acetone, mentioned previously. A negative $\mathbf{w}$ in mixtures of tetraethylmethane and *n*-octane has been attributed to the considerable difference in size and shape of these two molecular species. Of course, such a difference can take on catastrophic proportions for solutions of high-molecular-weight polymers in low-molecular-weight solvents. If we still want to describe *these* systems in terms of a quasi-crystalline lattice model, we must allow the polymer molecules to occupy more than one lattice

* For a very different theory, see M. L. McGlashan, pp. 252 ff. of reference 4 at the end of the chapter, where many additional references may be found.

† Prigogine and Defay, pp. 387–388 of reference 3 at the end of the chapter.

site. Needless to say, this brings about additional complications in the subsequent statistical mechanical treatment.

The temperature (and pressure) dependence of $\mathbf{w}$ has to be attributed, in the model under consideration, to the temperature (and pressure) dependence of both $\mathbf{z}$ and ε. We shall not attempt to give a detailed discussion of these matters.

Regular solutions, in the sense defined by Hildebrand, or better along the lines suggested by Prigogine and Defay [see the small print in subsection (d) above], arise when the A and B particles have very nearly the same size and shape, *and* ε (although not negligible) is very small compared to kT. For *athermal* solutions, $\Delta H_{\text{mix}} \approx 0$, hence ε is very small, so that the considerable magnitude of $\Delta S^{\text{E}}_{\text{mix}}$ is largely determined by the differences in geometry of the molecular species.

PROBLEMS

Problem 25-1

Consider a mixture of 100 g of benzene and 100 g of toluene at 20°C. At this temperature the vapor pressures of pure benzene and toluene are 75 torr and 22 torr, respectively.

Part I: Assume that the two components form an ideal solution, that its equilibrium vapor (of which there is a very small amount) is an ideal mixture of ideal gases, and that Raoult's Law is strictly valid.

(a) Compute x^{ℓ}, x^{g}, the partial pressures of benzene and toluene, and the total pressure.

(b) Compute ΔS_{mix} and ΔG_{mix} (for the liquid phase).

Part II: Assume that the two components form a *simple* solution with $\mathbf{w} = -10$ J mole^{-1} and $d\mathbf{w}/dT = -0.6$ J K^{-1} mole^{-1}, that its equilibrium vapor is still an ideal mixtures of ideal gases, and that Eqs. (25-37) are strictly valid.

(a) Repeat the calculations of Part I(a).

(b) Compute ΔH_{mix}, $\Delta S^{\text{E}}_{\text{mix}}$, ΔS_{mix}, $\Delta G^{\text{E}}_{\text{mix}}$, and ΔG_{mix}.

(c) Compute the activity coefficients of benzene and toluene in this liquid mixture. Use the answers to Parts I and II to ascertain whether the ideal solution is a good model for benzene–toluene mixtures.

Problem 25-2

We hinted in section 25.1 that substances differing only by isotopic substitution should approach ideal solutions as well as any set of compounds. In 1968, M. Lal and F. L. Swinton (*Physica* **40**, p. 446) found experimentally that for an equimolar mixture of benzene and benzene-d_6 at 30°C, $\Delta H_{\text{mix}} = 0.65$ J mole^{-1} and $\Delta V_{\text{mix}} = 0.0004 \pm 0.0002$ cm^3 mole^{-1}. No data on $\Delta S^{\text{E}}_{\text{mix}}$ were obtained.

(a) Assuming that $\Delta S^{\text{E}}_{\text{mix}}$ can be neglected entirely, obtain $\mathbf{w}$ at 30°C.

(b) Assuming the validity of the quasi-crystalline lattice model with a coordination number 12 (as in a closest-packed crystalline lattice of identical atoms), obtain ε as defined by Eq. (25-38).

(c) What can you conclude from a comparison of ε with kT?

(d) Obtain the order of magnitude of $(\partial \mathbf{w}/\partial P)_T$ at 30°C, in J atm^{-1} mole^{-1}.

REFERENCES

In the author's opinion, the following books on thermodynamics give a particularly good treatment of the subject matter in this chapter.

1. G. N. Lewis and M. Randall, as revised by K. S. Pitzer and L. Brewer, *Thermodynamics*, McGraw-Hill, New York, 1961, Chapter 21.
2. E. A. Guggenheim, *Thermodynamics*, 3rd ed., North-Holland, Amsterdam, 1957, Chapter V.
3. I. Prigogine and R. Defay, *Chemical Thermodynamics*, D. H. Everett, trans., Longmans, Green, London, 1952, Chapters XXIV, XXV, and XXVI.
4. M. L. McGlashan, *Chemical Thermodynamics*, Academic Press, London, 1979, Chapter XVI.

The last three books contain considerable excursions into statistical mechanics. Many additional references can be found in all four treatises. Among books dealing with various aspects of solutions, the following are particularly pertinent:

5. J. H. Hildebrand and R. L. Scott, *The Solubility of Nonelectrolytes*, 3rd ed., Reinhold, New York, 1950.
6. J. H. Hildebrand, J. M. Prausnitz, and R. L. Scott, *Regular and Related Solutions*, Van Nostrand Reinhold, New York, 1970.
7. E. A. Guggenheim, *Mixtures*, Clarendon Press, Oxford, 1952.
8. J. S. Rowlinson and F. L. Swinton, *Liquids and Liquid Mixtures*, 3rd ed., Butterworth, London, 1982.
9. H. C. Van Ness and M. M. Abbott, *Classical Thermodynamics of Nonelectrolyte Solutions*, McGraw-Hill, New York, 1982.
10. J. M. Prausnitz, *Molecular Thermodynamics of Fluid-Phase Equilibria*, Prentice-Hall, Inc., Englewood Cliffs, N. J., 1969.

Chapter 26

The (Ideally or Infinitely) Dilute Solution*

26.1 Phenomenological Definitions and the Corollary for the Solvent

(a) We shall first concentrate our attention on the simplest case, that of a solution of one or more *nondissociating nonelectrolytes*, i $(= 2, \ldots, r)$, in an appropriate solvent, 1. Even in this case, there are six different, though closely related, ways to define the model at hand, for the following two reasons:

1. Our choice to couch the discussion in terms of fugacities or in terms of chemical potentials.† This issue has already been raised in several earlier chapters, in which the author admitted his preference for the second option. Nevertheless, he feels that he would do potential readers of the vast literature concerned a disservice if he were not to familiarize them with the alternative scheme as well.
2. Our choice to work in terms of mole fractions, molalities, or molarities of the solute(s), dependent on what is more convenient in any given situation. (For the solvent, the mole fraction will be used as our concentration unit.) In this context readers are urged to reacquaint themselves with the contents of section 8.1.

In terms of fugacities, the phenomenological definition of an ideally dilute solution reads

$$f_i = k_i^{(\alpha)}(P, T, \text{solvent})\,\alpha_i \qquad (\alpha = X, m \text{ or } c;\ i = 2, \ldots, r), \qquad (26\text{-}1)$$

in which the $k_i^{(\alpha)}$ are as yet undetermined dimensionless constants. By Eqs. (8-8) and (8-9) and these defining equations (26-1), we readily obtain the

* For some comments on nomenclature and a few other pertinent issues, see section 25.1.

† See also the footnote on p. 304. Activities are a third alternative, less frequently used in *this* context.

relations

$$k_i^{(m)} = \frac{|M_1|}{1000} k_i^{(X)} \tag{26-2a}$$

and

$$k_i^{(c)} = \frac{|M_1|}{1000|\rho_1|} k_i^{(X)}, \tag{26-2b}$$

in which M_1 and ρ_1 are the weight of 1 mole and the density of the pure solvent, respectively. Of course, ρ_1 is a function of (P and) T. As mentioned in section 25.1, the corpuscular definition of this model requires all interactions between *solute* particles to be negligible. Strictly speaking, this requires these solutions to be *infinitely dilute*, in which case Eqs. (26-2) involve no approximations.

In this entire chapter, the equations pertaining to this model will be manipulated in the limiting case $\alpha_i \to 0$ ($\alpha = X, m$ or c; $i = 2, \ldots, r$), corresponding to *infinitely* dilute solutions. How useful the model turns out to be in the description of *reasonably* dilute *real* solutions depends on the specific system at hand, in the same way that the ideal gas model (strictly valid as $P \to 0$) is useful up to *reasonably* low pressures, as determined by the nature of the gas.

To obtain the three alternative definitions in terms of chemical potentials, we have to combine Eqs. (26-1) with

$$\mu_i = \mu_i^{\ominus(\alpha)} + RT \ln \frac{f_i}{f_i^{\ominus(\alpha)}}, \tag{19-63}$$

in which the addition of the superscript (α) anticipates that *our choice of standard state for the solute(s) depends on the concentration unit we decide to use.* In fact, we shall take

$$f_i^{\ominus(\alpha)} = k_i^{(\alpha)}[1_\alpha], \tag{26-3}$$

in which $[1_X] \equiv 1$, $[1_m] \equiv 1$ mole/kg solvent, and $[1_c] \equiv 1$ mole per liter of solution. We shall rationalize this choice in a moment. First we note that the substitution of Eqs. (26-1) and (26-3) into Eq. (19-63) gives us the desired alternative phenomenological definitions of an ideally dilute solution:

$$\mu_i = \mu_i^{\ominus(\alpha)}(P, T, \text{solvent}) + RT \ln |\alpha_i| \qquad (\alpha = X, m, \text{or } c, i = 2, \ldots, r), \tag{26-1'}$$

in which $|\alpha_i|$ is a shorthand for $\alpha_i/[1_\alpha]$. We see that in this scheme, the standard state for a solute, i, corresponds to the *hypothetical* solution, which would still be ideally dilute at $|\alpha_i| = 1$. First and foremost, the choice of standard state is a matter of *convenience*. Realizing this, the reader may very well wonder why it is more convenient to refer $\mu_i^\ominus$ to a *hypothetical* solution, as indicated, rather than to the pure solute. This alternative choice, embodied in the equalities $f_k^\ominus = f_k^\bullet(P, T)$ and $\mu_k^\ominus = \mu_k^\bullet(P, T)$, appeared to work quite well for all the mixtures discussed in Chapter 25. To understand why we now select the convention (26-3), we must recall that in Chapter 25 we were dealing with *liquid* mixtures over the entire possible range of composition variables. In the context of the present chapter, however, the pure stable solutes, at the temperature and pressure of interest, may very well be *solids* (e.g., sugars, salts) or *gases* (e.g., O_2, N_2, CO_2), which makes them very *in*convenient as reference states to be used in conjunction with *liquid* solutions. On the other hand, the *limiting law constants* $k_i^{(\alpha)}$ can, at least in principle, be determined by the appropriate extrapolation (of experimental data) to infinite dilution, in the same way as many properties of ideal gases can be obtained by the extrapolation (of experimental data) to zero pressure. For these reasons, our choice, Eq. (26-3), leading to Eq. (26-1′), is indeed the most convenient one.

The phenomenological definitions, given above, involved only the *solutes* in the dilute solution. The *solvent* entered only indirectly, through the important quantities $\mu_i^\ominus$. Since these potentials completely determine the equilibrium constants of chemical reactions, the solvent dependence of the $\mu_i^\ominus$ confirms the experimental fact that the equilibria for reactions in solution are often very strongly influenced by the choice of solvent.

For gas reactions, the $\mu_i^\ominus$ and the correct thermodynamic equilibrium constant were strictly functions of the temperature only. In solutions, they are also dependent on the pressure and the aforementioned choice of solvent. Since reactions in solution are usually carried out at atmospheric pressure, and the small variations concerned have little or no effect on chemical equilibria, the pressure dependence rarely concerns us in this context. However, pressure changes *do* affect chemical potentials in solution, as witnessed, for example, by the osmotic pressure phenomenon (see section 26.3).

Of course, in some instances, the solvent may enter the stoichiometric equation as a full-fledged reaction partner. In addition, solvent properties are vital in determining the well-known *colligative properties* (freezing-point lowering, boiling-point elevation, osmotic pressure, etc.) of solutions, to which we shall turn in the third section of this chapter. For these and other reasons it is important to obtain an expression for the chemical potential of the solvent, μ_1. We shall now show that such an expression can be *derived*, as a corollary to the definition(s) (26-1).

(b) We start with the Gibbs–Duhem equation, at constant P and T, in the form

$$\sum_{k=1}^{r} X_k \, d\mu_k = 0. \tag{8-28}$$

In order to differentiate between the solvent, 1, and the $(r-1)$ solutes, i $(=2, \ldots, r)$, we split the sum over k in two parts:

$$X_1\,d\mu_1 + \sum_{i=2}^{r} X_i\,d\mu_i = 0. \tag{26-4}$$

The μ_i in this equation are given by Eq. (26-1′), with $\alpha \equiv X$. Since P and T are constant and the solvent is fixed, the total differential of each μ_i reduces to $RT\,d\ln X_i$, so that Eq. (26-4) becomes

$$\begin{aligned} X_1\,d\mu_1 &= -RT\sum_{i=2}^{r} X_i\,d\ln X_i = -RT\sum_{i=2}^{r} dX_i = -RT\,d\sum_{i=2}^{r} X_i \\ &= -RT\,d(1-X_1) = RT\,dX_1; \end{aligned}$$

hence

$$d\mu_1 = RT\,d\ln X_1. \tag{26-5}$$

Integration at constant P and T yields

$$\mu_1 = RT\ln X_1 + \text{constant}. \tag{26-6}$$

Since for $X_1 = 1$, $\mu_1 = \mu_1^\bullet(P, T)$, our final result is

$$\mu_1 = \mu_1^\bullet(P, T) + RT\ln X_1, \tag{26-7′}$$

the desired solvent corollary to Eqs. (26-1′). As our standard state for the solvent we simply take the pure component at the same pressure and temperature:

$$\mu_1^\ominus(P, T) = \mu_1^\bullet(P, T). \tag{26-8}$$

Consequently, by Eqs. (19-63), (26-7′), and (26-8), we readily obtain the alternative corollary in terms of fugacities:

$$f_1 = X_1 f_1^\bullet(P, T). \tag{26-7}$$

Exercise 26-1

By writing the Gibbs–Duhem type of equation, at constant P and T, in terms of fugacities (rather than chemical potentials), derive Eq. (26-7) directly from Eq. (26-1), with $\alpha \equiv X$.

Two comments are in order regarding the surprising results Eqs. (26-7) and (26-7′). The first one is rather trivial: Since we never refer to the molality or molarity of the *solvent*, we need not consider analogies for μ_1 in terms of these concentration units. The more important comment concerns the formal identity of Eqs. (26-7) and (26-7′) with Eqs. (25-1) for the ideal solution of Chapter 25. This apparent identity has certainly contributed to the ambiguity of the attribute "ideal behavior," already mentioned in section 25.1. With regard to this issue, two points should be made:

1. The Eqs. (25-1) were valid *for all components* in an ideal (liquid) mixture; Eqs. (26-7) and (26-7′) are true *only for the solvent.*
2. The Eqs. (25-1) were valid *at all compositions*; Eqs. (26-7) and (26-7′) are true *only in the range where the solution is ideally dilute.* (Strictly speaking, the latter interval is an infinitely narrow band near $X_1 = 1$.)

(c) At this stage a digression on dissociating (or associating) solutes is in order. For our model to remain valid, we must still impose the strict requirement that the interactions between the solute particles, *as they are actually present* in the solution, are negligible.

This is hard to realize for the important systems known as *electrolyte solutions*, even if these are very dilute in a colloquial sense. The reason is, of course, that the (fairly long range) Coulomb forces between ions can rarely be neglected. Thus these solutions create special problems, to which we shall have to return later in the chapter. In the meantime, readers may safely assume that when in the rest of this section we talk about "dissociation," we shall always have *nonelectrolytic* dissociation in mind.

In this case the equations given above are still valid provided that all species present are properly considered as separate solutes. This also implies that Eqs. (26-7) retain their validity if X_1 is correctly interpreted as the *actual* mole fraction of the solvent. Occasionally, in some simple cases, it is possible to rewrite Eqs. (26-7) in terms of a *stoichiometric* mole fraction of the solvent, computed from the original number of moles of (undissociated) solute(s) dissolved in a given amount of solvent. For all practical purposes, this is useful only if we are dealing with a *single* solute, 2, since with *multiple* dissociating solutes too many additional variables are introduced. We end this section by considering this special case in some detail.

Let us take n_1 moles of solvent, 1, and n_2 moles of a solute, 2, which dissociates into v particles, with a degree of dissociation, d. This could represent, e.g., a solution of PCl_5 in benzene. In this example the dissociation takes place according to

$$PCl_5 \rightleftharpoons PCl_3 + Cl_2;$$

hence $v = 2$. Next, we define

$$\tilde{v} \equiv 1 - d + vd, \tag{26-9}$$

so that $\tilde{v} n_2$ is the total number of moles of solute particles actually present in the solution at equilibrium. The *actual* mole fraction of the solvent will be given by

$$X_1 = \frac{n_1}{n_1 + \tilde{v} n_2} = \left(1 + \tilde{v}\frac{n_2}{n_1}\right)^{-1}.$$

At infinite dilution, this yields

$$\ln X_1 = -\tilde{\nu}\frac{n_2}{n_1}. \tag{26-10a}$$

But the stoichiometric mole fraction of the solvent, X_1^{st}, is still given by $n_1/(n_1 + n_2) = [1 + (n_2/n_1)]^{-1}$, which yields, at infinite dilution,

$$\ln X_1^{st} = -\frac{n_2}{n_1}. \tag{26-10b}$$

Comparison of Eqs. (26-10a) and (26-10b) shows that

$$\ln X_1 = \tilde{\nu} \ln X_1^{st}; \tag{26-11a}$$

hence

$$X_1 = (X_1^{st})^{\tilde{\nu}}, \tag{26-11b}$$

so that Eqs. (26-7) and (26-7′) can be modified to read

$$f_1 = (X_1^{st})^{\tilde{\nu}} f_1^{\bullet}(P, T) \tag{26-12}$$

and

$$\mu_1 = \mu_1^{\bullet}(P, T) + \tilde{\nu}RT \ln X_1^{st}, \tag{26-12′}$$

respectively. From Eq. (26-9) we see that in the case of *complete dissociation* ($d = 1$) $\tilde{\nu}$ equals ν, and Eqs. (26-12) and (26-12′) can be simplified accordingly. For all *dissociative* equilibria, ν is larger than unity. In case we deal with *association* of solute molecules, ν will become fractional.

Equations (26-12) and (26-12′) are important because, as we shall see in section 26.3(d), they lead to simple modifications of the laws of the dilute solution, compared to those based on Eqs. (26-7) and (26-7′). Thus we can use experimental data on osmotic pressure, freezing-point lowering, etc., to obtain information on dissociative or associative equilibria. In this context, readers may recall some of their favorite "problems" in general chemistry courses.

Armed with equations for the chemical potentials of solute(s) and solvent, we shall discuss the most important properties of the ideally dilute solution in the next three sections. Some additional features will be incorporated in the problem set at the end of the chapter.

26.2 Vapor-Pressure Laws

(a) We shall again confine our attention to nondissociating nonelectrolytes. Many remarks made, techniques used, and equations derived in section 25.2 are very pertinent to the analysis below. As in Chapter 25, we shall assume that we are dealing with equilibrium vapors that can be considered as (ideal mixtures of) ideal gases. Once more we stress that this is an *additional* assumption, independent of any stipulations regarding the *liquid* phase. Of course, in the present chapter we have to distinguish between solvent and solute(s), and not all components need have finite vapor pressures. [For a (dilute) solution of sugar in water, at room temperature, we surely can neglect the partial pressure of the sugar in the equilibrium vapor!]

Let us first consider the *solvent*. The procedure outlined in section 25.2(e) can be used without modification to yield

$$X_1^g P \equiv p_1 = X_1^\ell P_1^\bullet, \tag{26-13}$$

which is *Raoult's Law*. Note that for the ideally dilute solution model, it is *valid for the solvent only* (and only in the narrow concentration range to which the model pertains). Since, in this case, P will always be very close to $P_1^\bullet$ (at the same temperature), the "correction term," corresponding to the one in Eq. (25-9), need not concern us in the derivation of Eq. (26-13).

Next we turn to the *solute(s)*. The basic equilibrium condition $\mu_i^g = \mu_i^\ell$, under the conditions at hand, becomes

$$\mu_i^{\ominus g}(T) + RT \ln |p_i| = \mu_i^{\ominus (X)}(P, T, \text{solvent}) + RT \ln X_i^\ell. \tag{26-14}$$

This can be written in the form

$$|p_i| = k_i^*(P, T, \text{solvent}) X_i^\ell, \tag{26-15}$$

with

$$k_i^* \equiv e^{-(\mu_i^{\ominus g} - \mu_i^{\ominus (X)})/RT}. \tag{26-16}$$

These equations should hold for every volatile solute in an ideally dilute solution when in contact with its equilibrium vapor, if the latter behaves as an ideal mixture of ideal gases.

(b) For the ideal solution (see Chapter 25) it was possible to obtain a fairly simple expression for the *total* equilibrium vapor pressure. Unfortunately, adding up p_1 from Eq. (26-13) and $(r - 1)$ different p_i, given by Eq. (26-15), leads to no particularly revealing relation for P. However, some interesting results can be written down by considering either of the following two *extreme cases*:

Case I. The solvent is much more volatile than any of the solutes.

This situation would pertain, for example, to a solution of sugar(s) in water. In this case $P = p_1 (X_1^g \approx 1)$; hence by Eq. (26-13),

$$P = X_1^\ell P_1^\bullet. \tag{26-17}$$

Call $\Delta P \equiv P_1^\bullet - P$ the *vapor-pressure lowering*, and $\Delta P / P_1^\bullet \equiv 1 - P/P_1^\bullet$ the *relative vapor-pressure lowering*. If Eq. (26-17) is valid, we can write

$$\frac{\Delta P}{P_1^\bullet} = 1 - X_1^\ell = \sum_{i=2}^{r} X_i^\ell,$$

and *in case there is only a single solute,*

$$\frac{\Delta P}{P_1^\bullet} = X_2^\ell = \frac{n_2^\ell}{n_1^\ell + n_2^\ell} \approx \frac{n_2^\ell}{n_1^\ell}.$$

The equation

$$\frac{\Delta P}{P_1^\bullet} = \frac{n_2^\ell}{n_1^\ell} \tag{26-18}$$

is known as the *law of the vapor-pressure lowering.* Although it is mentioned in many textbooks, it really conveys no more information than the "original" relation (26-17).

Since $n_1 = g_1/M_1$, it appears as if the law of the vapor-pressure lowering could provide us with a means of obtaining an experimental value of $|M_1|$, the molecular weight of the solvent. Even allowing for an *approximate* validity of the model at best, this should give us at least an *idea* about the degree of possible *association* of this solvent. If one carries out such experiments with water as the solvent, one *always* finds $|M_1| = 18$, in contradiction to generally accepted ideas about the structure of liquid water. This discrepancy is resolved* by observing that in the derivation of Eq. (26-13), from which Eqs. (26-17) and (26-18) were obtained, we used Eq. (25-7), with $k = 1$, as a starting point. In this identity we specify *1 mole of gas,* since the right-hand side contains the term $RT \ln |p_1|$. (Had we considered n_1 moles of gas, the term $n_1 RT \ln |p_1|$ would have appeared in its place.) In writing down the identity (25-7), the same amount of liquid must be involved. Therefore, in Eq. (26-18), n_1^ℓ is really the weight of the solvent divided by its molecular weight *in the equilibrium vapor*; the experimental results tell us only that the water *vapor* is not associated, and give no relevant information about the *liquid* phase at all.

Case II. There is only a single solute, much more volatile than the solvent.

An example would be a solution of a gas, such as O_2 or N_2, in water, at room temperature. In this case, by Eq. (26-15),

$$|p_2| = k_2^* X_2^\ell. \tag{26-19}$$

Call $(1/k_2^*) \equiv k_H$. Strictly speaking, k_H will depend on P, T, and the solvent. If we neglect the pressure dependence and consider a particular solvent, Eq. (26-19) yields, in effect,

$$X_2^\ell = k_H(T)|p_2| = k_H(T)(|P| - |p_1|). \tag{26-20}$$

This expresses the experimental discovery, made by William Henry in 1801, that *the solubility of a gas in water is linearly proportional to the gas pressure.* In the application of Eq. (26-20) it is usually permitted to approximate p_1 by $P_1^\bullet$, the vapor pressure of the pure solvent at the temperature of interest. It is rarely advisable to neglect p_1 altogether.

---

* J. M. Bijvoet and A. F. Peerdeman, *J. Chem. Educ.* **35**, 240 (1958).

In many publications, the designation *Henry's Law* is not confined to Eq. (26-20) or *obvious* equivalents such as Eq. (26-19). Sometimes this designation is given to an equation of the type Eq. (26-15), and even the defining equations (26-1) have been identified as such. In this terminology the *limiting law constants*, $k_i^{(\alpha)}$, are frequently called *Henry's Law constants.* In the same vein, the authors involved may refer to Eqs. (25-1a) and (26-7) as *Raoult's Law.* Logically, this leads to statements of the type: "A(n ideally) dilute solution is a mixture in which the solvent satisfies Raoult's Law and the solute(s) Henry's Law." It is left up to the readers to judge whether such liberties with history are justifiable because they illuminate certain unmistakable connections, or are undesirable because they obscure the more subtle extra conditions which have to be satisfied for the two relevant sets of equations to be *strictly* equivalent.

Henry's Law influenced Dalton in the development of his atomic theory. The solubility of O_2 in body fluids, as a function of pressure, is of importance in the study of respiration at high altitudes.

26.3 Colligative Properties; the van't Hoff Laws

(*a*) The *colligative properties* include *vapor-pressure lowering*, discussed in the preceding section, *osmotic pressure*, *boiling-point elevation*, and *freezing-point lowering.* In a direct sense, all refer to *solvent properties* and how these are influenced by the presence of solutes. More precisely, we have to know how μ_1 depends on the composition of the solution, and draw the proper conclusions. We already obtained the desired functional form of μ_1 in an ideally dilute solution, and the subsequent derivations, to be given below, will lead to the laws first obtained by J. H. van't Hoff in the 1880s. For simplicity, we shall deal with the case of a single solute, 2, in the solvent 1. It will become apparent anyhow that *as long as the model is valid, the effect of different solutes is simply additive*, i.e., the colligative properties are determined by the collection* of solute particles; specifically, these properties are proportional to the sum of all the n_i $(i = 2, \ldots, r)$. All derivations will initially assume that the solute(s) is (are) nondissociating; some generalizations will be given in subsection (d) below.

(*b*) *Osmotic Pressure*

Let us first look at this phenomenon qualitatively. From Eq. (26-7′) it is apparent that dissolving some 2 in 1 (reducing X_1 from unity to a fraction), at constant P and T, reduces μ_1. When such a solution is separated from the pure solvent by means of a semipermeable membrane, **M**, permeable to 1 but impermeable to 2, the substance 1 tends to flow from the pure solvent

* The latin *colligatus* means "collected together."

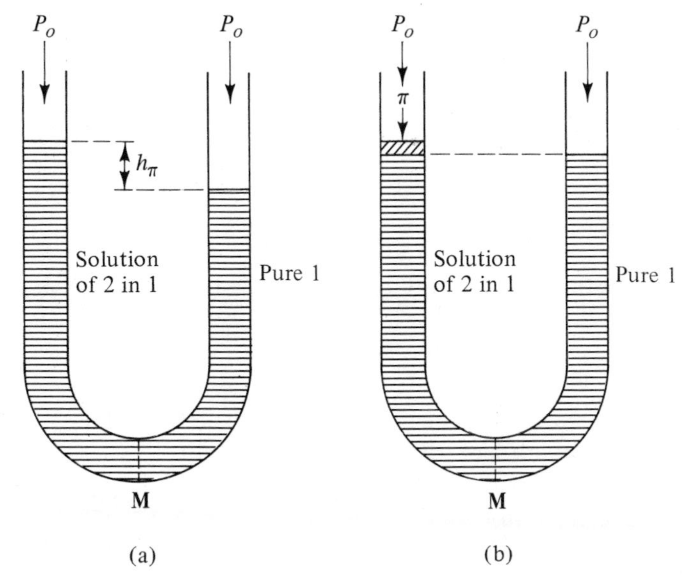

Fig. 42 Illustrating the osmotic phenomenon. (**M** is a semipermeable membrane: permeable to the solvent, 1, impermeable to the solute, 2.)

(region of highest μ_1) to the solution (with the lower μ_1). [Compare section 19.1(c).] In a typical classroom demonstration, no doubt familiar to many readers, one indeed observes that the liquid level rises on the solution side, and falls on the solvent side, of a U-shaped apparatus, as drawn in Fig. 42(a). When equilibrium is reached, the difference in these levels, h_π, is related to the *osmotic pressure*, π. For the purposes of actual measurements, it is faster and more accurate to obtain π as the extra pressure that must be applied to the solution side in order to prevent any net flow of solvent from taking place.* This is illustrated schematically in Fig. 42(b).

The thermodynamic equilibrium condition is

$$\mu_1^\bullet(P_o, T) = \mu_1(P_o + \pi, T). \tag{26-21}$$

At this stage we introduce our model by substituting Eq. (26-7′) for μ_1, to yield

$$\mu_1^\bullet(P_o, T) = \mu_1^\bullet(P_o + \pi, T) + RT \ln X_1$$

or

$$\mu_1^\bullet(P_o + \pi, T) - \mu_1^\bullet(P_o, T) = -RT \ln X_1. \tag{26-22}$$

* This is the principle on which the Hepp osmometer is based. For a summary of modern measuring techniques, see, e.g., D. P. Shoemaker and C. W. Garland, *Experiments in Physical Chemistry*, McGraw-Hill, New York, 1967, pp. 273ff.

Since we are dealing with an infinitely dilute solution, π is very small, so that we can write the truncated Taylor series:

$$\mu_1^\bullet(P_o + \pi, T) = \mu_1^\bullet(P_o, T) + \left(\frac{\partial \mu_1^\bullet}{\partial P}\right)_{T, P = P_o} \pi;$$

hence

$$\mu_1^\bullet(P_o + \pi, T) - \mu_1^\bullet(P_o, T) = V_1^\bullet \pi.$$

Substitution of this result into the left-hand side of Eq. (26-22) yields

$$\pi = -\frac{RT}{V_1^\bullet} \ln X_1, \tag{26-23}$$

the desired expression. As long as the model is strictly valid, Eq. (26-23) can be modified in a number of ways. First, we can write

$$-\ln X_1 \equiv -\ln \frac{n_1}{n_1 + n_2} = \ln \frac{n_1 + n_2}{n_1} = \ln\left(1 + \frac{n_2}{n_1}\right) \approx \frac{n_2}{n_1},$$

always remembering that $n_2 \ll n_1$. Hence

$$\pi = \frac{RT}{n_1 V_1^\bullet} n_2. \tag{26-24}$$

But

$$n_1 V_1^\bullet = \mathbf{V}_1 \approx \mathbf{V};$$

therefore,

$$\pi = n_2 \frac{RT}{\mathbf{V}}. \tag{26-25}$$

This form is striking, since it *formally* resembles the ideal gas law. The author joins those who consider this resemblance fortuitous. Finally, since $(n_2/\mathbf{V}) \equiv c_2$, the molarity of the solute, we obtain

$$\pi = c_2 RT, \tag{26-26}$$

the original formulation of van't Hoff (1885).

In case there are several noninteracting (nondissociating) solutes $i(= 2, \ldots, 4)$, *and the membrane* **M** *is still permeable to the solvent only*, Eqs. (26-25) and (26-26) must be replaced by

$$\pi = \frac{RT}{\mathbf{V}} \sum_{i=2}^{r} n_i = RT \sum_{i=2}^{r} c_i. \tag{26-27}$$

Exercise 26-2

Derive Eq. (26-27).

Note that π ultimately appears as an intrinsic property of the solution, whether or not it ever gets a chance to exhibit osmotic behavior. If, as we have assumed, the solution is ideally dilute, π is proportional to $\sum_i n_i$, hence is indeed a "colligative property" in the sense that this designation was introduced in subsection (a).

The osmotic phenomenon is of great importance in biology and physiology. In these sciences one often has to consider two solutions involving the same solvent but different sets of solutes, brought into contact through some particular natural membrane. If this membrane were *permeable to the solvent only*, a flow of this solvent would still take place from the solution with the higher π to the one with the lower π, each π being determined by Eq. (26-27). If these two π's would happen to be numerically the same, the solutions involved are said to be *iso-osmotic*, and nothing would happen if they were to be brought into contact through a membrane possessing the assumed characteristics. Under certain conditions, however, *some* biological membranes are permeable not only to the solvent, but also to *some* of the solutes. In this case it is impossible to decide whether an osmotic flow will take place or not *solely* on the basis of the numerical value of the two π's, as computed by Eq. (26-27).*

As an example, consider the following two solutions in the same solvent, 1:

1. Solution (I) is 0.01 molar in the solute 2 and 0.005 molar in the solute 3.
2. Solution (II) is 0.01 molar in the solute 2 and 0.002 molar in the solute 3.

These solutions have different π; they are *not iso-osmotic*. Suppose that they are brought into contact through a membrane, $\mathbf{M}_a$, permeable to 1 *and* 3 but impermeable to 2. Obviously, the 3 molecules will equilibrate themselves between the two solutions, and no osmotic phenomenon will take place at all. Biologists would call solutions (I) and (II) mutually *isotonic*, with respect to this particular membrane, $\mathbf{M}_a$. If these same solutions are brought into contact through a different membrane, $\mathbf{M}_b$, permeable to the solvent only, an osmotic phenomenon *does* occur: solvent will flow from (I) to (II).† In this situation biologists would call (I) *hypertonic* with respect to (II) [and (II) *hypotonic* with respect to (I)], in conjunction with the membrane $\mathbf{M}_b$.

If in an apparatus of the type illustrated schematically in Fig. 42b, the pressure applied to the solution is *larger than* $(P_o+)\pi$, pure solvent will flow out of this solution, through the semipermeable membrane **M**, into the pure solvent compartment. This process is called *reverse osmosis.* It can be used, in principle, to achieve such goals as the purification of brackish water and the desalinization of seawater, but there are serious technical problems, e.g., the clogging up or failing of the membranes involved. Consequently, it is

---

* For the purpose of this discussion it is assumed, all along, that the biological solutions are ideally dilute.

† In this case, always assuming the strict validity of the model, the relevant experiment will yield the same result as if (II) were pure solvent and (I) a 0.003 molar solution of 3 in 1.

not easy to apply reverse osmosis, on a large scale, in an economically competitive manner (see also Problem 26-8), although small home units are not uncommon in affluent areas with drinking water of poor quality.

(c) *Freezing-Point Lowering and Boiling-Point Elevation*

Next we return to the freezing-point-lowering phenomenon, discussed in section 22.3(d) (see also Fig. 36). As before, the freezing point of the pure solvent is denoted by T_f, that of the solution by $T_f + \delta T_f$. Since the solid phase consists solely of 1 (this is not true universally, but it *is* in the case illustrated by Figs. 33 and 36, on which we are still focusing our attention; see also Exercise 26-4), the general thermodynamic equilibrium condition tells us that

$$\mu_1^{\bullet s}(P, T_f + \delta T_f) = \mu_1^{\ell}(P, T_f + \delta T_f). \tag{26-28}$$

For an ideally dilute solution of 2 in 1, we have, by Eq. (26-7'),

$$\mu_1^{\ell}(P, T_f + \delta T_f) = \mu_1^{\bullet\ell}(P, T_f + \delta T_f) + R(T_f + \delta T_f) \ln X_1^{\ell}.$$

Since $\delta T_f \ln X_1^{\ell}$ $(\approx -(n_2^{\ell}/n_1^{\ell})\delta T_f)$ is an order of magnitude smaller than $T_f \ln X_1^{\ell}$ (the model corresponds to infinite dilution!), this can be written as

$$\mu_1^{\ell}(P, T_f + \delta T_f) = \mu_1^{\bullet\ell}(P, T_f + \delta T_f) + RT_f \ln X_1^{\ell},$$

which yields in Eq. (26-28),

$$-\ln X_1^{\ell} = \frac{1}{RT_f}\left[\mu_1^{\bullet\ell}(P, T_f + \delta T_f) - \mu_1^{\bullet s}(P, T_f + \delta T_f)\right]. \tag{26-29}$$

For each phase, $\mu_1^{\bullet}$ can be expanded in a Taylor series around T_f. If we discard terms in $(\delta T_f)^2$, etc., consistent with earlier approximations in this chapter, this expansion gives

$$\mu_1^{\bullet}(P, T_f + \delta T_f) = \mu_1^{\bullet}(P, T_f) + \left(\frac{\partial \mu_1^{\bullet}}{\partial T}\right)_{P,T_f} \delta T_f,$$

which can be written

$$\mu_1^{\bullet}(P, T_f + \delta T_f) = \mu_i^{\bullet}(P, T_f) - S_1^{\bullet}(P, T_f)\, \delta T_f. \tag{26-30}$$

Introducing Eq. (26-30) into the right-hand side of Eq. (26-29) for both phases yields

$$\frac{1}{RT_f}\left[\mu_1^{\bullet\ell}(P, T_f) - \mu_1^{\bullet s}(P, T_f)\right] - \frac{1}{RT_f}\left[S_1^{\bullet\ell}(P, T_f) - S_1^{\bullet s}(P, T_f)\right]\delta T_f = -\ln X_1^{\ell}. \tag{26-31}$$

The first term in brackets is clearly zero, since it corresponds to the freezing of the pure solvent under equilibrium conditions (at pressure P, normally atmospheric pressure). The second term in brackets is $-(1/RT_f)$. $\Delta H^{\bullet}_{\text{fus}}/T_f$, and finally $-\ln X_1$, can be equated to (n_2/n_1) as before. Hence Eq. (26-31) reduces to

$$\delta T_f = -\frac{n_2}{n_1}\frac{RT_f^2}{\Delta H^{\bullet}_{\text{fus}}}, \tag{26-32}$$

in which we have dropped the henceforth implied superscripts ℓ from the composition variables. This equation is entirely equivalent with the earlier result obtained as Eq. (22-22).

Exercise 26-3

In similar fashion, obtain an expression for the boiling-point elevation, δT_b, in terms of (n_2/n_1), the boiling point of pure 1 (at the same P), T_b, and $\Delta H^{\bullet}_{\text{vap}}$. You should assume that only the solvent is volatile. Carefully note at what stage in the derivation the sign of δT_b establishes itself as the opposite to that of δT_f.

Exercise 26-4

Derive an expression for δT_f in the case that a solid solution of 2 in 1 crystallizes out. Assume that this solid phase can *also* be treated as an ideally dilute solution. The result you obtain should leave the sign of δT_f undetermined in the sense that it depends on the relative magnitudes of X_1^{s} and X_1^{ℓ}.

(Optional for those familiar with phase diagrams: Draw a melting diagram for a binary system in which the solids 1 and 2 are completely miscible, to illustrate that δT_f can be either positive or negative. Confirm that this indeed correlates with the relative magnitudes of X_1^{s} and X_1^{ℓ}, as stipulated in the main part of this exercise. Does the phase rule allow us any flexibility with regard to the values of X_1^{s} and X_1^{ℓ}?)

Equation (26-32) embodies the essential result of the thermodynamic treatment. As in the case of the basic expression for the osmotic pressure, this equation can be modified to yield several equivalent forms. For example, since, by Eq. (8-8),

$$X_2 \approx \frac{n_2}{n_1} = \frac{m_2}{1000/M_1},$$

in which m_2 is the molality of the solute and M_1 the weight of 1 mole of the solvent, and also

$$\Delta H^{\bullet}_{\text{fus}} = \ell^{\bullet}_{\text{fus}}|M_1|,$$

in which ℓ refers to 1 *gram* (rather than 1 *mole*) of the solvent, the freezing-point *lowering*, $|\delta T_f|$, can be expressed as

$$|\delta T_f| = \mathbf{K}_f m_2, \tag{26-33}$$

in which

$$\mathbf{K}_f \equiv \frac{RT_f^2}{1000\ell^{\bullet}_{\mathrm{fus}}}. \tag{26-34}$$

Equation (26-33) represents the *Law of the Freezing-Point Lowering*, familiar from general chemistry. It is a particularly useful equation because (at a given pressure) the *freezing-point constant* $\mathbf{K}_f$ is a characteristic property of the solvent; e.g., for water it is usually taken to be 1.86 K kg mole^{-1} (see also Problem 26-1). Note that while the most concise expression for the osmotic pressure, Eq. (26-26), contained the molarity, the freezing-point lowering and the boiling-point elevation are most conveniently expressed in terms of the (temperature-independent) molality of the solute.*

(d) To end this section we return briefly to the case of dissociating (or associating) solutes, discussed in more general terms in section 26.1(c). We can retrace the derivations of the preceding two subsections with only one important change: Eq. (26-7′) must be replaced by Eq. (26-12′). Readers should verify that if only a single solute is present, all colligative properties simply acquire a factor $\tilde{v}$, as defined by Eq. (26-9). For multiple solutes, the strict additivity [compare section 26.3(a)] dictates that the term $n_2\tilde{v}_2$ should be generalized to $\sum_{i=2}^{r} n_i\tilde{v}_i$. Remembering that each $\tilde{v}_i$ depends, in general, on *two* parameters (v_i and d_i), readers will appreciate that this sum usually contains too many unknowns to be disentangled by comparison with measured π, δT_f, etc. In case only a *single* solute is present, the measurement of colligative properties *can* provide us with a good idea about its dissociative or associative properties. Relevant numerical "problems" are usually assigned in general chemistry courses. At all times the readers should be wary about the possible limitations to the validity of the ideally dilute solution model, so that the conclusions drawn may have qualitative or semiquantitative value at best.

The limitations of the model become particularly acute when we have to deal with electrolyte solutions, since forces between ions are usually non-negligible. For a single "strong electrolyte," $\tilde{v} = v$ is *known* (2 for a "1-1 electrolyte" such as NaCl, 3 for a "2-1 electrolyte" such as Na_2SO_4, and so on), and one can predict the values of colligative properties based on the modified formulas referred to above. Such predictions rarely match measured values, unless the solutions involved are extremely dilute (see Table 1). One can attempt to remedy the situation by introducing an empirical "fudge factor,"

* Remember that molarities are temperature dependent!

g. As an example, for a single strong electrolyte in a solvent such as water, Eq. (26-33) is modified to read

$$|\delta T_f| = g\nu\mathbf{K}_f m_2. \qquad (26\text{-}35)$$

Sometimes $g\nu$ is written as i, called the "van't Hoff factor." The present author prefers to be reminded explicitly that i is the product of the "obvious" factor ν and the "true" interaction correction, g.

Some pertinent data are summarized in Table 1. We see that everything else being equal, g deviates the more from its ideal value of unity, (1) the higher m_2, and (2) the higher the charge of the ions. Of course, this makes eminent sense in the light of a qualitative corpuscular picture. Quantitative corpuscular theories, such as those pioneered by Debye and Hückel, fall outside the scope of this book.

Table 1

g Factors, as Defined by Eq. (26-35), from Experimental $|\delta T_f|$ Values

	g		
Molality, m_2	NaCl ($\nu = 2$)	$MgSO_4$ ($\nu = 2$)	K_2SO_4 ($\nu = 3$)
0.100	0.94	0.61	0.77
0.0100	0.97	0.77	0.90
0.00100	0.99	0.91	0.95

While g values based on, say, osmotic pressure measurements should *roughly* parallel those of Table 1, there is no reason to assume that interionic interactions should affect such different properties in *exactly* the same way.

Not everyone enthusiastically supports the introduction of empirical correction factors such as g. Of course, when the model truly fails us, the proper but much more difficult way to proceed on a phenomenological level is to abandon Eqs. (26-7′) and (26-12′) altogether and express μ_1 in terms of a solvent *activity* (see section 26.5):

$$\mu_1 = \mu_1^\bullet(P, T) + RT \ln a_1. \qquad (26\text{-}36)$$

Exercise 26-5

In a "fairly dilute" solution of a single, nondissociating solute 2 in a solvent 1, it may be necessary to use Eq. (26-36) instead of Eq. (26-7′). Assume that the solution is dilute enough for the following three approximations to be valid:

(1) $$\bar{V}_1 \approx V_1^\bullet \quad \text{and} \quad \mathbf{V} \approx n_1 V_1^\bullet, \tag{26-37a}$$

(2) $$\ln X_1 \approx -\frac{n_2}{n_1}, \tag{26-37b}$$

and

(3) $$\frac{\ln a_1}{\ln X_1} = g, \textit{ a constant.} \tag{26-37c}$$

Show that in this situation, the osmotic pressure is given by

$$\pi = gn_2 \frac{RT}{\mathbf{V}}. \tag{26-38}$$

Thus the correction factor g can be introduced in an acceptable fashion, provided that we can justify the approximations (26-37).

26.4 Chemical Equilibrium in an Ideally Dilute Solution

If a solution of several solutes in an appropriate solvent were ideally dilute in a strict sense, there would be no interactions at all between the solute molecules, and no chemical reaction between these molecules could ever take place. The situation is entirely analogous to that already discussed in section 19.2 for chemical reactions between components in an ideal mixture of ideal gases. The readers should reacquaint themselves with the contents of the brief paragraph, in small print, just below Eq. (19-23).* The essential points are that although *some* interactions must be allowed for, they should be weak enough not to impair the validity of the phenomenological equations for the model, in particular the validity of Eq. (26-1′) for the chemical potential of the solutes. As long as this is the case, it is quite straightforward to derive expressions for the equilibrium constant of the reaction $\sum_i \nu_i i = 0$ in our solution. Starting with the basic equilibrium condition $\sum_i \nu_i \mu_i = 0$, we only have to retrace the steps in the type of derivation first given in section 19.2, with Eq. (26-1′) taking the place of Eq. (19-22). The equilibrium constant can be expressed in terms of molarities, molalities, or mole fractions. In the general notation introduced in section 26.1(a), the results can be summarized as follows. The "correct" equilibrium constant, in the sense that it is a function of P, T and the choice of solvent only, is K_α ($\alpha = c$, m or X), defined by

$$K_\alpha \equiv \frac{\mathfrak{Q}_\alpha^e}{\mathfrak{Q}_\alpha^\ominus} = \frac{\mathfrak{Q}_\alpha^e}{[1_\alpha]^{\Delta\nu}}, \tag{26-39}$$

---

* See p. 226.

in which the reaction quotient, $\mathfrak{Q}_\alpha$, is the usual shorthand:

$$\mathfrak{Q}_\alpha \equiv \prod_i (\alpha_i)^{\nu_i}. \tag{26-40}$$

As before, K_α is related to the standard free energy change of the reaction

$$\Delta \mathbf{G}^{\ominus(\alpha)}(P, T, \text{solvent}) = -RT \ln K_\alpha. \tag{26-41}$$

The superscript (α) to $\Delta \mathbf{G}^{\ominus}$ should remind us [compare the relevant discussion in section 26.1(a)] that the standard states of the reactants and products depend on our choice of concentration units.

Exercise 26-6

Derive the results embodied in the set of equations (26-39) through (26-41).

Usually, K_α is *markedly* influenced by T and the choice of solvent, while the pressure dependence is of *minor* importance.

In introductory chemistry courses, molarities are still the favorite way to express the concentration of solutes; hence K_c becomes "the" equilibrium constant of record and is often simply denoted by K. As we pointed out in section 8.1, physical chemists usually prefer not to work in terms of molarities, since these themselves, unlike molalities (and mole fractions), are temperature dependent. For an ideally dilute solution, by Eq. (8-9), $\mathfrak{Q}_c$ and $\mathfrak{Q}_m$ are related as follows:

$$\mathfrak{Q}_c = \mathfrak{Q}_m(\rho_1)^{\Delta\nu}, \tag{26-42}$$

in which ρ_1 is the density of the (pure) solvent. Equation (26-42) clearly shows that a change in $\mathfrak{Q}_c^e$ [which, by Eq. (26-39), must also affect K_c], caused by a raising or lowering of T, may not only be due to an equilibrium shift but also to a change in $\rho_1(T)$. The last factor does not come into play for reactions with $\Delta\nu = 0$, or in the exceptional case when $(d\rho_1/dT) = 0$ (as for water at 4°C).

26.5 *Appendix: Equilibria, Activities, and Activity Coefficients in Real Solutions*

In this section we obviously leave the realm of model systems. Therefore, the readers may very well wonder why this material is included as the last section of the present chapter, rather than as a true "appendix" at the very end of this book. As we hinted at the conclusion of Chapter 19, the reason is that several aspects of the treatment to follow are closely related to the corresponding features of the ideally dilute solution model. Consequently, we prefer to present this discussion as long as such matters are still fresh in our minds, before we turn to Chapter 27, which addresses entirely different topics.

From the point of view of the solution chemist, the contents of this appendix merely expose the tip of the iceberg of a very extensive range of applications. Entire monographs have been devoted to this subject, in particular with emphasis on electrolyte solutions, and some books on "chemical thermodynamics" treat the material concerned in many lengthy chapters. In accordance with the nature of this book, the author has set himself the more modest goal of illuminating a few of the more salient aspects of this branch of thermodynamics. In the process we aim at bridging the gap between the fundamental features involved and a vast specialized area of applications. For further details, the readers should consult the pertinent references at the end of the chapter.

For pedagogical purposes it seems advisable to split the ensuing treatment into two parts. In subsection (a) we confine ourselves to solutions of (nondissociating) nonelectrolytes, as we did in the main portion of this chapter. In subsection (b) we touch on some of the difficulties we encounter when attempting to extrapolate the discussion to dissociating solutes, electrolytes in particular.

(a) Solutions of (Nondissociating) Nonelectrolytes

(i) As soon as we leave model systems behind us, the thermodynamic properties of solvent and solutes can no longer be expressed directly in terms of composition variables, but must be given instead in terms of the activities, defined by

$$a_k \equiv \frac{f_k}{f_k^{\ominus}}. \tag{19-64}$$

This immediately reintroduces the problem of how to select the standard states. It turns out to be most convenient to let our choices concerned closely parallel those adopted for the ideally dilute solution model. We remind the readers [see section 26.1(a)] that these choices must take into account the way we wish to express the composition of the solution. In the context of the more accurate descriptions we aim at in this appendix, we shall confine ourselves to the use of either molalities or of mole fractions; too many complications arise in conjunction with molarities, as a result of their temperature dependence. Consequently, Eq. (19-64) takes on the following forms:

For the solvent:

$$a_1 = \frac{f_1}{f_1^{\bullet}}. \tag{26-43a}$$

For the solutes:

$$a_i^{(\alpha)} \equiv \frac{f_i}{f_i^{\ominus(\alpha)}} \qquad (i = 2, \ldots, r; \alpha = m \text{ or } X). \tag{26-43b}$$

By Eq. (26-3), these last equations can be written in the form

$$a_i^{(\alpha)} = \frac{f_i}{k_i^{(\alpha)}[1_\alpha]}. \tag{26-43b$'$}$$

Therefore,

$$\frac{a_i^{(X)}}{a_i^{(m)}} = \frac{k_i^{(m)}}{k_i^{(X)}}[1_m], \tag{26-44}$$

or by Eq. (26-2a),

$$\frac{a_i^{(X)}}{a_i^{(m)}} = \frac{|M_1|}{1000}, \tag{26-45}$$

in which $|M_1|$ is the usual (dimensionless) molecular weight of the solvent. Note that the right-hand side of Eq. (26-45) is independent of i. Therefore, for the general chemical reaction $\sum_i \nu_i i = 0$, the ratio of the two correct thermodynamic equilibrium constants becomes

$$\frac{K_a^{(m)}}{K_a^{(X)}} = \left(\frac{|M_1|}{1000}\right)^{\Delta\nu}; \tag{26-46}$$

$K_a^{(m)}$ and $K_a^{(X)}$ are equal only if $\Delta\nu = 0$.

(ii) Take the natural logarithm of Eq. (26-46) and multiply the resulting equation by $-RT$:

$$-RT \ln K_a^{(m)} = -RT \ln K_a^{(X)} - RT(\Delta\nu) \ln \frac{|M_1|}{1000}. \tag{26-47}$$

If we call

$$R \ln \frac{|M_1|}{1000} \equiv b_1 \tag{26-48}$$

a constant for a given solvent, and use the basic thermodynamic relation

$$\Delta\mathbf{G}^\ominus = -RT \ln K_a, \tag{19-69}$$

Eq. (26-47) becomes

$$\Delta\mathbf{G}^{\ominus(m)} = \Delta\mathbf{G}^{\ominus(X)} - b_1 T \Delta\nu. \tag{26-49}$$

This result clearly shows that in quoting $\Delta\mathbf{G}^\ominus$ values, or in obtaining these from tabulated standard free energy data, we must insist on specifying the way in which we express the composition of our solution. Should we take similar care in quoting $\Delta\mathbf{S}^\ominus$ and $\Delta\mathbf{H}^\ominus$ values for chemical reactions? Since

$\Delta\mathbf{S}^\ominus = -(\partial\,\Delta\mathbf{G}^\ominus/\partial T)_p$, Eq. (26-49) immediately yields

$$\Delta\mathbf{S}^{\ominus(m)} = \Delta\mathbf{S}^{\ominus(X)} + b_1\,\Delta\nu. \tag{26-50}$$

But, since $\Delta\mathbf{H}^\ominus = \Delta\mathbf{G}^\ominus + T\Delta\mathbf{S}^\ominus$, we also obtain

$$\Delta\mathbf{H}^{\ominus(m)} = \Delta\mathbf{H}^{\ominus(X)}; \tag{26-51}$$

$\Delta\mathbf{H}^\ominus$ is unaffected by our choice of composition variable. This last result is initially surprising, but perhaps it becomes more acceptable if we realize that all $\bar{H}_i^\ominus$ are equal to $\bar{H}_i^{\ominus\infty}$, the partial molar enthalpy of i at infinite dilution, irrespective of whether the composition is expressed in terms of molalities or in terms of mole fractions.

The proof of this contention is simple. For the chemical potential of any solute, i, we write

$$\mu_i = \mu_i^{\ominus(\alpha)} + RT\ln a_i^{(\alpha)}$$

Since

$$\left(\frac{\partial\mu_i/T}{\partial T}\right)_{P,\,\text{composition}} = -\frac{\bar{H}_i}{T^2},$$

we can conclude that

$$\left(\frac{\partial\ln a_i^{(\alpha)}}{\partial T}\right)_{P,\,\text{composition}} = -\frac{\bar{H}_i - \bar{H}_i^{\ominus(\alpha)}}{RT^2}. \tag{26-52}$$

For an infinitely dilute solution, Eq. (26-52) becomes

$$\left(\frac{\partial\ln a_i^{\infty(\alpha)}}{\partial T}\right)_{P,\,\text{composition}} = -\frac{\bar{H}_i^\infty - \bar{H}_i^{\ominus(\alpha)}}{RT^2}. \tag{26-53}$$

But $a_i^{\infty(m)} = m_i$ and $a_i^{\infty(X)} = X_i$ (we have returned to the ideally dilute solution model!), so that the left-hand side of Eq. (26-53) is zero regardless of whether α stands for m or X. Hence the right-hand side of this equation must also vanish for both $\alpha = m$ and $\alpha = X$, so that

$$\bar{H}_i^{\ominus(m)} = \bar{H}_i^{\ominus(X)} = \bar{H}_i^\infty. \tag{26-54}$$

This result confirms the validity of Eq. (26-51), since

$$\Delta\mathbf{H}^{\ominus(\alpha)} \equiv \sum_i \nu_i\bar{H}_i^{\ominus(\alpha)}.$$

Exercise 26-7

Show that

$$\bar{H}_i^{\ominus(c)} \neq \bar{H}_i^\infty. \tag{26-55}$$

The conclusion is that in enthalpy calculations we can use the infinitely dilute solution as our standard state, but in dealing with free energy or entropy changes there is no way around the complications first discussed in conjunction with Eq. (26-3).

(iii) In section 19.3 we pointed out that it is sometimes convenient to express the properties of real *gases* in terms of *fugacity coefficients* rather than in terms of the fugacities themselves. By the same token, the properties of the (solvents 1 and the) solutes, i, in a real *solution* can often be expressed conveniently in terms of the *activity coefficients*, $\gamma_i^{(\alpha)}$, defined by

$$\gamma_i^{(\alpha)} \equiv \frac{a_i^{(\alpha)}}{|\alpha_i|}. \tag{26-56}$$

In an ideally dilute solution, all $\gamma_i^{(\alpha)}$ are unity. In analogy with Eq. (19-61) for chemical reactions in gaseous mixtures, the equilibrium constants $K_a^{(\alpha)}$ in solutions can be factored as follows:

$$K_a^{(\alpha)} = \text{“}\mathfrak{Q}_\gamma^{e(\alpha)}\text{”}\,\frac{\mathfrak{Q}_\alpha^{e}}{\mathfrak{Q}_\alpha^{\ominus}}. \tag{26-57}$$

As before, $\mathfrak{Q}_\gamma^{e(\alpha)}$ is put between quotation marks, because it is a correction factor to the K_α of Eq. (26-39), not a reaction quotient at all.

(iv) To obtain numerical values of activities, we must judiciously interpret the appropriate experimental data by means of a thermodynamic formalism. The entire procedure is often lengthy and complex, which explains why chemists are so frequently tempted to use the ideally dilute solution model outside the range of its reliability, or to accept simple empirical corrections such as those mentioned at the end of section 19.3.

In principle, activities can be related to measured colligative properties by very similar techniques as were employed in the main part of this chapter. Let us illustrate these methods briefly, by means of two examples.

Example 1 Activities from equilibrium vapor pressures

Obviously, this makes sense only if the substance involved is sufficiently volatile. For the liquid–gas equilibrium of a volatile solute i, the most general thermodynamic equilibrium condition is

$$\mu_i^{\ominus g} + RT \ln |f_i^g| = \mu_i^{\ominus \ell(X)} + RT \ln a_i^{\ell(X)},$$

from which

$$a_i^{\ell(X)} = |f_i^g|\,e^{(\mu_i^{\ominus g} - \mu_i^{\ominus \ell(X)})/RT} \tag{26-58}$$

By Eq. (26-16), this can be written

$$a_i^{\ell(X)} = \frac{1}{k_i^*}\,|f_i^g|, \tag{26-59}$$

in which k_i^* is a limiting law constant for the ideally dilute solution, to be determined by extrapolating a series of experimental data to infinite dilution.

The procedure is most useful if pressures are low enough for the equilibrium *vapor* to be ideal. In that case,

$$a_i^{\ell(X)} = \frac{1}{k_i^*} p_i. \tag{26-60}$$

If this simplication *cannot* be made, we are faced with the additional task of relating the f_i^g to the measured p_i, a problem discussed in earlier chapters.

Example 2 Activities from freezing-point measurements

In a direct sense this can lead only to the solvent activity, a_1^ℓ. Proceeding as in section 26.3(a), we immediately write for the thermodynamic equilibrium condition at the freezing point of the solution:

$$\mu_1^{\bullet s}(P, T_f + \Delta T_f) = \mu_1^{\bullet \ell}(P, T_f + \Delta T_f) + R(T_f + \Delta T_f) \ln a_1^\ell. \tag{26-61}$$

Of course, ΔT_f is not necessarily very small. Call $T_f + \Delta T_f \equiv T'_f$, so that Eq. (26-61) can be written

$$\frac{\mu_1^{\bullet s}(P, T'_f)}{T'_f} = \frac{\mu_1^{\bullet \ell}(P, T'_f)}{T'_f} + R \ln a_1^\ell. \tag{26-62}$$

From this equation we subtract the equilibrium condition for the freezing of the pure solvent at the same pressure:

$$\frac{\mu_1^{\bullet s}(P, T_f)}{T_f} = \frac{\mu_1^{\bullet \ell}(P, T_f)}{T_f}, \tag{26-63}$$

to yield

$$\ln a_1^\ell = \frac{1}{R}\left[\frac{\mu_1^{\bullet s}(P, T'_f)}{T'_f} - \frac{\mu_1^{\bullet s}(P, T_f)}{T_f}\right] - \frac{1}{R}\left[\frac{\mu_1^{\bullet \ell}(P, T'_f)}{T'_f} - \frac{\mu_1^{\bullet \ell}(P, T_f)}{T_f}\right]. \tag{26-64}$$

Since, in general,

$$\left(\frac{\partial \mu_1^\bullet / T}{\partial T}\right)_P = -\frac{H_1^\bullet}{T^2}, \tag{26-65}$$

the result Eq. (26-64) can be written much simpler as

$$\ln a_1^\ell \equiv \ln a_1 = \int_{T_f \atop (P)}^{T_f + \Delta T_f} \frac{\Delta H_{\text{fus}}^\bullet}{RT^2}\, dT, \tag{26-66}$$

in which $\Delta H_{\text{fus}}^\bullet$ is the molar heat of fusion of the solvent, $H_1^{\bullet \ell} - H_1^{\bullet s}$, at a pressure P and a temperature T. The problem is that $\Delta H_{\text{fus}}^\bullet$ is not independent of the temperature. For example, for H_2O, this quantity is 6.008 kJ mole^{-1} at 0°C, and 5.619 kJ mole^{-1} at -10°C. It is possible to relate this temperature dependence to $\Delta C_p^\bullet$ values, but one way or another we must obtain calorimetric data involving supercooled liquids. For a few important solvents such data have indeed been determined.

Finally, we remind the readers that the activities of various components in a solution are related by a Gibbs–Duhem type of relation, which at given P and T can be written

$$\sum_{k=1}^{r} n_k \, d \ln a_k = 0 \tag{26-67a}$$

or

$$\sum_{k=1}^{r} X_k \, d \ln a_k = 0. \tag{26-67b}$$

This can be very useful, for example, for the determination of the activity of a single solute, a_2, in a solution for which the solvent activity, a_1, has been obtained by means of freezing-point measurements.

(*b*) *Electrolyte Solutions*

(i) In order to appreciate the special considerations that we must give to electrolytic dissociation, let us first return briefly to a *nonelectrolytic* dissociation such as

$$PCl_5 \rightleftharpoons PCl_3 + Cl_2,$$

e.g., in benzene solution. The general equilibrium condition

$$\mu_{PCl_5} = \mu_{PCl_3} + \mu_{Cl_2} \tag{26-68}$$

is valid whether the solution is ideally dilute or not. In a real solution the respective chemical potentials must be expressed in terms of activities rather than mole fractions or molalities, but this creates no new difficulties, at least not on a conceptual level.

Purely *formally*, electrolytic dissociation leads to relations analogous to Eq. (26-68). For example, the dissociation equilibrium for the *weak electrolyte* CH_3COOH in water is subject to the condition

$$\mu_{CH_3COOH} = \mu_{CH_3COO^-} + \mu_{H^+}, \tag{26-69a}$$

and for a *strong electrolyte* such as NaCl in water we can write

$$\mu_{NaCl} = \mu_{Na^+} + \mu_{Cl^-}. \tag{26-69b}$$

However, this formal resemblance of Eqs. (26-68) and (26-69) can be very misleading; there are at least two new features to which we must draw the reader's attention. First, in the case of CH_3COOH, we are still dealing with a "true" equilibrium condition, but since NaCl is usually considered totally dissociated into ions, we must look at both sides of Eq. (26-69b) as two equivalent ways of expressing the partial molar free energy of sodium chloride in water. Second, positive or negative ions cannot exist by themselves (unlike PCl_3 and Cl_2!), so that the question must be raised whether the chemical potentials of these *individual* ions have operational significance. We shall return briefly to the latter point below. Regardless of how this issue may be resolved, it is customary and convenient to adopt a formal scheme in which such potentials, as well as the corresponding individual ion activities,* are introduced. Thus, using sodium chloride as an example, we not only write

$$\mu_{NaCl} = \mu_{NaCl}^{\ominus} + RT \ln a_{NaCl}, \tag{26-70}$$

but also:

$$\mu_{Na^+} = \mu_{Na^+}^{\ominus} + RT \ln a_{Na^+} \tag{26-71a}$$

and

$$\mu_{Cl^-} = \mu_{Cl^-}^{\ominus} + RT \ln a_{Cl^-}. \tag{26-71b}$$

A (geometric) *mean ionic activity*, $a_\pm$, is defined by

$$a_\pm \equiv (a_{Na^+} a_{Cl^-})^{1/2}. \tag{26-72}$$

With the identification $\mu_{NaCl}^{\ominus} = \mu_{Na^+}^{\ominus} + \mu_{Cl^-}^{\ominus}$, the preceding set of equations, starting with Eq. (26-69b), is consistent if we write $a_{NaCl} = a_{Na^+} a_{Cl^-}$, so that

$$\mu_{NaCl} = \mu_{NaCl}^{\ominus} + RT \ln a_\pm^2. \tag{26-73}$$

As always, some prefer to work in terms of activity *coefficients.* Since the molality scale is the only suitable one in this context, its use will always be *implied* below. We define

$$\gamma_{Na^+} \equiv \frac{a_{Na^+}}{|m|_{Na^+}} = \frac{a_{Na^+}}{|m|_{NaCl}}, \tag{26-74a}$$

$$\gamma_{Cl^-} \equiv \frac{a_{Cl^-}}{|m|_{Cl^-}} = \frac{a_{Cl^-}}{|m|_{NaCl}}, \tag{26-74b}$$

and

$$\gamma_\pm \equiv (\gamma_{Na^+} \gamma_{Cl^-})^{1/2} = \frac{a_\pm}{|m|_{NaCl}} = \frac{a_{NaCl}^{1/2}}{|m|_{NaCl}}. \tag{26-74c}$$

(ii) The quantities μ_{NaCl}, a_{NaCl}, $a_\pm$, and $\gamma_\pm$ all have "obvious" *operational* meaning. By this we mean that several methods we can use in the determination of activities, etc., of *nonelectrolytes* [compare section 26.5(d), part (iv)] can, *at least in principle,*

---

* In order not to subject readers to a deluge of equations, this formalism will be presented exclusively in terms of chemical potentials and activities; fugacities will be bypassed.

still be used in the study of electrolytes, although some more severe difficulties manifest themselves in the latter case.

In contrast, the chemical potentials, activities, and activity coefficients of the individual ions cannot be obtained in similar fashion. The essential point is that we cannot study some property of a solution as a function of, say, the concentration of the *positive* ion of interest, because with any change in *this* concentration we must also change the concentration of *some* negative "counter ion." Nevertheless, there are at least two methods, one theoretical and one experimental, which can shed some light on the numerical values of these elusive properties. Theoretically, we can approach the problem on a *corpuscular* level by means of models of inter ionic interactions, such as the one introduced by Debye and Hückel in 1923, or its more refined descendants. In these theories, one calculates the extra free energy of an ionic species due to electrostatic attractions and repulsions involving the other ions. Success has been limited to relatively dilute solutions. The experimental method is based on the very difficult and controversial measurement of single-electrode potentials (absolute half-cell EMFs).* We remind readers of the relation between *ratios and/or products of* (ionic) activities and cell EMFs, given in section 19.4(a); see, in particular, Eq. (19-68) as well as Problems 19-3 and 26-7. In entirely similar fashion, absolute ionic activities are related to the aforementioned *half*-cell EMFs. For a recent discussion of the concept of absolute electrode potentials, we refer readers to a paper by Trasatti.† These matters fall outside the scope of this book.

(iii) Returning to the NaCl example, it is of interest to summarize the various limits in case we are dealing with an ideally dilute solution. As $m_{\text{NaCl}} \to 0$,

$$a_{\text{Na}^+} \to |m|_{\text{Na}^+} = |m|_{\text{NaCl}} \qquad \gamma_{\text{Na}^+} \to 1, \tag{26-75a}$$

$$a_{\text{Cl}^-} \to |m|_{\text{Cl}^-} = |m|_{\text{NaCl}} \qquad \gamma_{\text{Cl}^-} \to 1, \tag{26-75b}$$

$$a_{\pm} \to |m|_{\pm} = |m|_{\text{NaCl}} \qquad \gamma_{\pm} \to 1, \tag{26-75c}$$

$$a_{\text{NaCl}} \to |m|^2_{\text{NaCl}}. \tag{26-75d}$$

Apparently, the limiting formula for the chemical potential becomes

$$\mu_{\text{NaCl}} = \mu^{\ominus}_{\text{NaCl}} + RT \ln |m|^2_{\text{NaCl}}. \tag{26-76}$$

This $|m|$-*square* dependence can be verified by the appropriate extrapolation of experimental data. Such extrapolation procedures are much harder to carry out for electrolytes than for nonelectrolytes. The reason is that we have to go to extremely low concentrations (often $|m| \approx 0.0001$ or even lower) before ionic interactions can be neglected, hence before the ideally dilute solution model is approached with sufficient accuracy to this purpose.

* See, e.g., R. Gomer and G. Tyson, *J. Chem. Phys.* **66**, 4413 (1977).
† S. Trasatti, *J. Electroanal. Chem.* **139**, 1 (1982).

(iv) To illustrate the relevance of electrolyte activities, let us consider a familiar example from general and analytical chemistry, the solubility of AgCl in water, at room temperature and atmospheric pressure, in the possible presence of other electrolytes. This is of great importance in the gravimetric determination of silver or chloride. We are dealing with the process

$$AgCl(s) \rightleftharpoons AgCl(aq) \rightarrow Ag^{+}(aq) + Cl^{-}(aq),$$

for which the correct thermodynamic equilibrium constant is

$$K_a = \frac{a_{AgCl(aq)}}{a_{AgCl(s)}} = \frac{a_{Ag^+}a_{Cl^-}}{a_{AgCl(s)}}.$$

According to the discussion in section 19.4(b), $a_{AgCl(s)}$ can be put equal to unity. Expressing the product of the ion activities as the product of the corresponding molalities and the mean ionic activity coefficient (as in the NaCl example, discussed above), we obtain

$$K_a \equiv K_{sp} = (m_{Ag^+}m_{Cl^-})\gamma_{\pm}^2.$$

This is the *solubility product* of silver chloride in the solution. Under the conditions stipulated, K_{sp} is about 10^{-10}. When we teach our general chemistry students that $m_{Ag^+}m_{Cl^-}$* is *constant*, we are in effect assuming that $\gamma_{\pm}$ is unity, which is tantamount to assuming the validity of the ideally dilute solution model. This suffices to account for the *common-ion effect*, that is, for the *decrease* in solubility with the initial addition of an excess of precipitating agent, i.e., $AgNO_3$ or KCl.† Experimentally, it has been known for a long time, to analytical chemists in particular, that the solubility of AgCl in water *increases* in the presence of salts such as KNO_3. This *diverse ion effect* cannot be explained on the basis of the ideally dilute solution model; the presence of more or less KNO_3 cannot possibly alter m_{Ag^+} and m_{Cl^-}! The conclusion is that KNO_3 *must* affect $\gamma_{\pm}$. Qualitatively there is nothing mysterious about this; $\gamma_{\pm}$ is determined by the interactions of Ag^+ and Cl^- ions, not only with each other, but also with all other ions present. One of the early applications of the Debye–Hückel theory, mentioned in section 26.5(b), part (ii), clearly showed that $\gamma_{\pm}$ must in effect *decrease* as m_{KNO_3} *increases*,‡ so that the constancy of K_a, required by thermodynamics, indeed implies an increase in $m_{Ag^+}m_{Cl^-}$, as found experimentally. This example clearly illustrates the interplay of the correct thermodynamic analysis with corpuscular theories, in order to elucidate experimental results.

---

* Probably, $c_{Ag^+}c_{Cl^-}$ was considered instead.

† Continued addition of such electrolytes may cause the solubility of electrolytes to *increase* again for a variety of reasons, primarily the formation of complexes such as $AgCl_2^-$.

‡ See also Problem 26-6, in which $\gamma_{\pm}$ is given in terms of the *ionic strength*, $I \equiv \frac{1}{2}\sum_i m_i z_i^2$.

(v) In the preceding subsection we used NaCl and AgCl as examples. One of the reasons is that the formalism with which we wanted to acquaint readers is simplest for such a "1-1 electrolyte," i.e., an electrolyte that dissociates into *two* ions. For a ν_+-ν_- electrolyte, which dissociates into $(\nu_+ + \nu_-) \equiv \nu$ ions ($\nu > 2$), the scheme is generalized readily, but some of the equations concerned become quite clumsy and are much easier written down than "digested." They really become meaningful only if one becomes familiarized with this particular field. With this in mind, we conclude this appendix (and this chapter) with only a very brief summary of the formalism concerned. For details, interested readers must be referred to the literature on electrolyte solutions (see the references at the end of the chapter).

Consider a solution of a single strong electrolyte, 2, in a solvent, 1. Let 2 be ionized as indicated above, so that we can write for its chemical potential [compare the discussion following Eq. (26-69b)]:

$$\mu_2 = \nu_+\mu_+ + \nu_-\mu_-, \tag{26-77}$$

in which

$$\mu_2 = \mu_2^\ominus + RT \ln a_2, \tag{26-78a}$$

$$\mu_+ = \mu_+^\ominus + RT \ln a_+, \tag{26-78b}$$

and

$$\mu_- = \mu_-^\ominus + RT \ln a_-. \tag{26-78c}$$

With the identification $\mu_2^\ominus = \nu_+\mu_+^\ominus + \nu_-^\ominus\mu_-$, Eqs. (26-77) and (26-78) are consistent if

$$a_2 = a_+^{\nu_+} a_-^{\nu_-}. \tag{26-79}$$

The *mean ionic activity*, $a_\pm$, is defined by

$$a_\pm \equiv (a_+^{\nu_+} a_-^{\nu_-})^{1/\nu}, \tag{26-80}$$

so that μ_2 becomes

$$\mu_2 = \mu_2^\ominus + RT \ln (a_\pm)^\nu. \tag{26-81}$$

If one prefers to work in terms of activity *coefficients*, the following additional definitions are needed:

$$\gamma_+ \equiv \frac{a_+}{|m_+|}, \tag{26-82a}$$

$$\gamma_- \equiv \frac{a_-}{|m_-|}, \tag{26-82b}$$

and

$$\gamma_\pm \equiv \frac{a_\pm}{|m_\pm|}. \tag{26-82c}$$

In the last equation the *mean ionic molality*, $m_\pm$, is defined by

$$m_\pm \equiv (m_+^{\nu_+} m_-^{\nu_-})^{1/\nu} = [(\nu_+ m_2)^{\nu_+}(\nu_- m_2)^{\nu_-}]^{1/\nu} = m_2(\nu_+^{\nu_+}\nu_-^{\nu_-})^{1/\nu}. \qquad (26\text{-}83)$$

The readers can easily verify that for a 1-1 electrolyte $m_\pm = m_2$ [compare Eq. (26-75c)]. For other electrolytes the relation between $m_\pm$ and m_2 is more complicated; e.g., for a 1-2 electrolyte such as $BaCl_2$,

$$m_\pm = 4^{1/3} m_2.$$

The most important limiting laws for the ideally dilute solution are

$$\mu_+ = \mu_+^\ominus + RT \ln |m_+|, \qquad (26\text{-}84a)$$

$$\mu_- = \mu_-^\ominus + RT \ln |m_-|, \qquad (26\text{-}84b)$$

and

$$\mu_2 = \mu_2^\ominus + RT \ln [\nu_+^{\nu_+}\nu_-^{\nu_-}|m_2|^\nu] = \mu_2^\ominus + RT \ln |m_\pm|^\nu. \qquad (26\text{-}85)$$

Equations (26-81) and (26-85) show that the standard state of the electrolyte 2 is either the *real* solution with a mean ionic *activity* of 1, or the *hypothetical* solution that would still be ideally dilute at a mean ionic *molality* of 1. Just as for 1-1 electrolytes ($\nu = 2$), the m_2^ν dependence, embodied in Eq. (26-85), can be verified experimentally if the appropriate data can be extrapolated to very dilute solutions.

The two possible definitions of the standard state for 2 may remind the readers of a similar duality in fixing the standard state of a gas [see section 19.3(a)]. But while a gas with a fugacity of 1 atm usually has a pressure very close to this (a gas at 1 atm pressure is close to ideal), the situation is very different for electrolyte solutions. For example, it takes nearly exactly a 2 *m* solution of HCl in water to get $a_{HCl} = 1$! Hence such electrolyte solutions are very far from ideally dilute.

PROBLEMS

Problem 26-1

Empirical values usually given for $\mathbf{K}_f$ and $\mathbf{K}_b$ (the freezing- and boiling-point constants, respectively) of H_2O are 1.86 K kg mole^{-1} and 0.51 K kg mole^{-1}. Compute these constants from Eq. (26-34) and the corresponding equation derived via the result of Exercise 26-3, if $\ell_{fus}^\bullet$ and $\ell_{vap}^\bullet$ are 333 J g^{-1} and 2260 J g^{-1}, respectively.

Problem 26-2

Three grams of a high-molecular-weight polymer ($|M| = 30{,}000$) is dissolved in 1 kg of benzene at 25°C. The density of benzene, at this temperature, is 0.874 kg ℓ^{-1}, and its freezing-point constant, $\mathbf{K}_f$, is 5.12 K kg mole^{-1}.

(a) Compute the freezing-point lowering, $|\delta T_f|$, and the osmotic pressure, π, of this solution.

(b) Which of these two colligative properties appears to be more appropriate to determine M of this polymer (if it had not been known)? What difficulties may be created by increasing m to obtain a higher $|\delta T_f|$ (neglecting saturation problems)?

(c) What practical problems may arise in measuring π for a substance with *low* molecular weight?

Problem 26-3

Consider Henry's Law in the form (26-19)

$$|p_2| = k_2^* X_2^\ell \qquad \text{(only a single solute present).}$$

(a) Prove that

$$\left(\frac{\partial \ln k_2^*}{\partial T}\right)_P = \frac{-\Delta H^\ominus_{\text{soln}}}{RT^2},$$

in which $\Delta H^\ominus_{\text{soln}}$ refers to the dissolving of 1 mole of gas, 2, in the solvent, 1.

(b) If $2 \equiv H_2$ and $1 \equiv H_2O$, k_2^* equals 5.20×10^7 at 20°C and 5.34×10^7 at 25°C, provided that p_2 is expressed in torr. Compute $\Delta G^\ominus_{\text{soln}}$ at these two temperatures, and $\Delta H^\ominus_{\text{soln}}$ in the temperature range 20 to 25°C (assuming that it is constant in this interval).

Problem 26-4

In section 19.1(c) we discussed the distribution of a solute, i, between two immiscible solvents, I and II.

(a) If both solutions are ideally dilute, derive the *Nernst distribution law*:

$$\frac{\alpha_{i \text{ in I}}}{\alpha_{i \text{ in II}}} = K^{(\alpha)}, \tag{1}$$

in which $\alpha = X$, m, or c. As you carry out this derivation, carefully note the functional dependence of the *distribution coefficient*, $K^{(\alpha)}$.

(b) For $i \equiv I_2$, $\text{I} \equiv CCl_4$ and $\text{II} = H_2O$, $K^{(c)} = 85$, at a temperature of 300 K. If 0.15 g of I_2 is shaken up with a mixture of 1 ℓ of H_2O and 1 ℓ of CCl_4, at 300 K, compute the distribution of the iodine between the two solvents at equilibrium.

Problem 26-5

A solid substance, 2, is dissolved in a solvent, 1. Assume that the 2 molecules remain undissociated in this solution. Let a_2^{sat} and X_2^{sat} represent the activity and the mole fraction of 2, respectively, *upon saturation.*

(a) Prove that

$$\left(\frac{\partial \ln a_2^{\text{sat}}}{\partial T}\right)_P = \frac{\Delta H^\ominus_{\text{soln}}}{RT^2}, \tag{1}$$

and carefully note how the term $\Delta H^\ominus_{\text{soln}}$ arises.

(b) Under what assumptions can (1) be converted to

$$\left(\frac{\partial \ln X_2^{\text{sat}}}{\partial T}\right)_P = \frac{(\Delta H^\bullet_{\text{fus}})_2}{RT^2}? \tag{2}$$

(c) If T_{f2} is the melting point of pure 2, at the pressure of interest, P, show that (2) can be integrated to yield the solubility of 2 in 1 [at a given temperature T $(T < T_{f2})$], expressed in terms of its mole fraction:

$$(\ln X_2^{\text{sat}})_{\text{at } T} = \frac{(\Delta H_{\text{fus}}^{\bullet})_2}{R}\left(\frac{1}{T_{f2}} - \frac{1}{T}\right). \tag{3}$$

Note that Eqs. (2) and (3) do not involve any property of the solvent except in the indirect sense that our choice of solvent, in conjunction with the nature of the solute, must allow for some of the approximations inherent in the conversion of (1) to (2). Equation (1) *does* involve solvent properties. In what ways?

(d) At 25°C, $X_{\text{naphthalene}}^{\text{sat}}$ in benzene solvent is found to be 0.296, experimentally. If $T_{f,\,\text{naphthalene}} = 80.5°\text{C}$ and $(\Delta H_{\text{fus}}^{\bullet})_{\text{naphthalene}} = 19.3$ kJ mole^{-1}, compute $X_{\text{naphthalene}}^{\text{sat}}$ from Eq. (3). Is it sensible to assume that Eq. (3) is a good approximation for this particular solution? Try to answer this question *before* you carry out the computation.

Problem 26-6

The solubility product of AgCl in water is 1.78×10^{-10} mole^2 kg^{-2} at 25°C.

(a) Assuming that $\gamma_{\pm} = 1$, compute the solubility of AgCl in pure water (in mg of AgCl per kg of water) at 25°C.

(b) Compute the percent increase in solubility in a 0.100 molal aqueous KNO_3 solution at the same temperature. To calculate $\gamma_{\pm}$ for AgCl in such a fairly concentrated electrolyte solution, we use an equation given by C. W. Davies (*Ion Association*, Butterworth, London, 1962, pp. 39–43). For aqueous solutions at 25°C, this reads

$$\log \gamma_{\pm} = -0.51\, z_+|z_-|\left(\frac{|I|^{1/2}}{1 + |I|^{1/2}} - 0.30|I|\right),$$

in which I is the *ionic strength*, originally introduced by Lewis and Randall, of pivotal importance in the Debye–Hückel theory;

$$I \equiv \frac{1}{2}\sum_i m_i z_i^2.$$

(m_i is the molality of ion i, with charge $z_i|e|$.)

Problem 26-7

The cell

$$\text{Pt}|\text{H}_2(P = 1 \text{ atm})|\text{HCl}[\text{aq}(m)]|\text{AgCl, Ag}|\text{Pt}$$

has already been considered in Problem 15-6.

(a) Assuming that $a_{\text{H}_2} = |P_{\text{H}_2}| = 1$, show that at 25°C the EMF of this cell is given by

$$\mathscr{E} = \mathscr{E}^{\ominus} - 0.1183 \log [\gamma_{\pm} m_{\text{HCl}}].$$

(b) Obtain $\mathscr{E}^{\ominus}$ at 25°C from the empirical formula given in Problem 15-6.

(c) For $m_{\text{HCl}} = 0.01$ mole kg^{-1}, one finds experimentally $\mathscr{E} = 0.4642$ V. Compute $\gamma_{\pm}$ for HCl in water at this molarity.

Problem 26-8

For the purposes of this problem, we shall consider seawater simply as an aqueous solution containing 3.20% NaCl by weight (disregarding "minor" constituents, involving primarily the ions Ca^{2+}, Mg^{2+}, K^{+}, Cl^{-}, and SO_4^{2-}), having a density of 1.03 g/ml, at 20°C. The g factor [cf. the discussion in section 26.3(d)] for the calculation of the osmotic pressure of NaCl in this seawater may be taken as 0.91.

Based on these data:

(a) Compute the (approximate) osmotic pressure of seawater at 20°C.

(b) Compute the minimum amount of work, in kilojoules, needed to produce 1000 gal of pure water at this temperature, by *reverse osmosis*.

(c) Estimate the cost of such a desalinization, based on an energy price of 5.5 cents per kilowatt hour. (You may wish to substitute the current price in your area.)

(d) Compare the answers, obtained in part (b), with the amount of heat, in kilojoules needed to boil off 1000 gal of water at 100°C, 1 atm. (For $\ell^{\bullet}_{\text{vap}}$, see Problem 26-1.) Is this a fair comparison?

REFERENCES

The ideally dilute solution model is treated, to various degrees of sophistication, in all textbooks on physical chemistry and on (chemical) thermodynamics. The relevant chapters in the books listed as references 1 through 4 at the end of Chapter 25 are as good as any we can recommend. Different authors emphasize the use of fugacities (as compared to activities) to different degrees. It may be prudent to remind readers that there is no universally accepted nomenclature in denoting various models (see section 25.1).

For the material summarized in the Appendix (section 26.5), we strongly recommend Chapters 19 through 22 of I. M. Klotz and R. M. Rosenberg's *Chemical Thermodynamics*, 3rd ed. (W. A. Benjamin, Menlo Park, Calif., 1972). For electrolyte solutions, the "classic" book, *The Physical Chemistry of Electrolyte Solutions*, by H. S. Harned and B. B. Owen, 3rd ed. (Reinhold, New York, 1958), is probably still without equal.

Chapter 27

The van der Waals Model for Binary Fluids

This is an optional chapter, intended for those advanced readers with a special interest in phase equilibria. The theoretical development that follows will enable us to discuss the unusual, somewhat surprising, "retrograde" and "barotropic" phenomena. These are of interest both from a purely scientific and from a technical point of view. Prior to the study of this chapter, a rereading of section 18.4, which dealt with a *single*-component van der Waals type fluid, is strongly recommended.

27.1 Introduction; the Equation of State and the Helmholtz Free Energy

(a) There is a fair amount of overlap between the material of the present chapter and that given in Chapter 22, much of which also originated with van der Waals. On the whole, the contents of Chapter 22 is of more *general* validity, although, e.g., in Fig. 35 (top) we implicitly assumed a metastable van der Waals *type* liquid–gas continuation. In what follows we shall:

1. Confine ourselves to liquid–gas equilibria.
2. Explicitly incorporate the van der Waals equation of state for binary mixtures.
3. Following van der Waals, couch the discussion primarily in terms of $A(V, T, x)$, rather than $G(P, T, x)$, as our characteristic function.

This predominant use of A suggests that to avoid confusion we should label the two components 1 and 2 (rather than A and B, as in previous chapters), so that x must henceforth be identified with X_2 $(1 - x = X_1)$. In Chapter 22, superscripts **1** and **2** denoted *any* two phases, but in the present chapter, as already noted, we confine ourselves to liquid–gas equilibria. If necessary, the two phases will be specified by superscripts ℓ and g. There should be no difficulties in adapting to these changes in notation.

(b) Next, we remind the readers of the van der Waals equation of state for a *single*-component fluid:

$$\left(P + \frac{n^2 a}{\mathbf{V}^2}\right)(\mathbf{V} - nb) = nRT \tag{10-30}$$

or

$$P = \frac{nRT}{\mathbf{V} - nb} - \frac{n^2 a}{\mathbf{V}^2}. \tag{27-1}$$

The meaning of the positive van der Waals constants, a and b, has been discussed in section 10.2. The corresponding equation of state for a binary mixture, consisting of n_1 moles of 1 and n_2 moles of 2, is

$$P = \frac{(n_1 + n_2)RT}{\mathbf{V} - n_1 b_1 - n_2 b_2} - \frac{n_1^2 a_1 + 2n_1 n_2 a_{12} + n_2^2 a_2}{\mathbf{V}^2}, \tag{27-2}$$

in which a_i and b_i are the "ordinary" van der Waals constants for component $i\,(=1$ or $2)$ and a_{12} is a new (positive) constant characterizing the *interspecies* attractions.

The rationalization of this equation follows along essentially the same lines as that of Eq. (27-1), presented in section 10.2. If we assume, with van der Waals, that only *pair* interactions need be taken into account, there are $\frac{1}{2}N_1\,(N_1 - 1)$ 1-1 pairs, $\frac{1}{2}N_2\,(N_2 - 1)$ 2-2 pairs, and $N_1 N_2$ 1-2 pairs to consider. Since the $N_i\,(=n_i N_{\text{Av}})$ are very large, these numbers are in effect $\frac{1}{2}N_1^2, \frac{1}{2}N_2^2$, and $N_1 N_2$, respectively. This explains the appearance of the factor 2 with the interspecies attraction term, $2n_1 n_2 a_{12}$, in Eq. (27-2).

Divide the numerator and denominator of the first term on the right-hand side of Eq. (27-2) by $(n_1 + n_2)$, and the numerator and denominator of the second term by $(n_1 + n_2)^2$. With x as defined above and $V \equiv \mathbf{V}/(n_1 + n_2)$, the (mean) molar volume, we obtain

$$P = \frac{RT}{V - b_1(1 - x) - b_2 x} - \frac{(1 - x)^2 a_1 + 2x(1 - x)a_{12} + x^2 a_2}{V^2}. \tag{27-3}$$

After defining

$$a(x) \equiv (1 - x)^2 a_1 + 2x(1 - x)a_{12} + a_2 x^2 \tag{27-4a}$$

and

$$b(x) \equiv b_1(1 - x) + b_2 x, \tag{27-4b}$$

Eq. (27-3) for our *binary* mixture takes on the appealing simple form

$$P = \frac{RT}{V - b(x)} - \frac{a(x)}{V^2}, \tag{27-5}$$

in formal analogy with

$$P = \frac{RT}{V - b} - \frac{a}{V^2}, \tag{10-3\tilde{0}}$$

for 1 mole of a *single*-component fluid. While in the last equation, at least in van der Waals' approximation, a and b are true *constants* for a given substance, $a(x)$ and $b(x)$ of the *binary* equation of state, Eq. (27-5), are evidently *functions of the composition* (as well as of the nature of the two components).

(c) Let V^∞ indicate a very large molar volume corresponding to a very small P. We have the mathematical identity

$$A(V, T, x) = A(V^\infty, T, x) - \int_V^{V^\infty} \left(\frac{\partial A}{\partial V}\right)_{T,x} dV. \tag{27-6}$$

Since the integrand on the right-hand side is just $-P$, this equation becomes

$$A(V, T, x) = A(V^\infty, T, x) + \int_V^{V^\infty} P\, dV. \tag{27-7}$$

But for small enough P we can use Eq. (23-16) for 1 mole of an ideal mixture of ideal gases:

$$A(V^\infty, T, x) = (1 - x)A_1^*(T) + xA_2^*(T) - RT \ln V^\infty + RT[x \ln x + (1 - x) \ln (1 - x)], \tag{27-8}$$

with A^* defined by Eq. (23-9). Also, by Eq. (27-5),

$$\int_V^{V^\infty} P\, dV = RT \ln \frac{V^\infty - b(x)}{V - b(x)} + \frac{a(x)}{V^\infty} - \frac{a(x)}{V},$$

or, since V^∞ is very large,

$$\int_V^{V^\infty} P\, dV = RT \ln \frac{V^\infty}{V - b(x)} - \frac{a(x)}{V}. \tag{27-9}$$

If we substitute Eqs. (27-8) and (27-9) into Eq. (27-7), the undesirable terms $\ln V^\infty$ cancel out. With the obvious shorthand

$$A^*(T) \equiv (1 - x)A_1^*(T) + xA_2^*(T), \tag{27-10}$$

the final result reads

$$A(V, T, x) = A^*(T) - RT \ln [V - b(x)] - \frac{a(x)}{V} + RT[x \ln x + (1 - x) \ln (1 - x)]. \tag{27-11}$$

According to van der Waals, this equation should cover *both* the liquid *and* the gas phase of the system.

In the limiting case of a *single*-component van der Waals fluid ($x = 0$ or 1),

$$A(V, T) = A^*(T) - RT \ln (V - b) - \frac{a}{V}. \tag{27-12}$$

Note that $A(V, T, x)$ *cannot* be obtained from $A(V, T)$ if we simply replace the *constants* a and b by $a(x)$ and $b(x)$, respectively [compare the corresponding equations of state (27-5) and (10-30)]. $A(V, T, x)$ contains the *additional* term

$$RT[x \ln x + (1 - x) \ln (1 - x)], \tag{27-13}$$

our old friend, the "Gibbs paradox," first encountered in section 23.2 and discussed in section 25.3(c).

(d) In most basic physical chemistry textbooks it is shown that for a single-component van der Waals fluid, the critical temperature, T_{Cr}, is given by

$$T_{\mathrm{Cr}} = \frac{8}{27}\frac{a}{b}. \tag{27-14a}$$

In similar fashion we obtain for our binary system

$$T_{\mathrm{Cr}}(x) = \frac{8}{27}\frac{a(x)}{b(x)}. \tag{27-14b}$$

Exercise 27-1

Derive Eq. (27-14b) in the usual fashion, i.e., from Eq. (27-5) coupled to the vanishing of the first and second differentials of P with respect to V at constant T, and x.

The simplest shape of the $T_{\mathrm{Cr}}(x)$ curve is drawn in Fig. 43.

Let us consider the A,V,x surface at a constant temperature below the critical line [Fig. 44(a)]. The intersection of this surface with a plane at constant x, x' say, has the same general shape as the A,V curves for pure 1 and

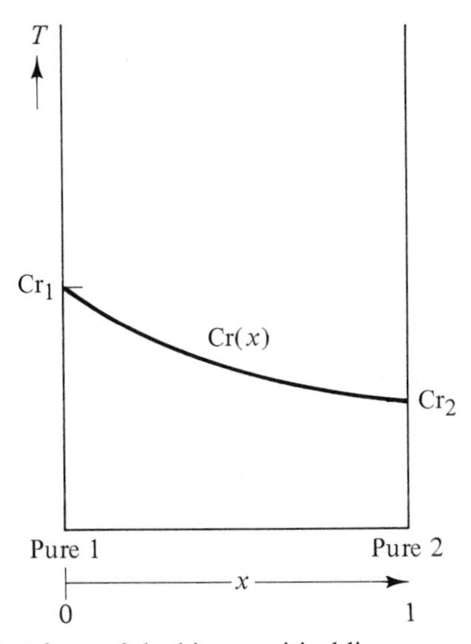

Fig. 43 Simplest form of the binary critical line, as projected on to the T,x plane.

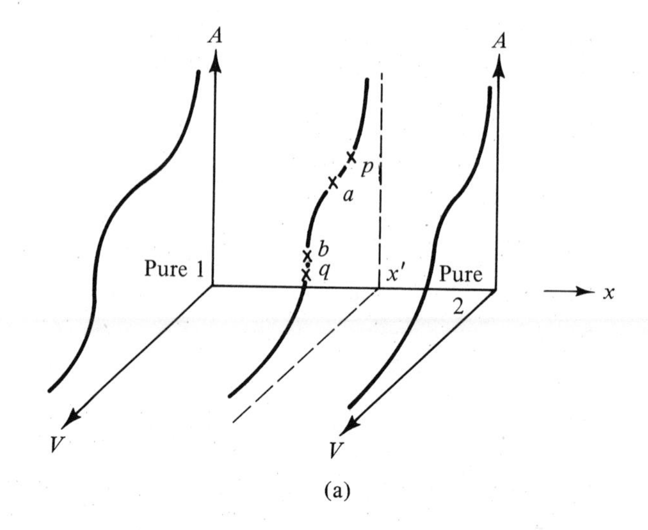

(a)

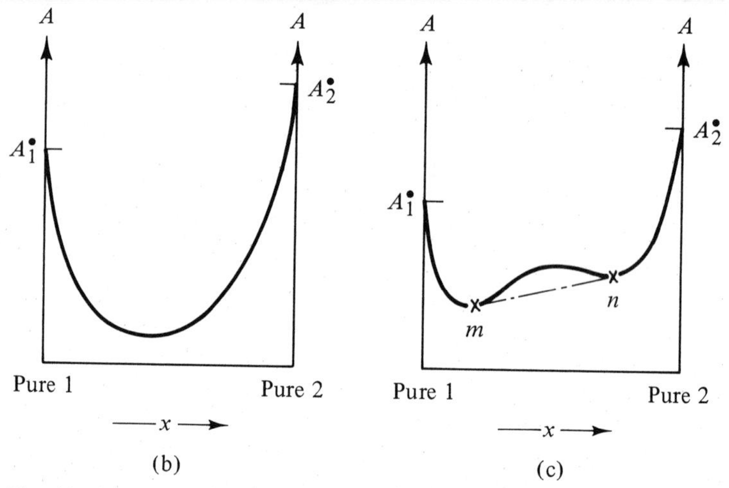

(b) (c)

Fig. 44 (a) A,V,x surface, at constant T, illustrating the *crossfold* below the critical line ($T < T_{Cr_2}$ of Fig. 43). (b) Simple form of the A,x curve at constant V and T. (c) Alternative form of the A,x curve, if a *lengthfold* is present as well.

pure 2. [Compare the top of Fig. 23 in Chapter 18. In Fig. 44(a), p and q (both at x') are *not* coexisting ℓ and g phases,* since the latter differ in composition!] In accordance with these shapes, the A,V,x surface exhibits what is called a *crossfold*. In order to visualize this surface in more detail, we should also consider intersections with planes at constant V. The resulting

* See section 27.2(b).

A,x curves [Figs. 44(b) and (c)] resemble, as far as their shapes are concerned, the G,x curves at constant P and T of Chapter 22 (see Fig. 34 and the top of Fig. 35, respectively).

To look at the shapes of the A,x curves (at constant V and T) more closely, we shall evaluate $(\partial A/\partial x)_{V,T}$ and $(\partial^2 A/\partial x^2)_{V,T}$. From Eq. (27-11),

$$\left(\frac{\partial A}{\partial x}\right)_{V,T} = RT \ln \frac{x}{1-x} - \frac{1}{V}\frac{da(x)}{dx} + \frac{RT}{V-b(x)}\frac{db(x)}{dx}. \tag{27-15}$$

From Eqs. (27-4a) and (27-4b),

$$\frac{da(x)}{dx} = 2[(x-1)a_1 + xa_2 - (2x-1)a_{12}] \tag{27-16a}$$

and

$$\frac{db(x)}{dx} = b_1 - b_2. \tag{27-16b}$$

Differentiate Eq. (27-15) anew with respect to x, at constant V and T, and use the results Eq. (27-16), to yield

$$\left(\frac{\partial^2 A}{\partial x^2}\right)_{V,T} = \frac{RT}{x(1-x)} - \frac{2(a_1 + a_2 - 2a_{12})}{V} + \frac{RT}{[V-b(x)]^2}(b_1 - b_2)^2 \tag{27-17}$$

Remember that a_1, a_2, and a_{12} are all positive. Near the pure component axes, where x is either close to zero or close to unity, the first term dominates in both Eqs. (27-15) and (27-17). Hence the A,x curves approach these axes with a vertical tangent, as indicated in Fig. 44(b) and (c). In the situations illustrated in Fig. 44(b), the A,x curve has a *positive* curvature everywhere. For a "bump" to occur as in Fig. 44(c), leading to an additional fold, a *lengthfold*, in the A,V,x surface, there must be points for which $(\partial^2 A/\partial x^2)_{V,T}$ is *negative.* Inspection of Eq. (27-17) reveals that this can only be the case if the second term dominates the sum of the positive first and third terms. In turn, this is favored by a small T, and requires the condition

$$a_{12} < \tfrac{1}{2}(a_1 + a_2), \tag{27-18}$$

preferably

$$a_{12} \ll \tfrac{1}{2}(a_1 + a_2). \tag{27-19}$$

The last two inequalities tell us that *interspecies* attractions should be smaller, preferably much smaller, then *intraspecies* attractions. This is exactly the situation in which phase separation may occur in the liquid phase. As in

Chapter 18, the "bump" between points m and n in Fig. 44(c) is associated with metastable states, since the fluid can lower its Helmholtz free energy by splitting into two phases.

Points m and n do *not* represent coexisting phases. To obtain the phase that can coexist with m, say, we must find the double tangent *plane*, through m, to the lengthfold of the A,V,x *surface*. This phase will, in general, be located at a different V.

While the appearance of a crossfold, at temperatures below the critical line, is universal, lengthfolds occur only in a limited number of systems which satisfy Eq. (27-19) at a sufficiently low temperature. If indeed both types of fold are present at a certain temperature, the corresponding A,V,x surface has a very complex appearance.

27.2 Equilibrium and Stability; the Spinodal and the Binodal

(a) The demarcation of stable, metastable (locally stable), and unstable portions on the A,V,x surface (at constant T) is much more difficult than the corresponding analysis associated with the A,V curve (at constant T) for a *single*-component van der Waals fluid (see section 18.4). We start by observing that the point (V_o, x_o) on our surface is metastable (locally stable) if all neighboring points *on* this surface lie above the plane tangent *to* this surface *in* this point (V_o, x_o). For, in this situation, any infinitesimal disturbance would lead to a phase separation with an accompanying *increase* in Helmholtz free energy. To put this criterion in a mathematical form, we first write down the equation for the tangent plane at (V_o, x_o):

$$A_t(V, x) = A_o + \left(\frac{\partial A}{\partial V}\right)_o (V - V_o) + \left(\frac{\partial A}{\partial x}\right)_o (x - x_o), \qquad (27\text{-}20)$$

in which A_o stands for $A(V_o, x_o)$, $(\partial A/\partial V)_o$ for $(\partial A/\partial V)_{T,x=x_o,\ \mathrm{at}\ V=V_o}$, and so on. Next, for points close to (V_o, x_o) on the A,V,x surface, we write the Taylor series expansion:

$$\begin{aligned} A(V, x) = A_o &+ \left(\frac{\partial A}{\partial V}\right)_o (V - V_o) + \left(\frac{\partial A}{\partial V}\right)_o (x - x_o) \\ &+ \left[\frac{1}{2}(x - x_o)^2 \left(\frac{\partial^2 A}{\partial x^2}\right)_o + (x - x_o)(V - V_o)\left(\frac{\partial^2 A}{\partial x\,\partial V}\right)_o \right. \\ &\left. + \frac{1}{2}(V - V_o)^2 \left(\frac{\partial^2 A}{\partial V^2}\right)_o\right] + \text{higher-order terms.} \end{aligned} \qquad (27\text{-}21)$$

Assuming that the higher-order terms can be neglected [for *neighboring* points, $(x - x_o)$ and $(V - V_o)$ are small!], the condition $A_t(V, x) < A(V, x)$ implies that the term in brackets in Eq. (27-21) must be positive definite. Hence we must have*

$$\frac{\partial^2 A}{\partial V^2} > 0 \tag{27-22}$$

and

$$\frac{\partial^2 A}{\partial x^2}\frac{\partial^2 A}{\partial V^2} - \left(\frac{\partial^2 A}{\partial x\,\partial V}\right)^2 > 0. \tag{27-23}$$

As in these inequalities, the indices specifying the point (V_o, x_o) are henceforth omitted.

In continuing the analysis, we shall assume that the A,V,x surface only has a crossfold, no lengthfold. Consider the A,V curve at constant $x = x'$ [Fig. 44(a)]. The portion between the two points of inflection, a and b, is unstable since $(\partial^2 A/\partial V^2)$ is negative. Beyond a and b this second partial differential is positive, as is $(\partial^2 A/\partial x^2)$ [see Fig. 44(b)], but this by itself does not ensure that the condition (27-23) is satisfied as well. In fact, *just beyond a* and b, $(\partial^2 A/\partial V^2)$ is not positive enough (it starts out from zero *at a* and b) to achieve this purpose. As we move farther out from a and b along the A,V curve, $(\partial^2 A/\partial V^2)$ continues to increase, until in points p and q we have

$$\frac{\partial^2 A}{\partial V^2}\frac{\partial^2 A}{\partial x^2} = \left(\frac{\partial^2 A}{\partial x\,\partial V}\right)^2. \tag{27-24}$$

From these points on outward, the states are no longer unstable, but become metastable [Eqs. (27-22) and (27-23) are *both* satisfied!].† As x goes from zero to unity, the collection of points p and q, all satisfying Eq. (27-24), forms a locus, known as the *spinodal* of the system. This locus separates the region of instability from that of local stability on the A,V,x surface, at the temperature chosen. Since this temperature has been taken below T_{Cr_2} (Fig. 43), the spinodal consists of two separate parts, corresponding to the locus of points p and q, respectively. A projection onto the V, x plane is drawn in Fig. 45.

(b) To find the stable coexisting phases, at a given T, we must find double tangent planes to the crossfold of the A,V,x surface. This may be proven

---

* For some brief comments on these inequalities and a reference to an appropriate linear algebra text, see the remarks in small print following Eqs. (17-9), with accompanying footnote, in section 17.1(b).

† Remember that for a *single*-component van der Waals fluid the points a and b *themselves* separate instability from metastability.

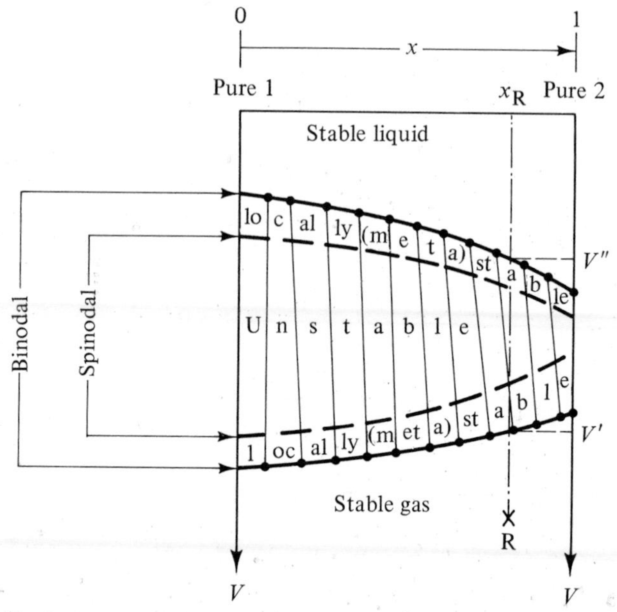

Fig. 45 Projection of the *spinodal* and the *binodal* onto the V,x plane. (T is constant, below the critical line of Fig. 43. Coexisting phases are connected by *tie lines*.)

analytically from the equilibrium conditions:

$$P^g = P^\ell, \tag{27-25a}$$

$$\mu_1^g = \mu_1^\ell, \tag{27-25b}$$

and

$$\mu_2^g = \mu_2^\ell, \tag{27-25c}$$

by techniques similar to those used in Chapters 18 and 22 [see also the small print following Eqs. (27-29)]. There are many such double tangent planes and as these "roll along" the A,V,x surface, the two tangent points form *another* two-part locus, which is called the *binodal.* Under the conditions stipulated so far (i.e., there is only a crossfold, and $T < T_{Cr_2}$), the projection of the binodal onto the V,x plane has the appearance shown in Fig. 45.* The coexisting phases are represented by heavy dots and are connected by *tie lines.* From such projections one may extract many features of the binary fluid

* Figures 45, 47a, and 48 are drawn in typical van der Waals fashion, "upside down"!

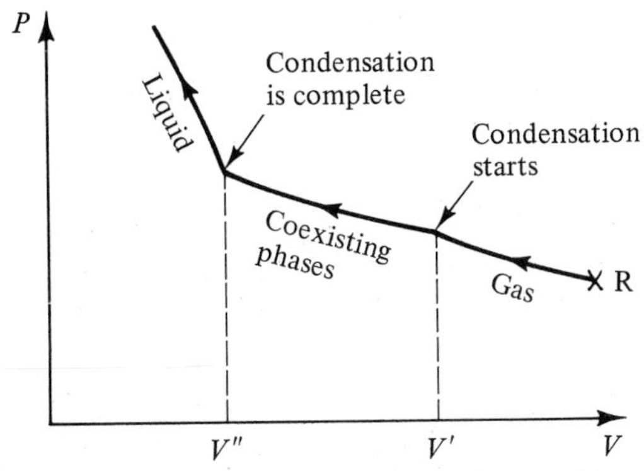

Fig. 46 Compressing the gas R at constant composition x_R (and at constant T); compare Fig. 45.

under consideration. For example, Fig. 46 shows us what happens if the gas R (Fig. 45) is compressed at constant (T and) composition, x_R. At V' the condensation starts, and at V'' it is complete. In between we have a mixture of stable coexisting phases, with net composition x_R, in a ratio given by the lever rule [section 18.4(c) and Fig. 24] as applied to the appropriate tie line. A metastable single phase is possible, in principle, between the binodal and the spinodal. Note (see Fig. 46) that as one might expect, the region of co-existing phases is *not* represented by a horizontal line, unlike the same region for a pure component.

Exercise 27-2

Verify that the statement made in the last sentence is in accordance with the Phase Rule.

27.3 Retrograde and Barotropic Phenomena

(a) As nearly all preceding chapters in Part IV of this book, Chapter 27 is very much "open ended." We shall close it by discussing two of the more striking phenomena that can be explained on the basis of the van der Waals theory. To this purpose we first need to find the shape of the spinodal and the binodal under the conditions that (1) the A,V,x surface still has a cross-fold only, but (2) the (constant) temperature T is between the critical points of the two pure components. It is evident that under these conditions, the crossfold cannot extend all the way to $x = 1$ but will end in some sort of binary critical point. In the V,x projection [Fig. 47(a)] this *folding point*, also called *umbilical point*, is represented by U. Here the binodal and spinodal, in this case both single closed curves, meet. We proceed to prove this.

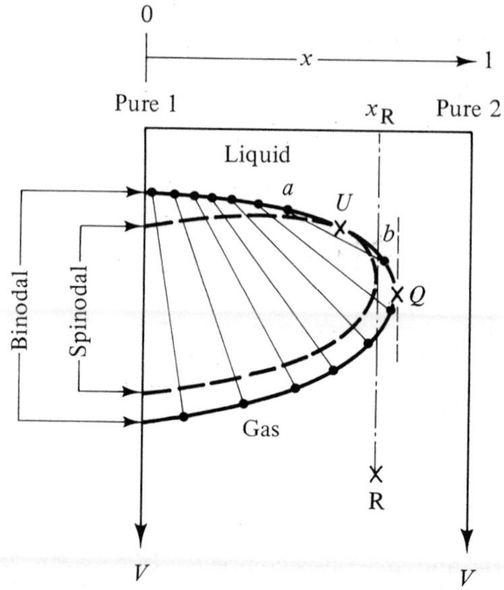

Fig. 47(a) As Fig. 45, but at a temperature T, such that $T_{Cr_1} > T > T_{Cr_2}$.

(b) First we apply Eqs. (27-25) to the liquid–gas equilibrium a-b, just "below" U [Fig. 47(a)]. Equation (27-25a) can immediately be put in the form

$$\left[\left(\frac{\partial A}{\partial V}\right)_{T,x}\right]_a^{\ell} = \left[\left(\frac{\partial A}{\partial V}\right)_{T,x}\right]_b^{g}. \tag{27-26a}$$

Equations (27-25b) and (27-25c) can be rewritten, if we use Eqs. (22-4) with an obvious change in notation, as follows:

$$\left[G - x\left(\frac{\partial G}{\partial x}\right)_{P,T}\right]_a^{\ell} = \left[G - x\left(\frac{\partial G}{\partial x}\right)_{P,T}\right]_b^{g} \tag{27-26b}$$

and

$$\left[G + (1 - x)\left(\frac{\partial G}{\partial x}\right)_{P,T}\right]_a^{\ell} = \left[G + (1 - x)\left(\frac{\partial G}{\partial x}\right)_{P,T}\right]_b^{g}. \tag{27-26c}$$

We can add and subtract Eqs. (27-26b) and (27-26c) to yield

$$\left[\left(\frac{\partial G}{\partial x}\right)_{P,T}\right]_a^{\ell} = \left[\left(\frac{\partial G}{\partial x}\right)_{P,T}\right]_b^{g} \tag{27-27a}$$

and

$$\left[G - x\left(\frac{\partial G}{\partial x}\right)_{P,T}\right]_a^{\ell} = \left[G - x\left(\frac{\partial G}{\partial x}\right)_{P,T}\right]_b^{g}. \tag{27-27b}$$

To convert the results, Eq. (27-27),* into equations involving A, rather than G, we observe that

$$G \equiv A + PV = A - V\left(\frac{\partial A}{\partial V}\right)_{T,x} \tag{27-28a}$$

and

$$\left(\frac{\partial G}{\partial x}\right)_{P,T} = \left(\frac{\partial A}{\partial x}\right)_{V,T} (= \mu_2 - \mu_1). \quad * \tag{27-28b}$$

With Eqs. (27-28), the results, Eq. (27-27), become

$$\left[\left(\frac{\partial A}{\partial x}\right)_{V,T}\right]_a^{\ell} = \left[\left(\frac{\partial A}{\partial x}\right)_{V,T}\right]_b^{g} \tag{27-29a}$$

and

$$\left[A - V\left(\frac{\partial A}{\partial V}\right)_{x,T} - x\left(\frac{\partial A}{\partial x}\right)_{V,T}\right]_a^{\ell} = \left[A - V\left(\frac{\partial A}{\partial V}\right)_{x,T} - x\left(\frac{\partial A}{\partial x}\right)_{V,T}\right]_b^{g}. \tag{27-29b}$$

Parenthetically, the last two equations, together with Eq. (27-26a), prove the double tangent plane construction referred to in section 27.2(b). Equations (27-26a) and (27-29a) *combined* show that the tangent planes in a and b are parallel. Equation (27-29b) then demonstrates that they also have the same intercept with the A-axis. The entire proof is reminiscent of the one given in section 18.3(d) for a *single-component* van der Waals fluid.

As the points a and b approach U, Eqs. (27-26a) and (27-29a) can be written as the vanishing of small variations accompanying the ℓ-g phase change:

$$\delta_{\ell \to g}\left(\frac{\partial A}{\partial V}\right)_{x,T} = 0 \tag{27-30a}$$

and

$$\delta_{\ell \to g}\left(\frac{\partial A}{\partial x}\right)_{V,T} = 0. \tag{27-30b}$$

Remembering that T is constant throughout, these two variations can be expanded in the form

$$\frac{\partial^2 A}{\partial V^2}\,\delta V + \frac{\partial^2 A}{\partial x\,\partial V}\,\delta x = 0 \tag{27-31a}$$

and

$$\frac{\partial^2 A}{\partial V\,\partial x}\,\delta V + \frac{\partial^2 A}{\partial x^2}\,\delta x = 0, \tag{27-31b}$$

* Compare the derivation of Eq. (22.3) in section 22.2(a).

respectively. Since A is taken to be a well-behaved function of V, T, and x, $\partial^2 A/\partial x\,\partial V = \partial^2 A/\partial V\,\partial x$. For Eqs. (27-31a) and (27-31b), homogeneous linear equations in δV and δx, to have a nontrivial solution,* the determinant of the coefficients has to vanish:

$$\begin{vmatrix} \dfrac{\partial^2 A}{\partial V^2} & \dfrac{\partial^2 A}{\partial x\,\partial V} \\ \dfrac{\partial^2 A}{\partial V\,\partial x} & \dfrac{\partial^2 A}{\partial x^2} \end{vmatrix} = \frac{\partial^2 A}{\partial V^2}\frac{\partial^2 A}{\partial x^2} - \left(\frac{\partial^2 A}{\partial x\,\partial V}\right)^2 = 0. \tag{27-32}$$

Equation (27-32) pertains to the ℓ-g equilibrium, hence is valid for the *binodal*, infinitely close to the umbilical point, U. But it is identical with Eq. (27-24) for the *spinodal*! Hence we have indeed proven that binodal and spinodal meet in this critical folding point, U, as indicated in Fig. 47(a).

(c) To observe retrograde condensation at constant T, we have to compress a gaseous mixture R with a composition x_R between x_U and x_Q (Fig. 47(a); Q is the only point in which the binodal has a vertical tangent). This compression, to be carried out at constant net composition, is somewhat easier to follow in the corresponding P,x diagram [Fig. 47(b)]. Remember

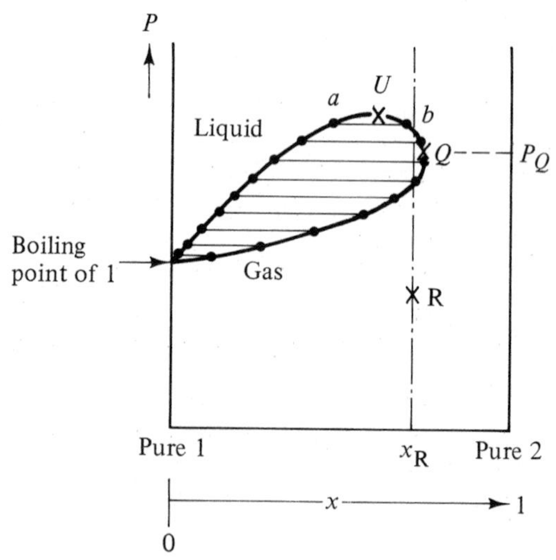

Fig. 47(b) P,x diagram of the same T of Fig. 45(a). [A P,x cut of this type is also drawn in Fig. 33, at T_2, with the difference that in Fig. 33 the "other" component (the one at $x = 0$) has the lower critical temperature.]

* That is, a solution other than the trivial one $\delta V = \delta x = 0$.

that, as we go up through the area of coexisting phases (indicated as heavy dots linked by, in this diagram *horizontal*, tie lines), their varying relative amounts are still determined by the lever rule as applied to *each* tie line, while the net composition of the mixture remains fixed at x_R. Hence as we first "hit" the coexistence area from below, a little liquid will form, and as we compress further, its amount will at first increase in "normal" fashion.

Exercise 27-3

Draw a P,x diagram at a constant temperature *below* the critical line of the binary system. Convince yourself that compressing a gas yields a monotonically increasing amount of liquid as we go through the coexistence area.

The remarkable feature is that *the amount of liquid reaches a maximum* (namely, at the pressure P_Q) *after which it diminishes again as the compression is continued.* This type of retrograde condensation, specified by some authors* as being *of the first kind,* was first observed by L. Cailletet in 1880 in studying the liquefaction of CO_2–air mixtures, and studied systematically by J. P. Kuenen in 1906. The latter, a pupil of van der Waals and Bakhuis-Roozeboom, used a mixture of 0.41 g of CH_3Cl and 0.5 g of CO_2 at 378 K. (The critical temperatures of these two substances are 536 and 304 K, respectively.) Retrograde condensation *of the second kind* is observed by reducing the *temperature* of an appropriate gaseous mixture at a *constant pressure* in between the critical pressures of the two pure components.

Exercise 27-4

Interpret this last phenomenon by means of a T,x diagram.

In recent decades there has been a renewed interest in retrograde condensation, since it is of importance in the high-pressure distillation of hydrocarbon mixtures in the petroleum industry.

(d) The *barotropic effects* are perhaps even more remarkable than the retrograde condensation phenomena, but they are not as "well known," insofar as this adjective is appropriate at all in this context. Just as retrograde condensation of the first kind, barotropic effects can be observed at a constant temperature between the critical ones of the two components, and only when these two temperatures are well separated. In that case the binodal *may* have *both* a horizontal *and* a vertical tangent, with the umbilical point U located "to the left," as illustrated in Fig. 48. [The spinodal, being irrelevant at this stage, is not drawn. It meets the binodal in U, as in Fig. 47(a).] In this V,x diagram, there is one *horizontal* tie line, c–d (the point d may be

---

* See, e.g., J. O. Hirschfelder, C. F. Curtis, and R. B. Bird, *The Molecular Theory of Gases and Liquids*, Wiley, New York, 1954, pp. 380ff.

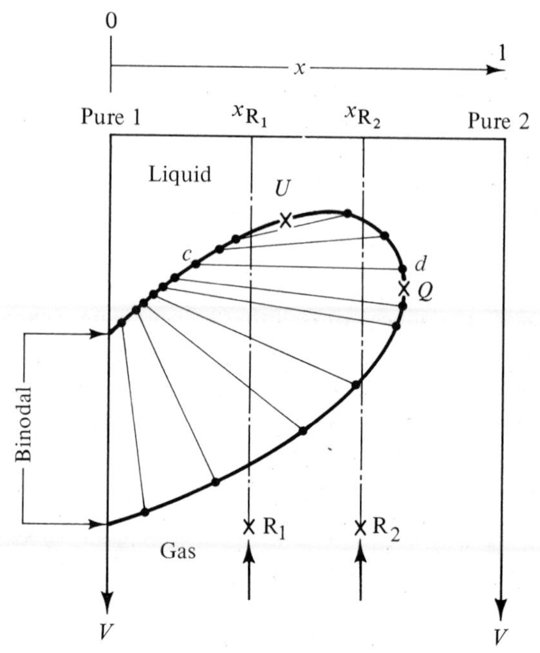

Fig. 48 Illustrating the projection of the binodal onto the V,x plane, at a constant temperature between T_{Cr_1} and T_{Cr_2}, allowing for *barotropic effects.*

close to but, in general, does not coincide with the vertical tangent point, Q). Barotropic phenomena are observed by compressing gaseous mixtures with composition anywhere between x_c and x_d, so that we cross c–d in the process. At this crossing, the coexisting vapor and liquid have the same density, and as the compression is continued from there, the *liquid will have the lower density.* Thus, as we cross c–d, we will see the vapor "sink into" the liquid. If we compress the gas R_1, we see *only this* effect [Fig. 49(a)], but if we compress the gas R_2, we observe a *combination* of barotropic and retrograde phenomena [Fig. 49(b)]. In the situation depicted in Fig. 49(a), we have crossed the tie line c–d somewhere between (1) and (2). In Fig. 49(b) we have crossed the tie line containing the vertical tangent point (for which the ℓ:g ratio is maximal) somewhere between (1) and (2), and the horizontal tie line between (2) and (3). As we pointed out already, these two tie lines may be close together, so that the sequence of Fig. 49(b) is usually very hard to observed in detail. In fact, both effects may be overlooked *jointly*, unless one is able to carry out the compression very slowly and pays particular attention to the changes in shape of the boundary meniscus. Barotropic effects, first studied experimentally by Kamerlingh Onnes, have been observed, e.g., in mixtures of argon and ammonia at a temperature of about 350 K. (The two critical temperatures of these components are 150 K and 406 K, respectively.)

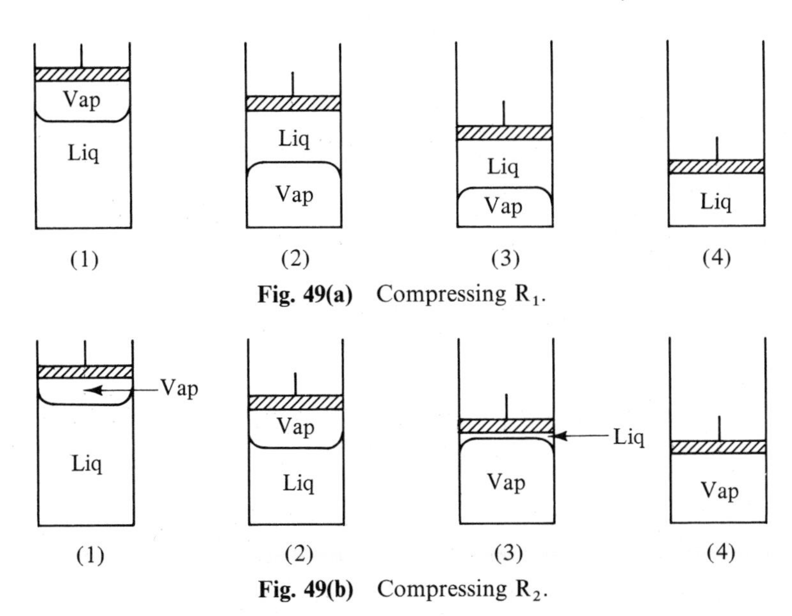

Fig. 49(a) Compressing R_1.

Fig. 49(b) Compressing R_2.

When these phenomena occur, it is sometimes hard to say whether a particular phase is a gas or a liquid. This has led to problems in the formulation of legal documents involving hydrocarbon mixtures obtained from oil wells.*

REFERENCES

Much of what has been said in the introductory comments to the references at the end of Chapter 22 is still very pertinent here. Some of the same books may be consulted, but retrograde condensation is alluded to in relatively few of these, and a discussion of barotropic phenomena is even harder to find. Much of Chapter 27 has been inspired by the lectures of F. E. C. Scheffer and G. E. Uhlenbeck. In turn, these two scientists acknowledged the direct influence of van der Waals and Lorentz respectively. Those who wish to go back to "the source" should consult the book by van der Waals and Kohnstamm (see the references of Chapter 22), and an article by H. A. Lorentz in *Ann. Physik. Chem.* **12**, 127–136 (1881).

* J. O. Hirschfelder, private communication, 1983.

List of Symbols Used

Ideally speaking, the use of all symbols should be totally consistent and completely unique, but unfortunately such an ideal cannot be realized. It is obvious that we should not use, e.g., p for *both* vapor pressure *and* number of phases in a description of the same phase equilibrium. It is equally obvious that it is hardly disturbing to use X_i for *both* the ith generalized force *and* the mole fraction of component i in a mixture, since these two meanings appear in entirely different sections of the book. In between such extremes, a judgment had to be made in each particular case whether a duplication of symbols could be permitted or had to be avoided at all cost by introducing some more "exotic" letter types. If a certain symbol occurs with a particular meaning in only one section of the book, this is indicated in parentheses in the list below. When a symbol is flanked by vertical bars, this may indicate either a dimensionless value (as in log $|\mathbf{V}|$) or an absolute value (as in the freezing-point lowering, $|\delta T_f|$); this dual use should cause no confusion. Standard states are indicated by the superscript $^{\ominus}$. Superscripts (rarely subscripts) ^{e} indicate that the entity concerned possesses its equilibrium value.

Starting in Chapter 8, the symbol **J** is introduced to denote any extensive thermodynamic property (**V**, **E**, $\mathbf{C}_p$, **S**, **G**, **A**, etc.). With each of these, we may associate the following:

$\mathbf{J}$	Total property.
$J \equiv \dfrac{\mathbf{J}}{n}$	Molar property (if we are dealing with a single-component system) or mean molar property (if we are dealing with a mixture).
$J_k^{\bullet}$	Property of 1 mole of pure k (usually given at specific P and T).
$\bar{J}_k$	Partial molar* property of component k, at given P and T, in a mixture of specific composition.
$\bar{J}_k^{\ominus}$	The same quantity in its standard state.
J'_k	Partial specific property of component k, at given P and T, in a mixture of specific composition.
${}^{\phi}J_2$	Apparent molar property of the solute, 2, in a solution of 2 in the solvent 1, at given P and T.
${}^{\phi}J'_2$	Apparent *specific* property.
ΔJ_{mix}	"**J** of mixing" (Chapter 25).
$\Delta J^{\mathbf{E}}_{\mathrm{mix}}$	Excess "**J** of mixing" (Chapter 25).

* Some authors prefer partial mola*l* property.

In the list to follow, only the relevant *total* properties are included; the rest of the symbolism is not repeated. This list is fairly complete, but not exhaustive in the sense that some obvious *ad hoc* symbols, frequently of a purely mathematical nature (e.g., summation indices), have been omitted.

A	Helmholtz free energy (work function).
$\mathscr{A}$	Cross-sectional area (of piston) (Section 4.4 only).
$\mathscr{A}$	Surface area (in general).
$\mathbb{A}$	Affinity (in a chemical reaction mixture).
a_k	Activity of k.
$a_\pm$	Mean ionic activity.
a	Van der Waals constant.
a	Constant in Debye Law (Chapter 16 only).
B, B	Volume correction in equation of state for hard-sphere model.
B	Second virial coefficient.
B_{jk}	Second virial coefficient for j-k interactions.
b	Van der Waals constant.
b	Constant determining contribution of free electrons to heat capacity (Chapter 16 only).
C	Heat capacity (a subscript such as p or v indicates that we refer to the heat capacity at constant p or v).
C	Third virial coefficient.
C, C_i	Constants acting as a parameter (Chapter 7 only).
c_k	Molarity of k.
$\mathbf{c}$	Number of components in the Gibbs sense.
d	Degree of dissociation.
E	Internal energy, total energy (see footnote, p. 78).
$\mathscr{E}$	EMF of galvanic cell.
e	Charge of an electron.
$\vec{F}$	Force, exerted by the system on its surroundings.
$\mathscr{F}$	The Faraday (unit of electric charge).
f	Fugacity
G	(Gibbs) free energy.
G	Weight of a body (section 4.5 only).
g	Acceleration of gravity.
g	Empirical correction factor for electrolyte solutions (Chapter 26 only).
H	Enthalpy.
$\mathscr{H}$	Magnetic field strength.
h	Height.
h_π	Height of a column of liquid, corresponding to the osmotic pressure, π (see Fig. 42).
I	Electric current.

I	Ionic strength.
i	Van't Hoff factor for electrolyte solutions.
$\mathbf{J}$	General (extensive) thermodynamic property (see the text above this list).
K	Degree Kelvin.
K	Equilibrium constant. [A subscript, such as p, indicates that the compositions are expressed in partial pressures. For superscripts, see under the symbol (α) below.]
K_{sp}	Solubility product.
K_p	Kinetic energy of the piston (section 4.4 only).
$\mathbf{K}_f$, $\mathbf{K}_b$	Freezing-point constant and boiling-point constant, respectively
k	Boltzmann constant.
k_H	Henry's Law constant [see Eq. (26-20)].
k_i^*	Constant determining the vapor pressure of solute i in ideally dilute solution.
k_i	Limiting law constant for solute i in ideally dilute solution (for superscripts, see under α below).
L	Surface in space (Chapter 7 only).
$\mathbf{L}_i$	See Eq. (22-32).
$\mathbf{L}$	See Eq. (22-33).
l	Length (of a wire).
ln	Natural logarithm.
log	Logarithm to the base 10.
M	Weight of 1 mole (in grams).
$\|M\|$	(Dimensionless) molecular weight.
M_p	Mass of piston (section 4.4 only).
$\mathcal{M}$	Magnetization.
$\mathbf{M}$	Semipermeable membrane.
m	Mass.
m_k	Molality of k (m without a subscript is also used for molality, if there is no ambiguity as to the species concerned).
$\mathbf{m}$	Number of special conditions (in the Wind Phase Rule; Chapters 21 and 22 only).
N_{Av}	Avogadro's number.
n_k	Number of moles of k.
n	Total number of moles.
$\mathbf{n}$	Number of conditions due to chemical reactions (in the Wind Phase Rule; Chapters 21 and 22 only).
P	Pressure.
p_k	Partial pressure of k.
$\mathbf{p}$	Number of phases.
Q	Heat flow from the environment into the system.
$\mathfrak{Q}$	Reaction quotient (for subscripts, see note under "Equilibrium constant" above).

"$\mathfrak{Q}$"	Correction factor to equilibrium constant for nonideal behavior (with the formal appearance of a reaction quotient).
q	Electrical charge.
q_{A-B}	Average bond energy (enthalpy) in a molecule with equivalent bonds.
R	Gas constant.
R	Electrical resistance.
r	Number of components or substances in a "generic" sense (cf. **c** and **s**).
$\vec{r}$	Displacement vector (of boundary between system and surroundings).
S	Entropy.
s	Number of chemical substances (in the Wind Phase Rule). (Not to be confused with **c**.)
T	Absolute temperature.
t	Empirical temperature.
	[Frequently used subscripts with temperatures: b, boiling; Cr, critical; f or fus, fusion (melting or freezing); i, inversion (of Joule–Kelvin effect).]
u	Speed (of gas molecules).
Δu	Electrostatic potential difference between electrodes. [For relation to $\mathscr{E}$, see Eq. (15-31).]
V	Volume.
$\mathscr{V}$	Potential energy in a gravitational field [section 6.2(b) only].
v	Variance.
W	Work done *by* the system *on* the environment.
w	Total mass of a system.
w_k	Mass of k.
w	Energy parameter in the phenomenological definition of a "simple" solution.
X_i	Generalized force.
X_k	Mole fraction of k.
x_i	Generalized coordinate.
x	Frequently used in lieu of X_2 for the mole fraction of a single solute, 2, in a solvent, 1, or in lieu of X_B in A–B systems.
Y_i, y_i	Pair of conjugate variables, used *ad hoc*, as in $dQ = \sum_i Y_i\, dy_i$ [Eq. (13-3)] or $d\mathbf{E} = \sum_i Y_i\, dy_i$ [Eq. (20-10)].
$y_1, \ldots, y_r$	State variables (section 6.2 only).
Z	Compressibility factor.
z	Displacement coordinate in a gravitational field.
z_k	Weight fraction of k.
z_+ and z_-	Integers giving, when multiplied by $\lvert e\rvert$ or $-\lvert e\rvert$, the ionic charge.
z	Coordination number (in quasi-crystalline model for liquid mixtures).

α	General symbol, frequently used in parentheses as a superscript, to denote X, m, or c. The symbol $[1_\alpha]$, in this context, stands for 1, 1 mole per kg solvent and 1 mole per liter, respectively.
α	Coefficient of thermal expansion.
Γ	Availability (section 15.5 only).
γ	Surface tension (very few places in Chapters 4 and 15).
γ_k	Fugacity coefficient of k or activity coefficient of k.
$\gamma_\pm$	Mean ionic activity coefficient.
δ	Very small change, as in $\delta\mathbf{E}$.
Δ	Is used in two meanings, to indicate a *general* macroscopic change ($\Delta\mathbf{E} \equiv \mathbf{E}_{\text{final}} - \mathbf{E}_{\text{initial}}$) and the *specific* operator $(\partial/\partial\xi)_{P,T}$ pertaining to a chemical reaction ($\Delta\mathbf{G} \equiv \sum_k \nu_k\mu_k$). See also under ξ below. For $\Delta\nu$ (and $\Delta\nu^g$) of a chemical reaction, see under ν below and section 8.4.
ε	*Ad hoc* symbol in Problem 10-2.
ε_i	Molecular energy level, i.
ε, ε_{AB}	Parameters in the quasi-crystalline lattice model of liquid mixtures.
θ	Ideal gas temperature.
κ	Isothermal compressibility.
Λ	Availability (section 15.5 only).
λ	Integrating factor (Chapter 7 only).
μ_k	Chemical potential of k (*ad hoc* symbol for $\bar{G}_k$).
$\mu_{J.T.}$	Joule–Thomson (or Joule–Kelvin) coefficient.
μ_M	Magnetic dipole moment (section 14.3 only).
ν	Number of particles in which a *single* solute dissociates in a given solvent.
ν_k	Stoichiometric coefficient of component k in a chemical reaction (taken *negative* for a "reactant" and *positive* for a "product").
$\Delta\nu$, $\Delta\nu^g$	Of a reaction denote $\sum_k \nu_k$ and $\sum_k \nu_k^g$, respectively (see section 8.4).
$\tilde{\nu}$	*Ad hoc* symbol in a solution [see Eq. (26-9)].
ξ	Degree of advancement of a reaction (progress variable of a reaction).
ξ	Efficiency (conversion factor) of a Carnot cycle.
$\prod$	Product of ....
π	Reduced pressure of a gas.
π	Osmotic pressure.
ρ	Density (of various types).
ρ_1	Density of the *solvent* associated with a solution.
Σ, σ	Denote a general surface (in Chapter 7 only).
$\sum$	Sum of ....
τ	Tension (to which, e.g., a wire is subjected).
τ	Integrating denominator (Chapter 7 only). (In Chapter 17, $d\tau$ labels a volume element.)

Φ	Potential energy associated with a conservative force field.
ϕ	As Φ, but per mole of fluid in such a field.
ϕ	Correction term in Gibbs Phase Rule (Chapters 21 and 22).
χ	Magnetic susceptibility.
Ω	Classically: state density in phase space. Quantum mechanically: the degeneracy of a system (Both used in Chapter 16 only.)
ω_i	Degeneracy of the molecular state i (Chapter 16).

Units and Conversion Factors

A. Some SI Units

Force	newton	N	$kg\ m\ s^{-2}$
Pressure	pascal	Pa	$kg\ m^{-1}\ s^{-2}$ or $N\ m^{-2}$
Energy	joule	J	$kg\ m^2\ s^{-2}$ or N m
Power	watt	W	$kg\ m^2\ s^{-3}$ or $J\ s^{-1}$
Electric charge	coulomb	C	A s (A ≡ ampere)
Electric potential difference	volt	V	$kg\ m^2\ s^{-3}\ A^{-1}$ or $J\ C^{-1}$
Electric resistance	ohm	Ω	$kg\ m^2\ s^{-3}\ A^{-2}$

B. Some Units Frequently Used in This Book, in Terms of SI

1 dyne = 10^{-5} N
1 atm = 760 torr = 101,325 Pa exactly
1 cal = 4.184 J exactly
1 erg = 10^{-7} J
1 ℓ atm = 101.325 J
1 ℓ = $10^{-3}\ m^3$
1 esu = 3.336×10^{-10} C

C. Some Physical Constants

Base of natural logarithms	2.71828
Zero of the Celsius scale in degrees Kelvin	273.15 K exactly
Avogadro's constant, N_{Av}	$6.0221 \times 10^{23}\ mole^{-1}$
Gas constant, R	$8.3148\ J\ K^{-1}\ mole^{-1}$
	$1.9873\ cal\ K^{-1}\ mole^{-1}$
	$0.082061\ \ell\ atm\ K^{-1}\ mole^{-1}$
Faraday's constant, $\mathscr{F}$	$96{,}486\ C\ mole^{-1}$
Electronic charge, e	-1.602×10^{-19} C
Standard acceleration of gravity, g	$9.80665\ m\ s^{-2}$ exactly

Answers to Numerical Problems

3-1. $t = 23.53$
3-2. (a) 820 kJ
(b) About 7 minutes
4-1. (b) $W = 9.81$ J
(c) 176 m
(d) About 6 minutes
4-2. (a) (i) $W = 12.2$ kJ
(ii) $W = 17.4$ kJ
(iii) $W = 22.5$ kJ
(iv) $W = 33.8$ kJ
8-2. (a) at $c = 1$, $\bar{V}_2 = 19.59$ ml
at $c = \frac{1}{2}$, $\bar{V}_2 = 18.63$ ml
$\bar{V}_1 \approx V_1^{\bullet} = 18.07$ ml
(b) $m = 0.5015$ mole kg^{-1}
$\bar{V}_2 = 18.63$ ml
8-3. $\xi_{max} = +2$

10-1.

	W	Q	ΔE	ΔH (in J)
(a) (i) Isothermal	1729	1729	0	0
(ii) Adiabatic	1384	0	−1384	−2307
(b) (i) Isothermal	1732	1732	0	−3
(ii) Adiabatic	1387	0	−1387	−2315

10-2. (c) $\varepsilon = 0.0086$ for He
$= 0.0610$ for H_2
$= 0.493$ for NO_2
10-3. (a) 3.98°C
(b) $C_p - C_v = 0.79$ J K^{-1}
(c) $C_p - C_v = 41.9$ J K^{-1}
10-4. (a) $Q_{500°C} = -393.3$ kJ
(b) $Q_{500°C} = -392.7$ kJ
10-5. $410\frac{1}{2}$ kJ mole^{-1}
10-6. Per mole of Pb reacting, the cell *absorbs* 8.3 kJ of heat
10-7. (a) $\mathbf{V} = 1.13\ \ell$
(b) $\Delta\mathbf{E} = 702$ J

12-2. (a) $|Q_1|/|W| = 29$
(b) $|Q_2|/|W| = 29$
12-3. (b) 42°C
14-1. $\Delta S = 23.05$ J K^{-1}
14-2. $\Delta S = 20.17$ J K^{-1} for *both* pathways.
14-3. (a) $\Delta S_{(1)} = -20.6$ J K^{-1}
(d) $P_\ell/P_s = 1.09^5$
(e) 0.8 J K^{-1}
14-4. (a) For the stretching, $\Delta \mathbf{S} = 6.67 \times 10^{-3}$ J K^{-1}
For the "snap back," $\Delta \mathbf{S} = -6.67 \times 10^{-3}$ J K^{-1}
(b) $W = -2$ J
14-5. (a) $T_i = 764$ K and 134 K
(b) $T_i = 752$ K and 132 K
(c) $T_i = 866$ K
(d) $P_{i,\max} = 301\frac{1}{2}$ atm
15-1. (b) $\Delta \mathbf{S} = 26$ J K^{-1}
15-2. (a) $\Delta \mathbf{G} = 0$
(b) $\Delta \mathbf{G} = -198.1$ J mole^{-1}
15-3. $Q = 40.58$ kJ mole^{-1}; $W = 3102\frac{1}{2}$ J mole^{-1};
$\Delta \mathbf{E} = 37.48$ kJ mole^{-1}; $\Delta \mathbf{H} = 40.58$ kJ mole^{-1};
$\Delta \mathbf{S} = 108.7^5$ J K^{-1} mole^{-1}; $\Delta \mathbf{A} = -3102\frac{1}{2}$ J mole^{-1};
$\Delta \mathbf{G} = 0$
15-4. (b) $\approx 15{,}000$ atm
15-5. $\Delta \mathbf{G} = -12.44$ kJ mole^{-1}; $\Delta \mathbf{S} = 29.72$ J K^{-1} mole^{-1};
$\Delta \mathbf{H} = -3.58$ kJ mole^{-1}
15-6. $\Delta \mathbf{G}^\ominus = -42.92$ kJ mole^{-1}; $\Delta \mathbf{S}^\ominus = 124.6$ J K mole^{-1};
$\Delta \mathbf{H}^\ominus = -80.07$ kJ mole^{-1}; $\Delta \mathbf{C}_p^\ominus = -343.0$ J K^{-1} mole^{-1}
16-1. 10%
16-2. $\Delta \mathbf{G}^\ominus_{1000} = 170.32$ kJ mole^{-1}
$K_p = 1.3 \times 10^{-9}$ (all pressures in atm)

16-3.

Reaction	$\Delta \mathbf{G}^\ominus$(kJ mole^{-1})	K at 25°C
A	−27.20	5.82×10^4
B	−10.88	8.06×10^1
C	−44.68	6.72×10^7
D	+18.46	5.83×10^{-4}
E	+8.74	2.94×10^{-2}
F	+8.16	3.72×10^{-2}

16-4. $S^\ominus = 259.66$ J K^{-1} mole^{-1}
17-1. (b) $P_{top}/P_{bottom} = 0.9999^5$
(c) 678 torr
(d) 632 torr (from barometric formula)
18-1. 95.3°C
18-2. 93.6°C
18-3. (a) $dP/dT = -130$ atm K^{-1}
(b) −0.76°C

19-1. Part I. (a) $K_p = 5.6 \times 10^5$
(b) (i) $K_p = 54$
(ii) $K_p = 45$
(d) (i) $K_p = 3.53 \times 10^{-5}$
(ii) 3.65%
(iii) 60%
Part II. 70%

19-2.

	P (atm)	f (atm)	γ
He	200	240	1.20
	400	579	1.45
H_2	200	232	1.16
	400	549	1.37

19-3. $a_{Zn^{2+}}/a_{Cu^{2+}} = K_a = 1.51 \times 10^{37}$
19-4. $p_{Cl_2} = 22.3$ torr; $p_{IC_1} = 28.3$ torr
23-1. Sample calculation for $X_A = X_B = \frac{1}{2}$; ΔS_{mix} has its maximum value of 5.76 J K^{-1} mole^{-1}
23-2. (a) $P = 0.70$ atm; $T = 452$ K
(b) $\Delta \mathbf{S}_{mix} = 0.286$ J K^{-1}

23-3.

	1 km above sea level		5 km above sea level	
i	p_i (atm)	X_i	p_i (atm)	X_i
N_2	0.697	0.783	0.444	0.795
O_2	0.185	0.208	0.110	0.197
Ar	0.008^5	0.009 +	0.004^5	0.008 +
	$P = 0.890$ atm	1.000	$P = 0.558$ atm	1.000

25-1. (bz ≡ benzene and tol ≡ toluene.)
Part I. (a) $X^{\ell}_{tol} \equiv x^{\ell} = 0.46$; $X^{\ell}_{bz} = 1 - x^{\ell} = 0.54$;
$p_{tol} = 10.1$ torr; $p_{bz} = 40.5$ torr;
$P = 50.6$ torr; $X^{g}_{tol} \equiv x^{g} = 0.20$
(b) $\Delta S_{mix} = 5.74$ J K^{-1} mole^{-1}
$\Delta G_{mix} = -1683$ J mole^{-1}
Part II. (a) No significant changes
(b) $\Delta H_{mix} = 41$ J mole^{-1}; $\Delta S^{E}_{mix} = 0.15$ J K^{-1} mole^{-1};
$\Delta S_{mix} = 5.89$ J K^{-1} mole^{-1}; $\Delta G^{E}_{mix} = -3$ J mole^{-1};
$\Delta G_{mix} = -1686$ J mole^{-1}
(c) $\gamma_{bz} = \gamma_{tol} = 0.999$
25-2. (a) $w = 2.60$ J mole^{-1}
(b) $\varepsilon = 3.6 \times 10^{-25}$ J molecule^{-1}
(c) $(\partial w/\partial P)_T \approx 0.00016$ J atm^{-1} mole^{-1}
26-1. $\mathbf{K}_f = 1.86$ K kg mole^{-1}
$\mathbf{K}_b = 0.512$ K kg mole^{-1}
26-2. (a) $|\delta T_f| = 5.12 \times 10^{-4}$ K; $\pi = 1.63$ torr

26-3. (b) $\Delta G^{\ominus}_{soln} = 27.14$ kJ mole^{-1} at 20°C, and 27.60 kJ mole^{-1} at 25°C; $\Delta H^{\ominus}_{soln} = -3.86$ kJ mole^{-1} between 20 and 25°C

26-4. (b) $w_{I_2 \text{ in } CCl_4} = 0.148$ g; $w_{I_2 \text{ in } H_2O} = 0.002$ g

26-5. (d) $X^{sat}_{naphthalene} = 0.295$

26-6. (a) 1.91 mg of AgCl per kg of water
(b) 13.2%

26-7. (b) $\mathscr{E}^{\ominus} = 0.2224$ V
(c) $\gamma_{\pm} = 0.904$

26-8. (a) 24.7 atm
(b) 9460 kJ
(c) 15 cents
(d) ($Q_{vap} \approx 1{,}260{,}000$ kJ)

Author Index

R

S

T

U

V

W

Z

Subject Index

F

J

K